AF783716

Micromechanics Modelling of Ductile Fracture

SOLID MECHANICS AND ITS APPLICATIONS
Volume 195

Series Editors: G.M.L. GLADWELL
Department of Civil Engineering
University of Waterloo
Waterloo, Ontario, Canada N2L 3GI

Aims and Scope of the Series

The fundamental questions arising in mechanics are: *Why?*, *How?*, and *How much?* The aim of this series is to provide lucid accounts written by authoritative researchers giving vision and insight in answering these questions on the subject of mechanics as it relates to solids.

The scope of the series covers the entire spectrum of solid mechanics. Thus it includes the foundation of mechanics; variational formulations; computational mechanics; statics, kinematics and dynamics of rigid and elastic bodies: vibrations of solids and structures; dynamical systems and chaos; the theories of elasticity, plasticity and viscoelasticity; composite materials; rods, beams, shells and membranes; structural control and stability; soils, rocks and geomechanics; fracture; tribology; experimental mechanics; biomechanics and machine design.

The median level of presentation is the first year graduate student. Some texts are monographs defining the current state of the field; others are accessible to final year undergraduates; but essentially the emphasis is on readability and clarity.

For further volumes:
http://www.springer.com/series/6557

Zengtao Chen • Cliff Butcher

Micromechanics Modelling of Ductile Fracture

Springer

Zengtao Chen
Department of Mechanical Engineering
University of New Brunswick
Fredericton, New Brunswick
Canada

Cliff Butcher
Department of Mechanical
and Mechatronics Engineering
University of Waterloo
Waterloo, Ontario
Canada

ISSN 0925-0042
ISBN 978-94-007-6097-4 ISBN 978-94-007-6098-1 (eBook)
DOI 10.1007/978-94-007-6098-1
Springer Dordrecht Heidelberg New York London

Library of Congress Control Number: 2013932705

© Springer Science+Business Media Dordrecht 2013
This work is subject to copyright. All rights are reserved by the Publisher, whether the whole or part of the material is concerned, specifically the rights of translation, reprinting, reuse of illustrations, recitation, broadcasting, reproduction on microfilms or in any other physical way, and transmission or information storage and retrieval, electronic adaptation, computer software, or by similar or dissimilar methodology now known or hereafter developed. Exempted from this legal reservation are brief excerpts in connection with reviews or scholarly analysis or material supplied specifically for the purpose of being entered and executed on a computer system, for exclusive use by the purchaser of the work. Duplication of this publication or parts thereof is permitted only under the provisions of the Copyright Law of the Publisher's location, in its current version, and permission for use must always be obtained from Springer. Permissions for use may be obtained through RightsLink at the Copyright Clearance Center. Violations are liable to prosecution under the respective Copyright Law.
The use of general descriptive names, registered names, trademarks, service marks, etc. in this publication does not imply, even in the absence of a specific statement, that such names are exempt from the relevant protective laws and regulations and therefore free for general use.
While the advice and information in this book are believed to be true and accurate at the date of publication, neither the authors nor the editors nor the publisher can accept any legal responsibility for any errors or omissions that may be made. The publisher makes no warranty, express or implied, with respect to the material contained herein.

Printed on acid-free paper

Springer is part of Springer Science+Business Media (www.springer.com)

Contents

List of Figures

Chapter 1
Introduction to Ductile Fracture Modelling

Plastic deformation is widely employed in metal shaping for various applications. Ideally, metals will be formed to a permanent shape as the final product or for subsequent processes. However, due to material heterogeneity, metals sometimes will fail prematurely through ductile fracture. During metal forming processes, the existence of initial second-phase particles and inclusions in metal alloys offers sites where damage can nucleate in the form of microvoids (Fig. 1.1b). With continued deformation, nucleation of damage will continue, accompanied by the growth of the existing voids, as shown in Fig. 1.1c. At a certain stage of deformation, the interaction of neighbouring voids triggers void coalescence which eventually leads to the formation of macro-cracking and failure through ductile fracture. While easily categorized into independent regimes, the reality is that these mechanisms are tightly interwoven and are related to many additional factors such as the second-phase particle/void distribution, void geometry, stress state, strain rate, material hardening and temperature (Horstemeyer et al. 2003). Furthermore, the highly localized and dynamic nature of ductile fracture makes observing void initiation and evolution extremely difficult. Due to the complexity of ductile fracture, damage-based constitutive models must employ many approximations and simplifications to reach a tractable analytical or numerical solution.

An extensive collection of modelling research exists in the literature that deals with ductile fracture. The following literature review will summarize the work done by other researchers that relates to the current monograph. The following review of the literature pertinent to the current research will focus on the three stages of damage-induced ductile fracture: void nucleation, growth and coalescence. An emphasis is placed upon the various micromechanical modeling techniques used to describe damage evolution within the constitutive models. The influence of microstructural heterogeneity and the limitations of the traditional approach to modelling ductile fracture are discussed, followed by a summary of the damage percolation model used in the present work.

Z. Chen and C. Butcher, *Micromechanics Modelling of Ductile Fracture*, Solid Mechanics and Its Applications 195, DOI 10.1007/978-94-007-6098-1_1,

© Springer Science+Business Media Dordrecht 2013

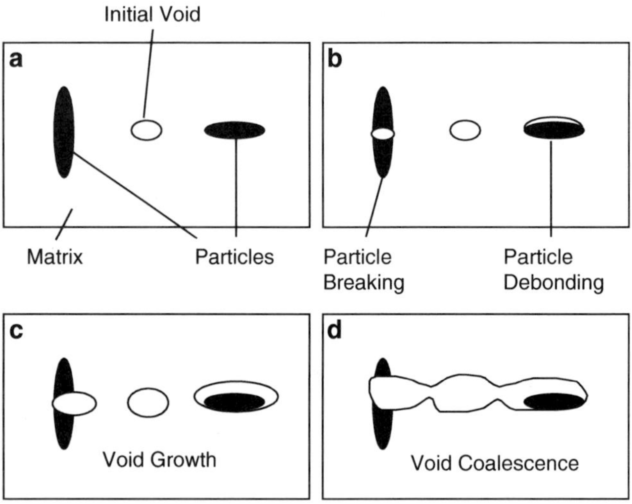

Fig. 1.1 Ductile damage evolution, (**a**) initial state; (**b**) void nucleation; (**c**) void growth; (**d**) void coalescence (Chen 2004)

1.1 Role of Material Damage

1.1.1 Void Nucleation

Void nucleation occurs as second-phase particles or inclusions crack or debond from the matrix material during plastic deformation as shown in Fig. 1.2. Void nucleation is difficult to model since the particle may crack, or debond (partially or completely) depending on the particle size, shape, distribution, stress state, temperature and strength of the particle and interface (Gurland and Plateau 1963; Gurland 1972; Cox and Low 1974; Tanaka et al. 1970a; Argon et al. 1975a, b, c; Goods and Brown 1979; Fisher and Gurland 1981a, b; Beremin 1981; Horstemeyer and Gokhale 1999; Horstemeyer et al. 2000a, 2003).

The size and shape of the second-phase particles/inclusions play a major role in the resulting nucleation mechanism. Large particles (radius greater than 1 μm) are prone to cracking at lower strains than smaller sub-micron particles which exhibit debonding (Thomason 1990). Nucleation occurs earlier for large particles because they contain more internal surface defects, have a larger interface and are irregularly shaped. An elongated particle aligned in the principal loading direction is more likely to crack while debonding is more probable if the elongated particle is loaded transversely. Small particles tend to be spherical or equiaxed and favour interface fracture due to the pile-up of dislocations at the particle-matrix interface. Equiaxed particles are less sensitive to the loading direction and nucleation is a competition between cracking and debonding (Lassance et al. 2006). Lower temperatures promote both cracking and debonding due to increased work hardening rates (Horstemeyer et al. 2003).

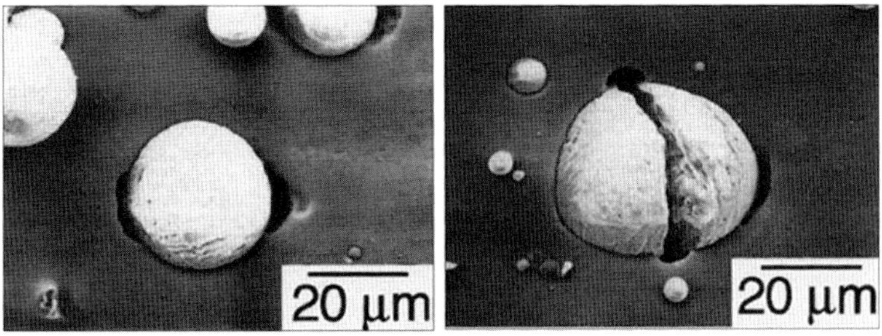

Fig. 1.2 Particle debonding (*left*) and cracking (*right*) in aluminum 6,061 reinforced with Al_2O_3 particles. The loading direction is horizontal (Reprinted with permission from Kanetake et al. 1995. Copyright 1995 W. S. Maney & Son Ltd.)

To further complicate matters, the nucleation mechanisms can be considered as either stress- or strain-controlled. The debonding of small particles is related to dislocation kinetics and can thus be considered strain-dependent (McClintock 1968). Particle debonding can also be described as stress-controlled since the separation of the interface is dependent upon the hydrostatic stress (Needleman 1987; Shabrov and Needleman 2002). Alternatively, the nucleation of large particles due to cracking can be described using a stress-controlled model since this is a brittle-type fracture (Thomason 1990). Overall, nucleation is a complex and material specific phenomenon which is difficult to model and observe experimentally. Consequently, void nucleation models typically follow a continuum-based approach where the objective is to model the average nucleation response of an alloy system. The shortcomings of these nucleation models are corrected (ideally) through the introduction of material specific calibration parameters.

1.1.2 Modeling Void Nucleation

Gurland and Plateau (1963) proposed one of the first nucleation models and adopted an energy approach for particle cracking. It was assumed that a particle fractures when the strain energy in the particle exceeds the surface energy of the newly formed crack surface. Tanaka et al. (1970a, b) performed a micromechanical analysis of a spherical inclusion subjected to uniaxial tension and found that the energy criterion of Gurland and Plateau (1963) is indeed a necessary condition for nucleation. Argon et al. (1975a, b, c) proved that the energy requirement is a necessary but not sufficient condition; the strength of the particle or interface must also be exceeded. The stress-based model of Argon et al. (1975a, b, c) assumed that the critical interface stress, σ_n, could be approximated as the sum of the flow and hydrostatic stress components

$$\sigma_n = \bar{\sigma} + \Sigma_{hyd} \tag{1.1}$$

where $\bar{\sigma}$ is the matrix flow stress and Σ_{hyd} is the macroscopic hydrostatic stress defined as

$$\Sigma_{\mathrm{hyd}} = \frac{\Sigma_1 + \Sigma_2 + \Sigma_3}{3} \tag{1.2}$$

Brown and Stobbs (1976) also developed a stress-based nucleation rule by relating the stress at the particle-matrix interface with dislocation density

$$\sigma_{\mathrm{n}} = \frac{\mu b_v \sqrt{p_d}}{2\pi} \tag{1.3}$$

where μ is the shear modulus, p_d is the dislocation density and b_v is the Burgers vector. The dislocation density can be related to the critical strain required to nucleate a particle based on particle size, r, and interface stress. The resulting nucleation criterion is

$$\varepsilon_{\mathrm{c}} = \frac{8\pi^2 {\sigma_{\mathrm{n}}}^2 r}{5.1 G^2 b_v} \tag{1.4}$$

The Brown and Stobbs (1976) model incorporated the effect of void size and observed that large particles tend to nucleate at smaller strains. Gurson (1977) suggested that nucleation can be expressed in terms of equivalent plastic strain based on Gurland's (1972) experimental results for nucleation in spheroidized steel. Following Gurson's result, Chu and Needleman (1980) theorized that the stress or strain required to nucleate a void follows a normal distribution. The strain-controlled nucleation model is expressed as

$$\dot{f}_{\mathrm{nucleation}} = \frac{f_{\mathrm{n}}}{s_{\mathrm{N}}\sqrt{2\pi}} \exp\left[-\frac{1}{2}\left(\frac{\bar{\varepsilon}^{\mathrm{p}} - \varepsilon_{\mathrm{N}}}{s_{\mathrm{N}}}\right)^2\right] \dot{\bar{\varepsilon}}^{\mathrm{p}} \tag{1.5}$$

where $f_{\mathrm{nucleation}}$ is the porosity contributed by void nucleation and an over-dot symbol indicates the time derivative; $\bar{\varepsilon}^{\mathrm{p}}$ is the equivalent plastic strain, f_{n} is the volume fraction of void nucleating particles and ε_{N} and s_{N} are the average and standard deviation of the strain at which void nucleation occurs. Traditionally, f_n is determined from metallurgical analysis of the material from the volume fraction of the second phase particles and inclusions. Theoretically, if plastic deformation could continue indefinitely, f_n would become the volume fraction of the second-phase particles because each particle would crack or debond to form a void. The volume fraction of nucleating particles can be estimated as

$$f_{\mathrm{n}} = W_{\mathrm{n}} \xi_{\mathrm{n}} f_p \tag{1.6}$$

where f_p is the volume fraction of the particles in the material, W_n is the aspect ratio of the nucleated void and ξ_n is the fraction of particles that nucleate voids. In the standard GT model, $W_n = 1$ and it is common to assume $\xi_n = 1$ so that all of the particles are available for nucleation. If an extended Gurson-based model that accounts for void shape is considered, $W_n = W_p$ for particle debonding while for particle cracking, W_n can be assumed to be a small value to represent a penny-shaped crack with $W_n \leq 0.01$ (Pardoen 2006). In most steels, the volume fraction of nucleating particles can be taken as the total fraction of manganese sulfide (MnS) particles that can be estimated from the sulfur content (Chen and Lambert 2003).

For stress controlled nucleation

$$\dot{f}_{\text{nucleation}} = \frac{f_n}{s_n \sigma_y \sqrt{2\pi}} \exp\left[-\frac{1}{2}\left(\frac{\dot{\sigma}_n - \sigma_N}{s_n \sigma_y}\right)^2\right] \dot{\sigma}_n \tag{1.7a}$$

where σ_y is the initial yield stress. An important distinction between the stress-controlled and strain-controlled nucleation models is that the nucleation stress, σ_n, cannot be considered to be a monotonically increasing quantity like the equivalent plastic strain. The hydrostatic stress may vary during a deformation process and lead to a nucleation stress that is lower than the previous nucleation stress. Therefore, the nucleation stress must be tracked during deformation with nucleation only occurring when the present nucleation stress exceeds its previous maximum value so that the nucleation stress increment is defined as

$$\dot{\sigma}_n = \left(\bar{\sigma} + \Sigma_{\text{hyd}}\right) - \sigma_n{}^{\max} \tag{1.7b}$$

It is also interesting to note than in a simple shearing operation, the stress-based nucleation model is equivalent to the strain-based nucleation model since the hydrostatic stress is zero and the nucleation stress becomes, $\sigma_n = \bar{\sigma} = h\bar{\varepsilon}^p$, where h is the hardening modulus of the matrix flow stress relation.

Beremin (1981) also proposed a stress-based nucleation criterion by employing the homogenization method of Bervellier and Zaoui (1979) to estimate the stress within the inclusions. The current stress within the particle is related to the stress in the matrix material as

$$\sigma_n = \Sigma_1 + k(\bar{\sigma} - \sigma_y) \tag{1.8}$$

where k is a parameter related to the composition of the particles and the loading direction. The Beremin (1981) model can be coupled into the normal distribution model in Eq. (1.7) by replacing the term σ_n with the stress obtained in Eq. (1.8).

The nucleation burst stress or strain, σ_N or ε_N, is the stress/strain where the maximum number of voids are nucleated according to the normal distribution and s_N is the standard deviation. The nucleation burst parameter is the dominant parameter and the standard deviation can be expressed as a function of the burst parameter. It is important to note that once the nucleation value is reached, only 50 % of the voids have nucleated as shown in Fig. 1.3.

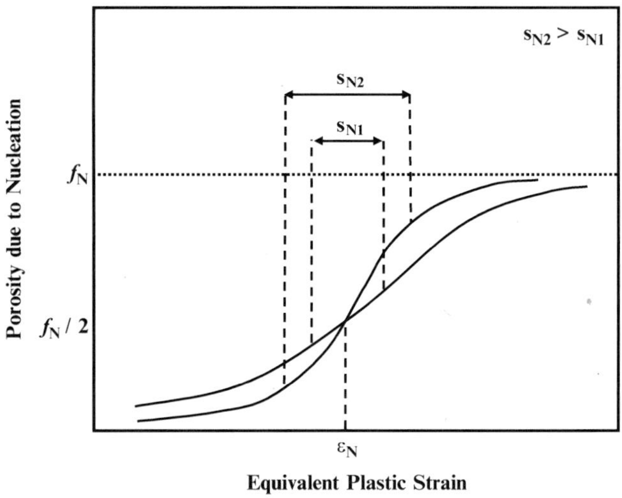

Fig. 1.3 Qualitative sketch of void nucleation in the strain-controlled statistical nucleation model of Chu and Needleman (1980)

The strain-controlled criterion in Eq. (1.5) has been widely used in the literature since it is easily implemented into finite-element codes. For aluminum-magnesium alloys, Chen (2004) suggests taking the standard deviation as 15–20 % the nucleation burst parameter. In the literature, strain-controlled nucleation is typically used with $\varepsilon_N = 0.3$ and $s_N = 0.1$ for many materials but these values are somewhat arbitrarily assumed and should be used with caution. Each alloy system will possess different nucleation mechanisms and characteristics, so the adoption of these universal nucleation parameters is not encouraged.

The stress-controlled nucleation model has not received much attention as it can cause non-normality in the flow rule and promote early flow localization (Saje et al. 1982). Recent works by Butcher et al. (2006, 2009) and Butcher and Chen (2011) have shown that the stress-controlled nucleation model in Eq. (1.7) can give good predictions of nucleation in aluminum-magnesium and advanced high strength steels. The advantage of using a stress-based nucleation model is that it is not sensitive to the load-path like the strain-controlled model and can be transferable to general stress states. Needleman (1987) numerically investigated the debonding of spherical particles within a characteristic unit cell and observed that only a fraction of the hydrostatic stress was present at the particle-matrix interface. A reduction factor, c, was introduced into the model of Argon et al. (1975a, b, c) in Eq. (1.1) as

$$\sigma_n = \bar{\sigma} + c\Sigma_{\text{hyd}} \tag{1.9}$$

where for a periodic particle distribution, $c \approx 0.35$ for spherical particles and $c \approx 0.44$ for cubic particles (Needleman 1987; Shabrov and Needleman 2002). Thomson (2001) extended the work of Needleman (1987) to investigate debonding within particle clusters and observed nucleation to be very sensitive to the cluster arrangement and density. This study highlighted the localized nature of the nucleation mechanism and the importance of accounting for the particle distribution on void nucleation, growth and coalescence. It is important to mention that the commonly used nucleation model of Chu and Needleman (1980) in Eqs. (1.5) and (1.7) does not account for the particle size or shape. None of the nucleation models described above account for the particle distribution.

For alloys that contain predominantly large particles that crack early during the deformation process, it can be assumed that nucleated voids can be treated as initial voids. This nucleation model is termed the "cluster" nucleation model by Zhang (1998). A significant advantage of this model is that the initial porosity is not required to be predetermined since the effective initial porosity will be calibrated to account for void nucleation. To determine the initial porosity, f_o, is parametrically varied until the finite-element and experimental results show good agreement. This model is physically realistic as long as the calibrated f_o is less than or equal to the sum of the actual initial porosity and volume fraction of particles in the material.

An alternative nucleation model to the cluster nucleation model is the so-called continuous nucleation model that assumes a constant fraction of voids are nucleated per stress or strain increment. This type of nucleation has been observed in steels (Gurland 1972) where the void nucleation trend is approximately linear with the equivalent plastic strain. The continuous model also has only one parameter, A_o, which is the volume fraction of nucleated voids per stress/strain increment. The stress and strain-controlled continuous nucleation models are expressed as

$$\dot{f}_{\mathrm{nucleation}} = A_o \dot{\sigma}_n \tag{1.10}$$

$$\dot{f}_{\mathrm{nucleation}} = A_o \dot{\varepsilon}^p \tag{1.11}$$

Although these models are quite simple, they have been shown to give good results for steel and aluminum alloys (Zhang et al. 2000; Brunet et al. 2005). As seen in Fig. 1.3, the continuous nucleation model emerges as a special case of the statistical nucleation model for materials with a large standard deviation for the nucleation stress/strain.

1.1.3 Void Growth

Void growth is the most understood stage of ductile fracture. Void growth is a continuum plastic deformation process and therefore easier to model than void nucleation or coalescence (Zhang and Skallerud 2010). While void nucleation and coalescence often occur suddenly, void growth is a relatively stable phase of deformation and easier to observe experimentally.

1.1.3.1 Modeling Void Growth

Voids are typically assumed to be initially cylindrical, spherical or spheroidal and change shape during loading. McClintock (1968) first analyzed a cylindrical void in an infinite work-hardening matrix and expressed void growth as a function of remote strain. Rice and Tracey (1969) analyzed a spherical void in an infinite, perfectly plastic material and demonstrated the exponential dependence of void growth upon the tensile hydrostatic stress (negative pressure). The intensity of the hydrostatic stress is defined by the stress triaxiality, T, which is the ratio of the hydrostatic to effective stress. Stress triaxialities of 0, 1/3 and 2/3 correspond to pure shear, uniaxial tension and equal-biaxial tension, respectively. Huang (1991) reworked the numerics in the Rice and Tracey (1969) model to better reflect growth at lower stress triaxiality. The resulting void growth rate is expressed as

$$\dot{f}_{\text{growth}} = 1.28 \exp\left(\frac{3}{2}T\right)\dot{\bar{\varepsilon}}^{\text{p}} \quad \text{for } T > 1 \tag{1.12a}$$

$$\dot{f}_{\text{growth}} = 1.28 T^{\frac{1}{4}} \exp\left(\frac{3}{2}T\right)\dot{\bar{\varepsilon}}^{\text{p}} \quad \text{for } \frac{1}{3} \leq T \leq 1 \tag{1.12b}$$

The void growth model of Rice and Tracey (1969) has been continually improved to yield results in good agreement with finite-element simulations (Le Roy et al. 1981; Budiansky et al. 1982; Huang 1991; Thomason 1993). More recently, Liu et al. (2003) extended the Rice-Tracey model using strain-gradient plasticity theory to account for the void size effect because small voids tend to grow more slowly than large voids. The growth of sub-micron sized voids is impeded by the toughening of the material at the void surfaces due to the pile up of dislocations. It is important to note that all of these void growth models are only valid for isolated voids as interactions with nearby voids are not included. The void growth rate is significantly larger within void clusters (Thomson et al. 1999; Bandstra and Koss 2008) where severe stress- and strain-gradients accelerate both void growth and coalescence.

Overall, the analytical solutions for void growth can provide decent agreement with the numerical results (Ragab 2004a) but are limited by their ability to account for void shape effects. To provide a void growth rule more suited for engineering applications, Ragab (2004a) compiled an exhaustive summary of the analytical and finite-element results for void growth and shape evolution in the literature to develop a set of semi-empirical equations for a wide range of materials, void shapes and stress states. Void growth in the Ragab (2004a) model is derived from a reduced form of the growth law in the Gurson (1977) model and will be discussed at length in Chap. 2.

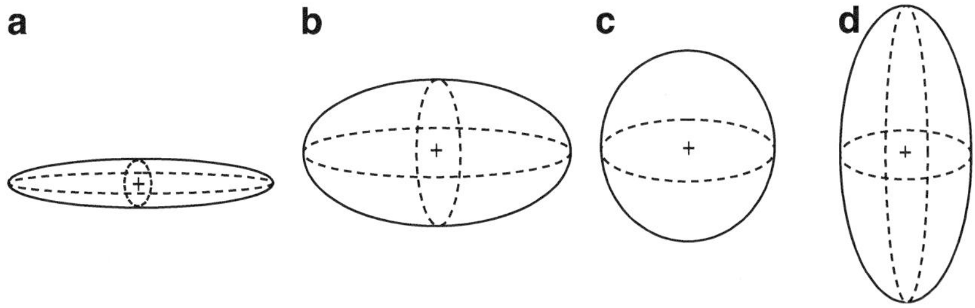

Fig. 1.4 Idealized spheroidal void shapes. (**a**) Penny-shaped spheroid: $W \to 0$, (**b**) oblate spheroid: $W < 1$, (**c**) sphere: $W = 1$, (**d**) prolate spheroid: $W > 1$

1.1.4 Void Shape Evolution

Naturally, the void can be expected to change its shape during deformation as the voids are not internally pressurized and must deform according to the plastic flow of the surrounding matrix material. The void growth rate and shape evolution are intrinsically linked as the void shape induces anisotropy, altering the stress state and the growth rate in a non-linear fashion. To simplify the modeling procedure, it is often assumed that the voids can be approximated as axisymmetric ellipsoids (spheroid) as shown in Fig. 1.4. In uniaxial tension, the void shape evolution is significant because initially penny-shaped voids can rapidly elongate in the principal loading direction into prolate voids. Alternatively, at high stress triaxialities (like found ahead of a crack tip), the voids will grow laterally and become flat and oblate, regardless of the principal loading direction. Penny-shaped voids are the exception to this trend and only appreciably grow in the opening direction even at high stress triaxiality. The mechanics for modeling void shape evolution are complex and no analytical model has yet been developed that can account for the void shape evolution in different stress states. As such, the adoption of heuristic parameters in these models is currently unavoidable (Keralavarma and Benzerga 2010).

Along with his study of semi-empirical rules for void growth, Ragab (2004a) consolidated the void shape evolution studies in the literature to develop a set of semi-empirical evolution laws using a total strain formulation. The semi-empirical evolution laws are functions of the nucleation strain, ε_{N}, equivalent plastic strain, initial void aspect ratio, W_i, hardening exponent, n, and stress triaxiality in the general form

$$W = g(W_{\mathrm{i}}, T, n, \bar{\varepsilon}^p, \varepsilon_{N}) \tag{1.13}$$

The void shape evolution laws of Ragab (2004a) in Eq. (1.11) will be described in detail in Chap. 2. Void growth and shape evolution in the Ragab (2004a) equations assume an isolated void and do not account for interactions with neighboring voids. The advantages of these semi-empirical expressions for void growth and shape evolution are that they are efficient to compute and ideally suited for implementation into a percolation-type model that may contain thousands of

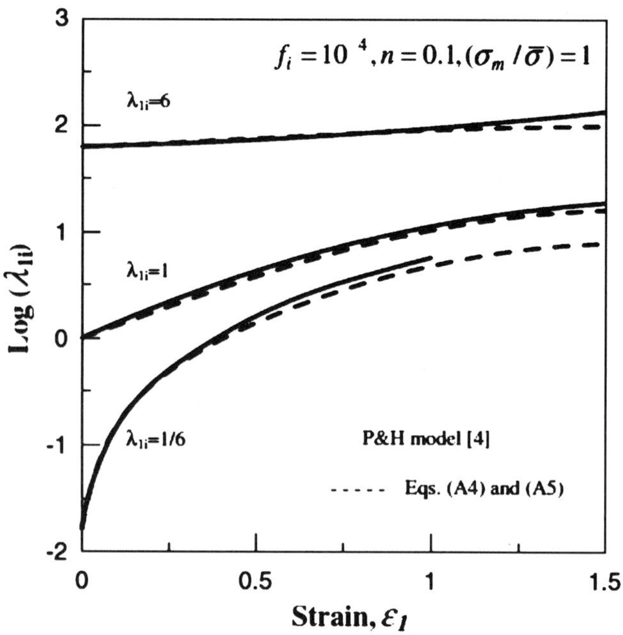

Fig. 1.5 Comparison of the semi-empirical equations for void growth and aspect ratio evolution with the results of Pardoen and Hutchinson (2000) at a stress triaxiality of unity (Ragab 2004b). The voids aspect ratio in this figure is denoted by λ_1

isolated voids of different shapes. A rate-based void evolution rule has also been widely used in the Gologanu-Leblond-Devaux (GLD) model (Gologanu et al. 1993, 1994) constitutive model with good accuracy and this rule could be used instead of the Ragab (2004a) model. The Ragab (2004a) model has been shown to give very good agreement with the numerical results of Pardoen and Hutchinson (2000) who employed the sophisticated GLD constitutive model (Fig. 1.5).

1.1.5 Void Coalescence

Void coalescence is a sudden and rapidly occurring phenomenon where the microvoids in a material rapidly link up to form microcracks that propagate throughout the material causing sudden failure. Void coalescence depends on many factors such as the initial porosity, stress triaxiality, void size, shape, spacing and material hardening. There are three idealized mechanisms of void coalescence: (i) primary void impingement, and failure of the inter-void ligament due to (ii) shearing, or (iii) necking. It is generally agreed that void coalescence is a combination of inter-ligament necking and shearing (Barsoum and Faleskog 2007a, b) at low to moderate stress triaxialities with necking coalescence occurring exclusively at high triaxialities. Coalescence by primary void impingement is theoretically possible but instability and failure of the ligament would

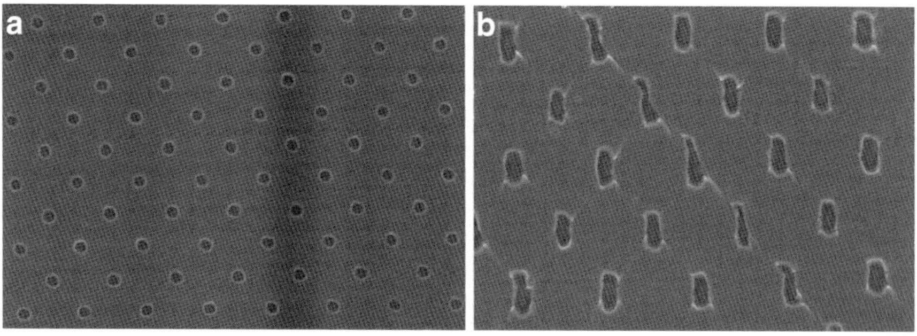

Fig. 1.6 (**a**) Triangular array of cylindrical holes in AA-5052 sheet and (**b**) coalescence due to shear localization under tensile loading. The loading direction is vertical and the holes have an initial diameter of 10 μm (Reprinted with permission from Weck 2006. Copyright 2006 Weck)

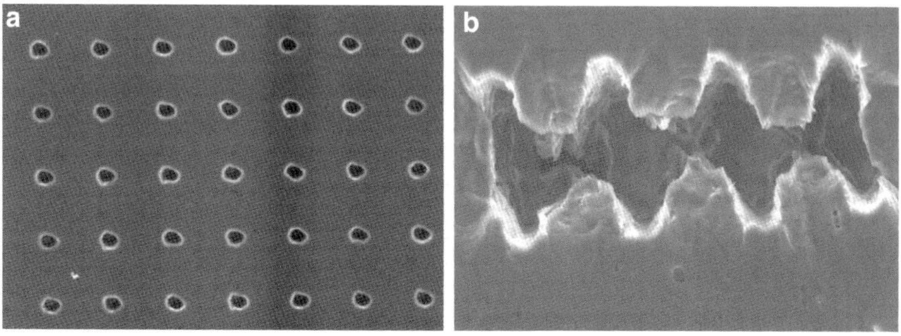

Fig. 1.7 (**a**) Rectangular array of cylindrical holes in AA-5052 sheet and (**b**) coalescence due to localized necking under tensile loading. The loading direction is vertical and the holes have an initial diameter of 10 μm (Reprinted with permission from Weck 2006. Copyright 2006 Weck)

occur before the voids would actually come into contact. Void coalescence due to inter-ligament shearing and necking are shown in Figs. 1.6 and 1.7, respectively.

The fracture surface for failure due to ligament necking (Fig. 1.8a) exhibits the classic dimpled structure associated with ductile fracture while shearing exhibits a smoother surface (Fig. 1.8b) as the strong shear stress smears out the voids. Inspection of the fracture surfaces reveals the presence of many void sizes indicating that coalescence involves secondary populations of smaller voids (Tvergaard 1982; Sun 1995b; Brocks et al. 1996; Faleskog and Shih 1997; Perrin and Leblond 2000; Fabregue and Pardoen 2008). These smaller voids nucleate between larger voids and link up to cause rapid failure by compromising the integrity of the ligament and hastening its collapse as seen in Fig. 1.9. The investigation of the role of secondary voids on ductile fracture has become an active area of research since the coalescence within void clusters can no longer be ignored (Fabregue and Pardoen 2008). Aluminum-magnesium alloys are particularly

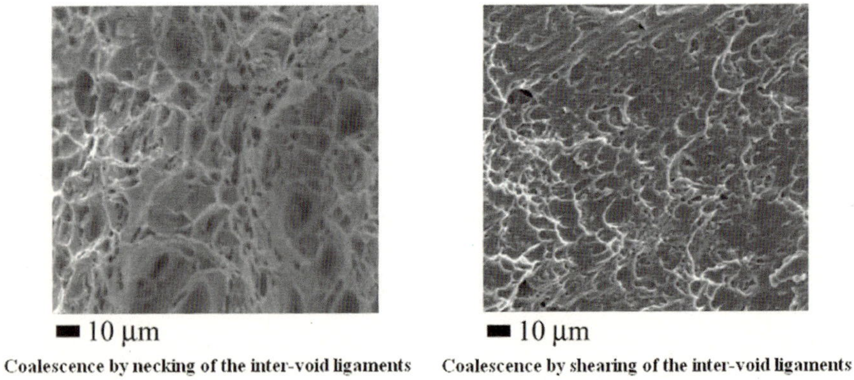

Fig. 1.8 Scanning electron microscope fractographs of Weldox 420 steel illustrating two different rupture mechanisms (Reprinted with permission from Barsoum and Faleskog 2007a, b. Copyright 2007 Elsevier)

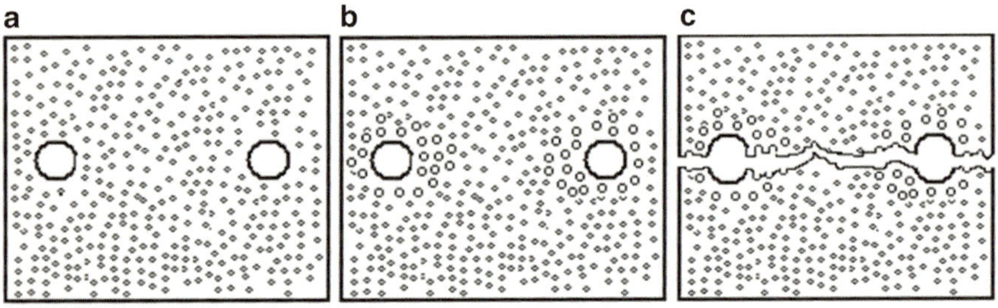

Fig. 1.9 (**a**) Dual population of large and small voids at beginning of deformation; (**b**) activation and growth of small voids in a region of high stress surrounding the large voids; (**c**) fracture as the voids link up throughout the material (Butcher 2011)

sensitive to this mechanism since fracture can be triggered by the coalescence between two or three primary void clusters (Worswick et al. 2001; Chen 2004).

Whether failure occurs due to necking or shearing is largely dependant upon the orientation of the void ligament relative to the principal loading direction. Tensile void coalescence occurs as a transition to uniaxial straining and gives rise to the flat, dimpled fracture surface commonly found in ductile materials (Faleskog and Shih 1997). Shear coalescence (Tvergaard 1982; Faleskog and Shih 1997; Barsoum and Faleskog 2007a, b) has been attributed to the formation of shear bands at the microscale and is favored at low stress triaxiality, low strain biaxiality and low hardening.

1.1.5.1 Modelling Void Coalescence

Coalescence models have typically been proposed based upon geometrical considerations where coalescence occurs once a material specific value has been reached. The critical geometric parameter could be the void radius, spacing and/or porosity (McClintock 1968; Rice and Tracey 1969; Brown and Embury 1973; Le Roy et al. 1981; Tvergaard and Needleman 1984; Rousselier 1987). The majority of work on void coalescence has been performed by assuming the existence of a critical porosity. Tvergaard and Needleman (1984) theorized void coalescence as the termination of stable void growth and nucleation where the voids suddenly link-up throughout the material resulting in failure. A critical porosity at coalescence, f_c, and at failure, f_f, were theorized as intrinsic material parameters and used to identify the transition to coalescence and subsequent loss of load bearing capacity.

Subsequent investigations by Becker (1987) and Koplik and Needlman (1988) observed that the critical and failure porosities were dependent upon the material, stress state and work hardening. Recent works have shown that critical porosity cannot be considered a material constant because it is related to the initial porosity, void shape, spacing, clustering, stress state, and strain (Zhang and Niemi 1994a, b; Pardoen and Delannay 1998a; Zhang et al. 2000; Pardoen and Hutchinson 2000; Bandstra and Koss 2008). While attractive for its simplicity, the assumption of a critical porosity is unrealistic because it contains no physical foundation.

To account for the physical mechanism of void coalescence, Thomason (1985a, b, 1990) theorized coalescence as the competition between stable, homogeneous and unstable, localized deformation modes. Initially, the presence of voids in the material is negligible and the deformation mode is homogeneous. The stress required to achieve a localized deformation mode decreases with continued plastic deformation as voids nucleate and grow. The onset of coalescence is identified when the stress required for both homogeneous and localized deformation modes becomes equal. At this point, the plastic limit-load of the ligament has been reached and deformation becomes localized in the ligament resulting in necking failure. Unlike the previous phenomenological coalescence models, the plastic limit-load criterion was rigorously derived from the void geometry and stress states associated with coalescence.

The pioneering work of Thomason (1990) has greatly contributed to the understanding and modeling of coalescence due to internal necking. The plastic limit-load model has been validated by numerous researchers through comparison with finite-element models (Pardoen and Hutchinson 2000; Zhang et al. 2000; Benzerga 2002; Zhang and Skallerud 2010). The plastic limit-load model has been extended to account for material hardening (Pardoen and Hutchinson 2000) and oblate or flat voids (Benzerga 2002). The modified plastic limit-load criterion of Benzerga (2002) predicts necking failure of the inter-void ligament when the following condition is satisfied

$$\frac{\Sigma_1}{\bar{\sigma}(1 - \kappa_{uc}\chi^2)} \geq C_f(W,\chi) \tag{1.14}$$

where $\kappa_{uc} = 1$ for an axisymmetric unit cell and $\kappa_{uc} = \pi/4$ for a cubic cell. The plastic constraint factor, C_f, is a function of the void aspect ratio, W, and the void spacing ratio, χ. Several variants of C_f, have been proposed in the literature to account for different void geometries. The constraint factor of Pardoen and Hutchinson (2000) is widely used and is expressed as

$$C_{f-PH} = \left(\alpha \left(\frac{1-\chi}{W\chi} \right)^2 + \frac{\beta}{\sqrt{\chi}} \right) \tag{1.15}$$

where $\alpha = 0.1 + 0.22n + 4.8n^2$ and $\beta = 1.24$.

A limitation of the Thomason (1990) and Pardoen and Hutchinson (2000) models is that infinite ductilities are predicted for flat voids as $W \to 0$. To remove this limitation, Benzerga (2002) compiled the unit cell results of Golaganu (1997) who employed richer velocity fields in the limit analysis than Thomason (1990) in analyzing the cell behaviour. An heuristic extension was introduced into the model to obtain a plastic constraint factor of the form

$$C_{f-B} = 0.1 \left(\frac{\chi^{-1} - 1}{W^2 + 0.1\chi^{-1} + 0.02\chi^{-2}} \right)^2 + \frac{1.3}{\sqrt{\chi}} \tag{1.16}$$

The ligament size ratio or void spacing ratio is defined as the ratio of the lateral void radius, R_x, to the lateral void spacing, L_x, for a periodic arrangement of 3-D unit cells as shown in Fig. 1.10. The ligament size ratio evolves with the microstructure and is related to the unit cell geometry as

$$\chi = \frac{R_x}{L_x} = \left(\frac{f}{\gamma_{\text{cell}}} \frac{\lambda}{W} \right)^3 \tag{1.17}$$

where λ is the aspect ratio of the unit cell and $\gamma_{\text{cell}} = 2/3$ for an axisymmetric unit cell and $\gamma_{\text{cell}} = \pi/6$ for a cubic unit cell (Fig. 1.10). Assuming the matrix material of the unit cell is incompressible, the evolution of the unit cell aspect ratio, $\lambda = L_y/L_x$, can be expressed in rate form as

$$\dot{\lambda} = \frac{3}{2} \lambda \dot{E}_1 \tag{1.18}$$

where $\dot{E}_1$ is the first principal strain rate.

While void coalescence due to internal necking of the inter-void ligament can be characterized using the plastic limit-load criterion, no such model exists for ligament shearing or combined necking and shearing. Shear coalescence has often been observed experimentally but few criteria have been developed aside from the work of McClintock (1968) and Richelsen and Tvergaard (1994). The importance of shear coalescence has been overlooked since the numerical investigations of void growth and coalescence focused on a range of triaxiality of one to three where coalescence occurs due to necking failure (Barsoum and Faleskog 2007a, b). The development of void evolution and coalescence models in shear has gained traction

Fig. 1.10 Cubic unit cell containing an axisymmetric elliptical void ($R_x = R_z$) subjected to uniaxial loading

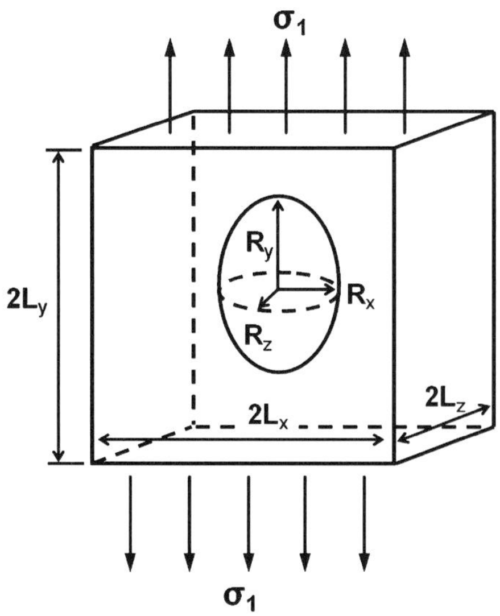

in recent years and is now an active area research area. Xue (2007) has modified the McClintock (1968) coalescence model to introduce a shear damage parameter for coalescence. Alternatively, Butcher and Chen (2009a) coupled the McClintock (1968) shear model with the plastic limit-load model of Benzerga (2002) to account for combined shear and necking coalescence. Most recently, Schyvaerts et al. (2011) have modelled internal necking coalescence in combined tension and shear, and Neilson and Tvergaard (2011) have considered shearing between both primary and secondary voids.

Recently, a physically-motivated post-coalescence model has been developed from unit-cell simulations by Scheyvaerts et al. (2010) for coalescence due to internal necking. This model was developed for implementation into the GLD (1997) damage model that accounts for the void shape, spacing and aspect ratio of the unit cell. Since the post-coalescence model of Tvergaard and Needleman (1984) can effectively mimic the loss of the load-bearing capacity, the model of Scheyvaerts et al. (2010) has been re-worked here so that the failure porosity can be estimated from the microstructure at coalescence as

$$f_f = W_f \frac{f_c}{W_c \chi_c^3} \tag{1.19}$$

where χ_c is the void spacing ratio at coalescence, W_c is the void aspect ratio at coalescence and the W_f is the aspect ratio at failure, which depends upon the geometric parameters at coalescence defined as $\gamma_1 = 0.5/\chi_c W_c$ and $\gamma_2 = 4\chi_c$, leading to

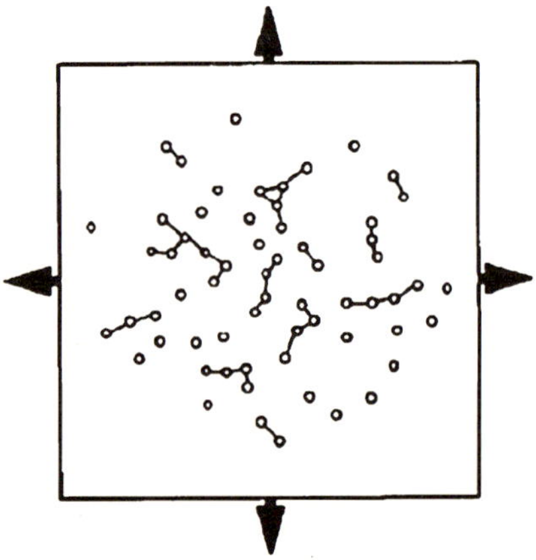

Fig. 1.13 Experimentally observed flow localization paths of Geltmacher et al. (1996) in test specimens with a minimum hole spacing of two hole diameters and loaded in uniaxial tension (*left*) and equal-biaxial tension (*right*) (Reprinted with permission from Geltmacher et al. 1996. Copyright 1996 Elsevier)

1998; Bandstra and Koss 2008; Toi and Kang 2005; Weck 2006; Jones et al. 2007; Hu et al. 2008). The influence of the distribution of voids and the stress state is clearly shown in Fig. 1.13 from Geltmacher et al. (1996) where the fracture path is different in biaxial tension because it activates coalescence paths in two directions. In uniaxial tension, coalescence occurs only in ligaments that are transverse to the loading direction. These studies are informative and can provide insight into the behaviour within void clusters. However, experimental tests are limited to approximating cylindrical holes as voids and these holes are significantly larger than actual voids. A notable exception to this limitation is the work of Weck (2006) who used laser-drilled holes to obtain holes comparable to actual void size with a diameter of 10 μm.

Another avenue to characterize void evolution within clusters is by creating unit cells that contain multiple voids as shown in Fig. 1.14 (Thomson et al. 1999; Horstemeyer et al. 2000a, b; Zhang and Chen 2007). These simulations are valuable for determining the influence of the cluster geometry on void growth and coalescence but are not representative of the true microstructure as it assumes a periodic arrangement of void clusters. As Fig. 1.13 has demonstrated, coalescence in a random distribution of voids does not resemble that of a unit cell. These types of simulations are only useful to visualize deformation within clusters or to obtain benchmarks for the determination and/or calibration of void evolution models.

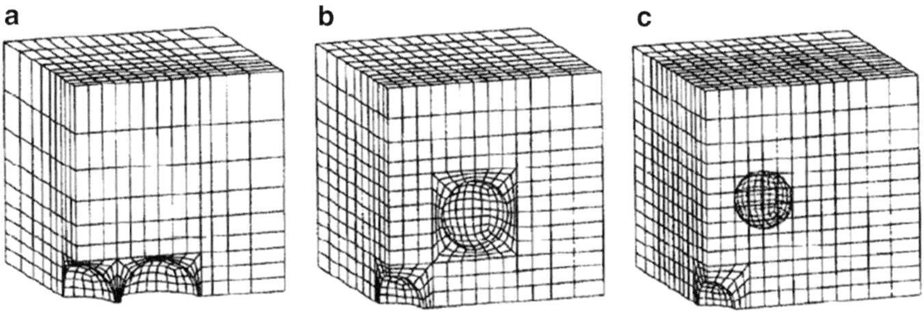

Fig. 1.14 Thomson's 3D unit cell model used to model the three void cluster types, (**a**) linear, (**b**) planar, and (**c**) quasi-spherical (Reprinted with permission from Thomson et al. 1999. Copyright 1999 Elsevier)

1.4 Damage Percolation Modeling

1.4.1 Role of the Void Distribution

The over-arching theme of this literature review has been to emphasize that the assumption of a periodic void distribution is employed in all facets of the ductile fracture modelling research area. Significant efforts have gone into refining these models and much progress has been made. However, there is a natural limit to the accuracy of these models because they cannot properly account for the physical mechanism of the failure: void initiation and evolution within heterogeneous clusters. A cynical perspective is that these models are no better than a phenomenological continuum damage model that has the same number of adjustable parameters. In addition, the continuum damage model is also not constrained by ensuring the parameters are realistic, just that they work. Certainly a void-based constitutive model is more realistic but it is debatable that is more useful. It is difficult to defend the advantages of a void-based constitutive model compared to a continuum damage model except from the academic argument of scientific purity. To obtain physically realistic fracture predictions one must first have a physically realistic model.

While rather unattractive from a modeling perspective, the role of the heterogeneous particle distribution cannot be neglected when formulating a ductile fracture model. Unfortunately, the assumption of a periodic void/particle distribution is a necessary evil in order to obtain a tractable geometry from which to formulate a yield criterion and void evolution models. Even if a unit cell contained a void cluster as studied by Thomson et al. (1999), a myriad of simulations would be required to accommodate the vast number cluster types, voids sizes, shapes, and spacings found in a typical microstructure. With this in mind, a modeling technique is required that can be applied to a general microstructure with a random distribution of voids and second-phase particles or inclusions.

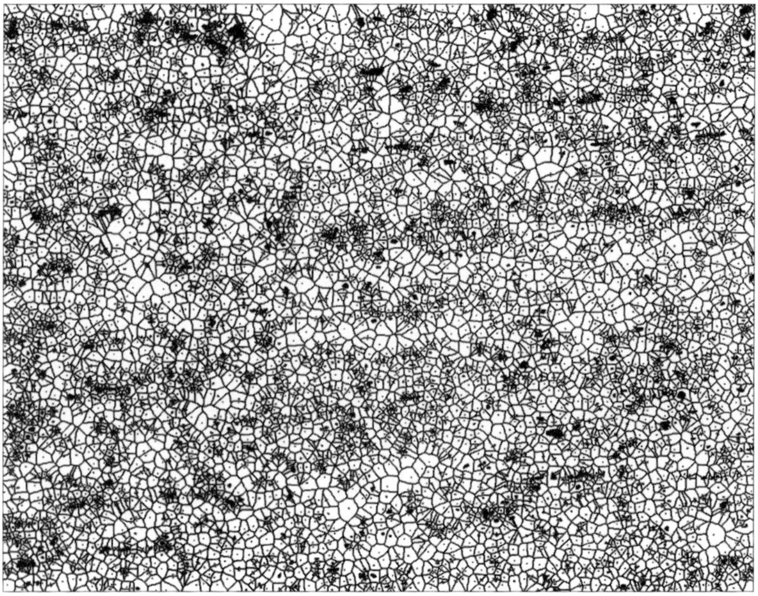

Fig. 1.15 5,500 × 4,250 pixel large-scale tessellated second phase particle field (corresponding to a physical size of 2.0 × 1.6 mm) of Al-Mg alloy AA5182 used in the RVE study. The rolling direction and loading directions are horizontal (Chen 2004)

1.4.2 Damage Percolation Modeling

The limitation of homogenous damage models can be overcome using a technique known as "damage percolation modeling" developed by Worswick et al. (2001) and extended by Chen (2004) and Butcher and Chen (2009a, b, c). In the damage percolation model, digital imaging techniques are used to obtain the actual particle distribution in a material (Fig. 1.15). The image tessellation algorithm approximates the void/particle/inclusion as an ellipse and extracts the particle size, shape, location and cluster, and nearest neighbour lists. Using this information, micromechanical models are applied to characterize void and crack formation leading to failure at the individual particle scale (Figs. 1.16 and 1.17). In this approach, all three stages of ductile fracture are captured with damage originating within heterogeneous particle clusters. Furthermore, ductile fracture in the percolation model naturally accounts for the presence of multiple void sizes, shapes and clusters which cannot be included using a traditional damage model. The damage percolation model is a powerful technique in material modeling as it directly relates changes in the local microstructure to the overall material behaviour.

Previous percolation models of Worswick et al. (2001) and Chen (2004) show great promise in predicting damage-induced ductile fracture but the framework requires further development and validation before it can make deterministic

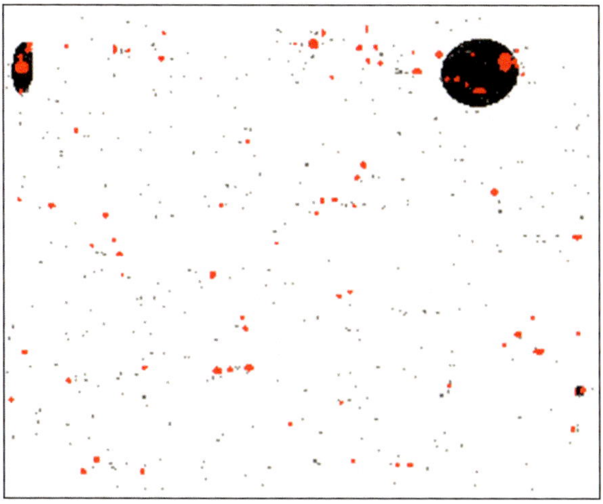

Fig. 1.16 Predicted damage prior to profuse coalescence within a particle field of AA5182 sheet under plane strain loading at 23.9 % major strain. The *grey* ellipses are second phase particles, the *red* ellipses are nucleated voids and the *black* ellipses represent voids or 'cracks' formed by coalescence. The rolling and loading directions are horizontal (Reprinted with permission from Butcher and Chen 2009a, b, c. Copyright 2009 IOP Publishing)

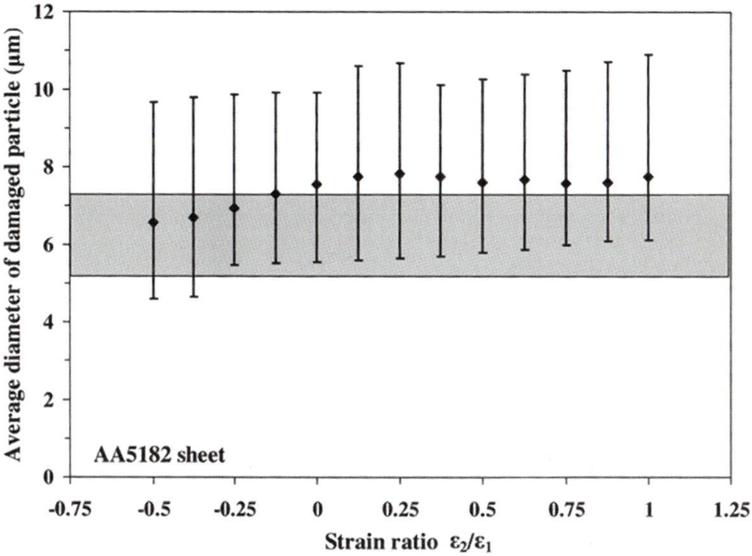

Fig. 1.17 Comparison of the predicted average diameter of the broken particles with the experiment data of Hadianfard et al. (2008) using the 2-D percolation model of Butcher and Chen (2009a, b, c). The experimental range for the broken particles is presented as a shaded band and three particle fields were evaluated in the model for each load ratio

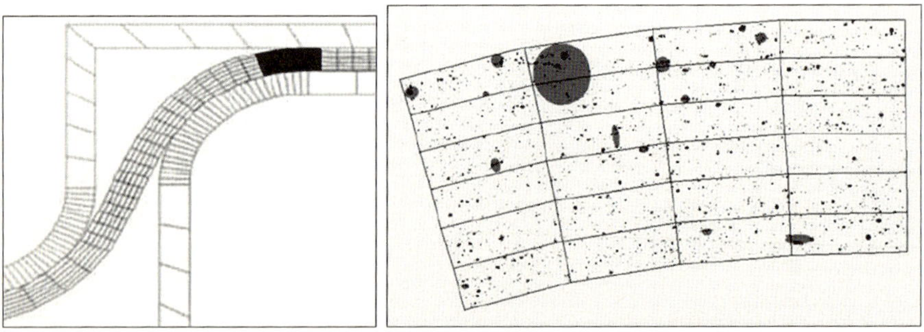

Fig. 1.18 Predicted damage within an AA5182 aluminum particle field during a stretch flange forming operation. The stretch flange finite-element model is axisymmetric and a two-dimensional particle field was used (Reprinted with permission from Chen et al. 2003. Copyright: Elsevier)

predictions of fracture in a metal forming operation. To date, the percolation models have been restricted to two-dimensional particle fields and do not account for void orientations and neglect the stress state which is critical for modeling void nucleation, growth and coalescence. Furthermore, the stress state evolves considerably during deformation as a consequence of both (i) material hardening and (ii) softening due to the presence of voids and geometric considerations. Since the stress state is not considered, all aspects of these percolation models are strain-controlled. This assumption significantly impedes the ability of the model to quantitatively predict fracture under general loading conditions because strain-based measures cannot uniquely define the stress state.

To apply the strain-based percolation model to a metal forming operation, Chen (2004) extracted the nodal displacements from a finite-element simulation and mapped them onto a large particle field for use in the percolation model. In this manner, the strain gradients throughout the material could be re-created within the particle field with the percolation model acting as post-processor for the finite-element model. This approach demonstrated the vast potential of the percolation model as localized damage within heterogeneous clusters was reflected in a metal forming operation (Fig. 1.18). The approach of Chen (2004) represents a loose-coupling between the micro- and macro-mechanical behaviour of the material and serves as the motivation for this work as we seek to develop a fully-coupled damage percolation model.

The percolation model of Chen (2004) was extended by Orlov (2006) to include a three-dimensional particle field obtained using x-ray micro-tomography. The experimental tomography data was used to control damage evolution in the percolation model to obtain predictions consistent with the experiment. The percolation model of Orlov (2006) should be considered a proof-of-concept rather than a predictive model since simple micromechanical models were employed and its strong performance is a function of using the experiment data to recreate the damage distribution which is an effective calibration that offsets its limitations. A damage percolation model that employs a rigorous treatment of each stage of void evolution is required before the model can be considered predictive.

Chapter 2
Averaging Methods for Computational Micromechanics

The averaging or homogenization process is the foundation of all unit cell models and any yield criterion derived from them. In porous materials, the presence of a void gives rise to an overall response of the bulk or aggregate material that is different than that of a damage-free material. The averaging process is employed to transition from the micro-scale (unit-cell) to the macro-scale to quantify the overall response of the material and these average quantities are frequently referred to as "macroscopic" quantities. The study of homogenization techniques is a very rich field and a proper treatment is outside the scope of this book and the interested reader is referred to basic textbooks on plasticity as well as the work of Eshelby (1957), Mori and Tanaka (1973), Nemat-Nasser (1993a, b) as well as Ponte Casteneda and Suquet (1998). Only a brief explanation of the extremum theory of plasticity is provided here since a great deal of attention will be paid to the application of upper and lower bound-based yield criteria for porous ductile materials.

2.1 Defination of Average Stress and Strain

Consider an arbitrary representative volume element (RVE) or unit cell that is large enough to statistically represent the material behaviour of the aggregate material. For an arbitrary unit cell geometry, the macroscopic stresses are obtained as the volume average of the microscopic stresses as

$$\Sigma_{ij} = \frac{1}{V}\int_V \sigma_{ij}(\boldsymbol{x})dV \tag{2.1}$$

where $\sigma_{ij}(x)$ are the micro-stress fields in a unit cell at a point, $\boldsymbol{x}$; Σ_{ij} is the macroscopic stress tensor and V is the volume of the unit cell. This integral can be converted to an integration of the applied surface traction vector, $T_i(x)$, over the outer boundary of the unit cell, S, using the Gauss theorem

Z. Chen and C. Butcher, *Micromechanics Modelling of Ductile Fracture*, Solid Mechanics and Its Applications 195, DOI 10.1007/978-94-007-6098-1_2,
© Springer Science+Business Media Dordrecht 2013

$$\Sigma_{ij} = \frac{1}{V}\int_V \sigma_{ij}(x)dV = \frac{1}{S}\int_S T_i(x)n_j dS \tag{2.2}$$

where n_j are the components of surface normal direction vector corresponding to $T_i(x)$. The average strain rate fields are similarly defined as

$$\dot{E}_{ij} = \frac{1}{V}\int_V \dot{\varepsilon}_{ij}(x)dV = \frac{1}{V}\int_S \frac{1}{2}\left[v_i(x)n_j + v_j(x)n_i\right]dS \tag{2.3}$$

where $\dot{\varepsilon}_{ij}$ and $\dot{E}_{ij}$ are the respective microscopic and macroscopic strain rates and v_i are the components of the velocity vector on the outer surfaces of the unit cell. The average work rate is thus defined by

$$\dot{W} = \frac{1}{V}\int_V \sigma_{ij}\dot{\varepsilon}_{ij}dV = \frac{1}{V}\int_S T_i v_i dS = \Sigma_{ij}\dot{E}_{ij} \tag{2.4}$$

Based on the energy Eq. (2.4) and the variational method, the stress strain relation can be obtained using the following equations

$$\dot{E}_{ij} = \frac{\partial \dot{W}}{\partial \Sigma_{ij}} \tag{2.5}$$

$$\Sigma_{ij} = \frac{\partial \dot{W}}{\partial \dot{E}_{ij}} \tag{2.6}$$

2.2 Fundamentals of a Constitutive Model for Plasticity

The constitutive equation of a material is used to relate the material response (stress) to an applied deformation (strain). The constitutive equation, or plastic potential, ϕ, is a scalar function of the stress tensor along with some internal variables arranged in a vector, $\boldsymbol{\alpha}$. The general form of the plastic potential for a perfectly plastic material is expressed as

$$\phi\left(\sigma_{ij}, \alpha_i\right) = \sigma_0 \tag{2.7}$$

From the concept of the plastic potential (Chakrabarty 1987), the microscopic plastic strain increment, $d\varepsilon_{ij}$, is related to the normal of the yield surface with a scalar factor known as the plastic multiplier, $d\lambda$, and thus defines the "flow rule" of the material as

$$d\varepsilon_{ij}^{p} = d\lambda \frac{\partial \phi(\sigma_{ij}, \alpha_i)}{\partial \sigma_{ij}} \tag{2.8}$$

This is known as an associated flow rule because the plastic potential is the same as the yield criterion. Non-associated flow rules can also be implemented but are not considered in this book.

2.3 Normality and Convexity of the Yield Surface

The fundamentals of the mathematical theory of plasticity require that the yield surface be convex since the work dissipation for a deforming material is always positive. They also constitute the basis for many minimum/maximum theorems in plasticity, including the upper- and lower-bound theorems to be discussed and utilized throughout this book. The condition of normality is fundamental to the development of the plastic flow rule since the vector of the plastic strain increment is always normal to the yield surface. The convexity condition forms the basis of the principle of maximum plastic work. Taking the yield locus as an example, convexity can be formulated as

$$(\Sigma_1 - \Sigma_2) : \left(\frac{\partial \Phi}{\partial \Sigma}\right)_{\Sigma_1} \geq \Phi(\Sigma_1) - \Phi(\Sigma_2) \tag{2.9}$$

where Σ_1 and Σ_2 indicate two different macroscopic stress tensors and $\Phi(\Sigma, \alpha)$ is the macroscopic plastic potential.

2.4 Principle of Virtual Work

A stress field can be said to be statically admissible if it satisfies the equilibrium equations

$$\frac{\partial \sigma_{ij}}{\partial x_i} = 0 \tag{2.10}$$

If we consider an admissible velocity field, v_i, to be independent of the equilibrium stress field, the work rate done by the surface tractions and its volume-based equivalent are

$$\int T_i v_i dS = \int n_i \sigma_{ij} v_i dS \tag{2.11}$$

$$\int n_i\sigma_{ij}v_i dS = \int \sigma_{ij}\frac{\partial v_i}{\partial x_j}dV = \int \sigma_{ij}\dot{\varepsilon}_{ij}dV \tag{2.12}$$

Equations (2.11) and (2.12) define the principle of virtual work which states that the rate of work done by the surface tractions with any virtual velocity field is equal to the rate of dissipation of internal energy by the stress field corresponding to the surface tractions

$$\int T_i v_i dS = \int \sigma_{ij}\dot{\varepsilon}_{ij}dV \tag{2.13}$$

2.5 Principle of Maximum Plastic Work

The work rate of plastic deformation is

$$dW = \sigma_{ij}d\varepsilon_{ij} \tag{2.14}$$

For a given plastic strain increment, $d\varepsilon_{ij}^p$, the corresponding stress σ_{ij}, can be determined from the normality rule and the yield function and denoted as point P in stress space. Now consider an arbitrary stress, σ_{ij}^*, that is statically admissible and denoted by a point P^* that lies on or inside the yield surface. From the principle of virtual work, the difference between the incremental plastic works done by the two stresses can be determined as follows

$$dW = \left(\sigma_{ij} - \sigma_{ij}^*\right)d\varepsilon_{ij}^p \tag{2.15}$$

since the yield surface is strictly convex the scalar product is positive. Hence,

$$dW = \left(\sigma_{ij} - \sigma_{ij}^*\right)d\varepsilon_{ij}^p \geq 0 \tag{2.16}$$

Equation (2.16) represents the principle of maximum plastic work: the actual work done in a given plastic strain increment is greater than or equal to the work done by an arbitrary stress that is less than or equal to the yield limit.

2.6 Extremum Theorems in Plasticity

The extremum principles in plasticity arise from the comparison of the work dissipations (or work rates) associated with actual stress fields and velocity fields. A lower bound estimate can be obtained by applying a statically admissible stress field at the cell boundary while an upper bound estimate is obtained by applying a kinematically admissible velocity field. The principle of virtual work is used to obtain the difference between the actual and possible fields.

For a mass of material with a volume, V and bounded by a surface S, the equation of energy conservation is

$$\int T_i v_i dS = \int \sigma_{ij}\dot{\varepsilon}_{ij} dV + \int \tau[v] dS_D \tag{2.17}$$

where S_D is the surface on which a discontinuity of velocity occurs and τ and $[v]$ are the shear stress and relative velocity on the dislocation surface S_D (Kachanov 1971). Equation (2.17) holds for any continuous medium in equilibrium and the velocities and stresses are, in general, not related. In other words, Eq. (2.17) is applicable to both the actual stress distribution σ_{ij} and to any kinematically possible velocity field v_i^*; it is also applicable to both to the actual velocity distribution v_i and a statically admissable stress field, σ_{ij}^*.

2.6.1 Upper Bound Solution

According to Kachanov (1974), the upper bound principle states that the total rate of work attains an absolute minimum for the actual velocity field when

$$\int T_i^* v_i dS \geq \int T_i v_i dS \tag{2.18}$$

where T_i^* are surface tractions solutions corresponding to any kinematically possible velocity fields v_i. Upper bound yield functions for porous materials can be obtained by constructing a uniform velocity field on the outer surface of a unit cell. The upper bound theorem is used to solve for the forces that arise due to deformation from a kinematically possible velocity field. In general, upper bound solutions are easier to obtain analytically because they represent solving for the stress (load) from an applied strain (deformation).

2.6.2 Lower Bound Solution

As stated by Kachanov (1974): the rate of work done by the actual surface tractions on prescribed velocities is greater than or equal to the rate of work developed by surface tractions corresponding to any statically admissible stress

$$\int T_i^* v_i dS \leq \int T_i v_i dS \tag{2.19}$$

where T_i^* are obtained from any statically admissible microscopic stress fields. Lower bound yield limits for porous materials can be obtained by constructing stress fields instead of velocity fields. Theoretically, the yield strength obtained using a yield criterion derived using the lower bound formulation will always be lower than the real yield stress.

2.7 Gurson's Upper Bound Solution for a Porous Ductile Material

Some of the earliest work in modelling damage-induced ductile fracture was performed by McClintock (1968) who studied the axisymmetric deformation of an infinitely long circular-cylindrical void in an infinite, perfectly plastic matrix material. Rice and Tracey (1969) considered the same problem but for the growth of an initially spherical void. It is important to mention that while these studies related the void growth and evolution to the far-field loading, the bulk behaviour of the material was independent of the void damage. Building upon the work of Green (1972) and the numerical results of Needleman (1972) for a cylindrical unit cell, Gurson (1977) derived a damage-based yield criterion and flow rules for both cylindrical and spherical unit cells. In this model, the macroscopic response of the material is directly linked to the porosity and its evolution. The Gurson model is based on upper bound plasticity theory and thus the model will overestimate the material strength and underestimate the porosity in the material by restricting void growth.

To simplify the analysis, the Gurson-based material is assumed to have a periodic distribution of voids with each void located at the centre of a cylindrical or spherical unit cell. In this manner, the randomly distributed voids in the material are replaced with an equivalent single void. In reality, the void distribution is random with voids and particles of many sizes as shown in Fig. 1.11. The unit cells used by Gurson are only approximations to a periodic microstructure because the assembling of the unit cells to represent the bulk material will neglect the material between the cells (Fig. 2.1). These approximations of the unit cell geometries are required to simplify the problem so that a closed-form yield criterion can be derived. Twenty years later, Gologanu et al. (1997) followed a similar approach as Gurson but considered ellipsoidal voids embedded in an ellipsoidal unit cell so that the evolution of the void shape and shape-induced anisotropy could be captured. To maintain the integrity of the presentation, the definition of porosity in the Gurson unit cells is repeated here

$$\text{Spherical: } f = \frac{a^3}{b^3} \qquad \text{Cylindrical: } f = \frac{a^2}{b^2} \qquad (1.21\text{a, b})$$

where a is the radius of the void and b the radius of the unit cell.

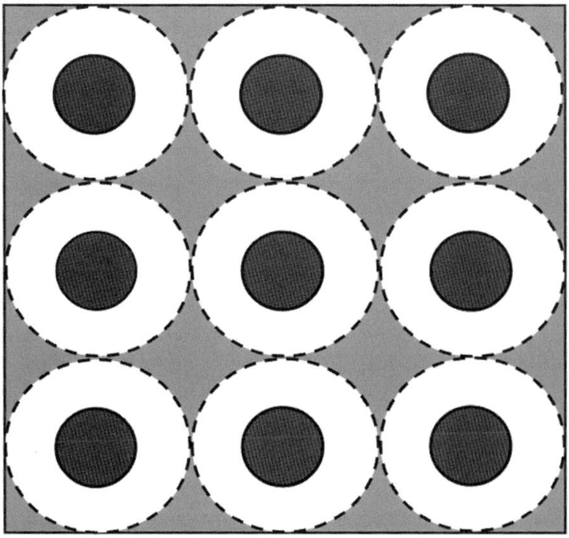

Fig. 2.1 Top-view of a material composed of a periodic array of spherical or cylindrical unit cells. Note the material that is not included in the homogenization process for these cell geometries

Another advantage of Gurson's model is its implicit accounting for material isotropy as the voids are assumed to remain spherical or cylindrical. The spherical model is geometrically isotropic while the cylindrical void experiences transverse isotropy. Gurson's yield criteria for spherical and cylindrical voids are

$$\text{Spherical void: } \Phi = \frac{\Sigma_{eq}^2}{\bar{\sigma}^2} + 2f\cosh\left(\frac{3}{2}\frac{\Sigma_{hyd}}{\bar{\sigma}}\right) - 1 - f^2 = 0 \qquad (2.20a)$$

$$\text{Cylindrical void: } \Phi = \frac{\Sigma_{eq}^2}{\bar{\sigma}^2} + 2f\cosh\left(\frac{3\sqrt{3}}{2}\frac{\Sigma_{hyd}}{\bar{\sigma}}\right) - 1 - f^2 = 0 \qquad (2.20b)$$

where $\bar{\sigma}$ is the equivalent tensile flow stress in the matrix material, neglecting variations in local stress, and Σ_{eq} and Σ_{hyd} are the macroscopic effective and hydrostatic stresses, respectively. Gurson's formulation reduces to the von Mises yield criterion for a damage free material by setting $f = 0$ in Eq. (2.20). The relationship between Gurson's damage model and the von Mises criterion with hydrostatic stress is presented in Fig. 2.2. The von Mises criterion is independent of the hydrostatic stress because it is assumed to be free of interal defects (voids, particles, inclusions) and because a hydrostatic stress will not induce any shear stresses. Damage-based materials are sensitive to the hydrostatic stress because a tensile hydrostatic stress will expand the voids, softening the material and resulting in earlier necking and failure. Conversely, a compressive hydrostatic stress will increase formability by shrinking the size of the voids.

The spherical void variant of the Gurson model was well-received but it was soon realized that it provided an overly stiff response when compared to the numerical

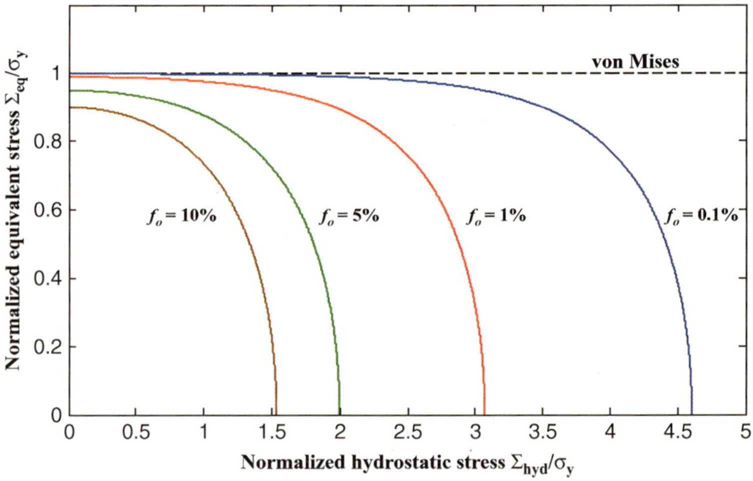

Fig. 2.2 Gurson (1977) yield function showing the reduction in the macroscopic equivalent stress with increasing hydrostatic stress and porosity

solutions of porous materials. Tvergaard (1981) extended Gurson's model to account for shear band instabilities and introduced three calibration parameters, q_1, q_2 and q_3 to better match the effects of voids during plastic deformation. The q_i parameters serve to artificially soften the material response and encourage void growth. Traditionally, the parameters are assumed to be material constants with $q_1 \approx 1.25 - 1.5$, $q_2 \approx 1$ and $q_3 = {q_1}^2$ based on the numerical work of Koplik and Needleman (1988). Other researchers such as Faleskog and Shih (1997) and Ragab (2004a) have proposed correlations for the q_i parameters as functions of the stress state, void shape and material properties and generally $q_1 \geq 1$ and $q_2 \leq 1$. The variation of the yield surfaces with the q_1 and q_2 parameters are shown in Figs. 2.3 and 2.4. The modification of Tvergaard forms the Gurson-Tvergaard (GT) model and has become the standard formulation of the Gurson yield surface.

2.7.1 Void Growth and Nucleation

The effects of void nucleation and void growth must be included to model damage evolution in a material. Void nucleation and growth are considered independently and damage development is expressed as the sum of these effects

$$\dot{f} = \dot{f}_{\text{growth}} + \dot{f}_{\text{nucleation}} \tag{2.21}$$

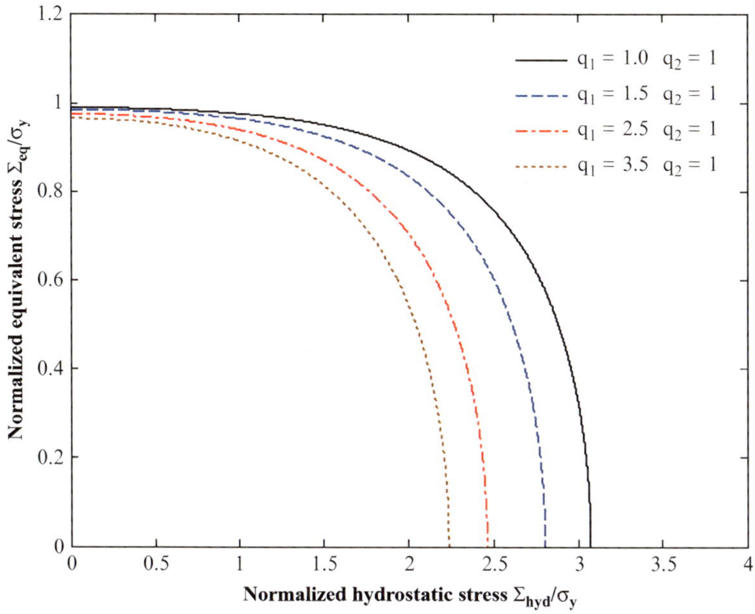

Fig. 2.3 Varation of the Gurson-Tvergaard yield surface with the q_1 parameter for a porosity of 1 %

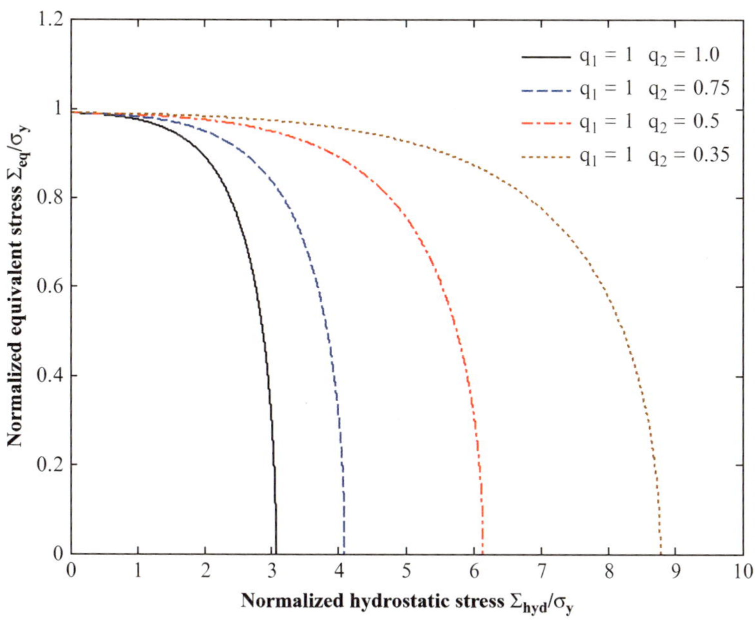

Fig. 2.4 Variation of the Gurson-Tvergaard yield surface with the q_2 parameter for a porosity of 1 %

The growth rate of the voids is proportional to the plastic volume dilatation rate

$$\dot{f}_{\text{growth}} = 3(1-f)\dot{\varepsilon}^{p}_{\text{hyd}} \tag{2.22}$$

where $\dot{\varepsilon}^{p}_{\text{hyd}}$ is the hydrostatic component of the plastic strain increment. The void growth equation in Eq. (2.22) is valid for all other damage-based constitutive models but the void growth rate will be different since $\dot{\varepsilon}^{p}_{\text{hyd}}$ is related to the shape of the yield surface through the associated flow rule:

$$\dot{\varepsilon}^{p}_{\text{hyd}} = d\lambda \cdot \frac{\partial \Phi}{\partial \sigma_{\text{hyd}}} \tag{2.23}$$

As discussed previously, void nucleation can be stress or strain-controlled. A review of the commonly used nucleation models can be found in Chap. 1.

2.7.2 *Void Coalescence*

The majority of work on void coalescence has been investigated using Gurson-based constitutive models because void growth, nucleation and material softening are included. However, the Gurson (1977) model is not particularly useful as a fracture criterion since material softening is a continuous process with a complete loss of material strength occurring when the porosity reaches 100 %. Obviously, this is unrealistic and a complete loss of load carrying capacity in a material occurs at porosities on the order of several percent. The introduction of the q_i parameters reduced the porosity at fracture to $1/q_1$ or between 66.7 and 100 %, depending on the value of q_1, but it was not sufficient to bring the model prediction closer to the reality.

2.7.2.1 Critical Porosity Coalescence Model

To address the unrealistic fracture porosities in the GT model, Tvergaard and Needleman (1984) proposed a phenomenological coalescence model that relies upon a critical porosity to identify the onset of coalescence and failure. In this approach, once the specified "critical porosity" has been reached, the porosity is rapidly increased to simulate the sudden drop in load carrying capacity associated with ductile fracture (see Fig. 2.5). The resulting criterion uses an effective porosity term, f^*, that replaces the original porosity term in the GT yield criterion:

$$f^* = \begin{cases} f & \text{if } f \leq f_c \\ f_c + \dfrac{f_u^* - f}{f_f - f_c}(f - f_c) & \text{if } f \geq f_c \end{cases} \tag{2.24}$$

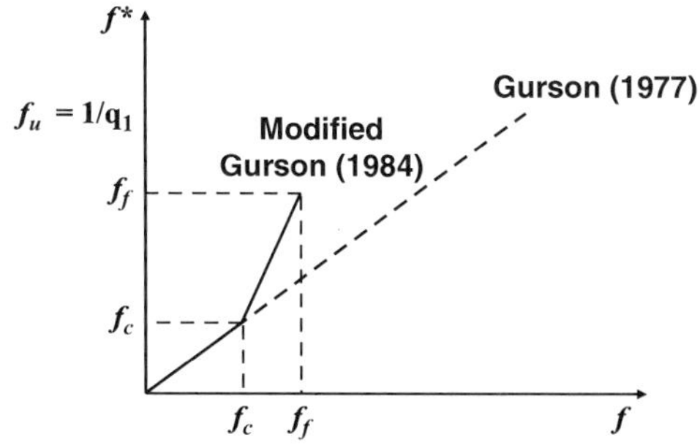

Fig. 2.5 Qualitative sketch of Tvergaard and Needleman's modification to simulate coalescence and rapid loss of material strength (Reprinted with permission from Zhang (1998). Copyright 1998 WIT Press Southampton)

where f_c is the critical porosity. Void nucleation, growth and coalescence continue to until the failure porosity is reached, f_f, where all material strength vanishes. The ultimate porosity, f_u^*, has no physical significance and is equal to $1/q_1$. The critical void volume fraction and porosity at failure were initially assumed to be universal constants with values of 0.15 and 0.25, respectively. This variant of the model is commonly referred to as the GTN model (Gurson-Tvergaard-Needleman) and is the most widely used variant of the Gurson model.

The value of f_f controls the rate that the load carrying capacity is lost with a smaller failure porosity corresponding to a steeper load drop prior to fracture. One method to identify the value of f_f is to best match the experimental drop in load carrying capacity from a tension test. It is important to note that the value of f_f does not play a significant role in numerical modeling once the critical porosity, f_c is determined and if f_f is kept the same during fitting (Zhang and Niemi 1994a, b). For practical sheet metal forming operations where the triaxiality (ratio of the hydrostatic-to-effective stress) is low (less than 1), the post-coalescence regime is not significant with a negligible difference between the coalescence and fracture strains (Scheyvaerts et al. 2010).

Numerical simulations of voided unit cells show that the phemenological post-coalescence response in Eq. (2.24) can accurately describe the material response since the porosity and the corresponding load drop behave in a linear fashion as shown in Figs. 2.6 and 2.7.

Becker (1987) applied the GTN model to investigate the effect of porosity distribution on the ductile fracture of a porous iron compact. Becker observed that the critical and final porosities were not universal constants among different materials and suggested values of $f_c = 0.075$ and $f_f = 0.09$ for the model material. Koplik and Needleman (1988) agreed with Becker's conclusion that the critical porosity is related to initial porosity and stress triaxiality and suggested that f_c should instead be specified as a material constant. To determine f_c for a given material, the value of f_c is taken that best fits the load drop point in a tension test (Sun et al. 1989) or from microscopy of the fracture surface. The critical porosity can also be determined using a unit cell model (Koplik and Needleman 1988) but it will still be challenging to match the experimental value.

Nowadays, it is universally recoginized that the critical porosity is not a material constant since it is a function of the initial porosity, void shape, stress state and

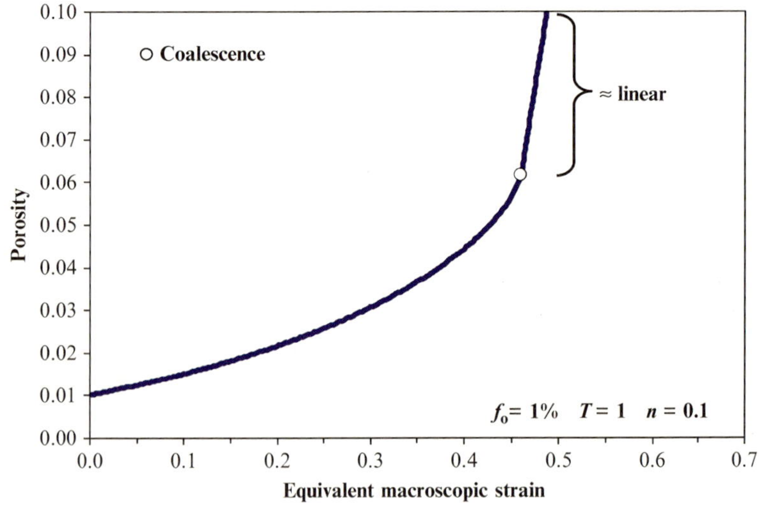

Fig. 2.6 Porosity evolution in an axisymmetric unit cell containing an initially spherical void

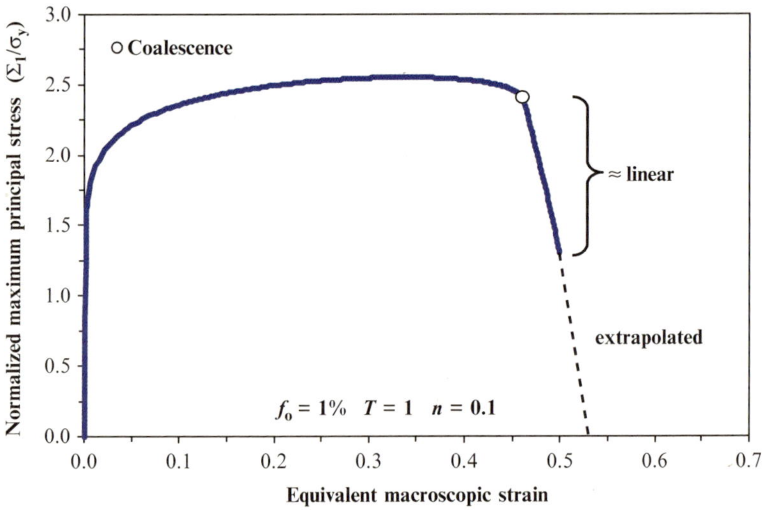

Fig. 2.7 Macroscopic stress response in an axisymmetric unit cell containing an initially spherical void

material properties (Zhang and Niemi 1994a; Pardoen and Delannay 1998b; Pardoen and Hutchinson 2000; Zhang et al. 2000). The variation of the critical porosity with the void shape and stress triaxiality obtained from unit cell simulations is presented in Fig. 2.8. From this figure, it is clearly shown that the

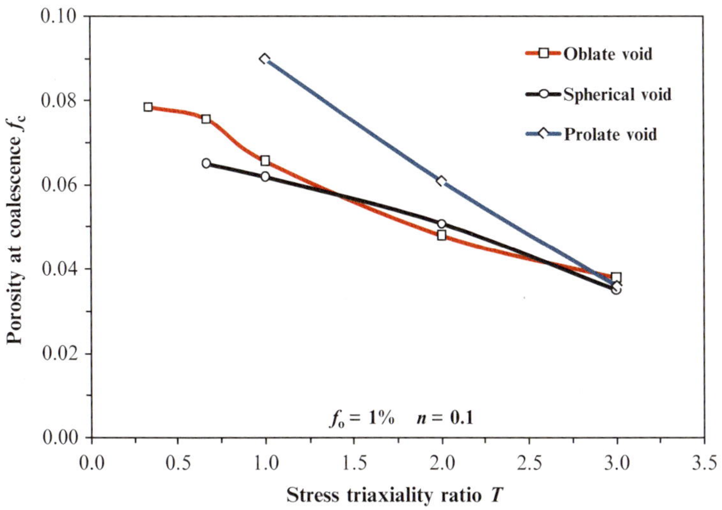

Fig. 2.8 Variation of the porosity at coalescence with the stress triaxiality and initial void shape. The aspect ratio of the oblate void is 1/6 and the aspect ratio of the prolate void is 6. The coalescence porosities were obtained from finite-element simulations of an axisymmetric unit cell

critical porosity model is reasonable at high stress triaxialities which is the intended regime for the Gurson model. The GTN model is well suited for fracture mechanics and applications related to crack propagation since the stress triaxiality ahead of a crack tip is high enough for the voids to grow in an approximately spherical manner. The critical porosity coalescence model is still in use today because it is available in commercial finite-element codes and some researchers have previously characterized the f_c value for their materials of interest. However, the reader is cautioned that this coalescene model should only be used for a well-defined material and that the critical porosity determined in one stress state will likely not be valid in a different application.

Fortunately, the critical porosity coalescence model has largely been supplanted by the plastic limit-load criterion of Thomason (1990) that describes the necking failure of the inter-void ligaments. In this physically-motivated model, void coalescence is a function of the stress state and the microstructure geometry with f_c becoming a field quantity. By removing f_c as a material constant, the nucleation parameters can be determined from tensile tests, instead of being pre-assumed in order to determine f_c by calibration with experiments. This is a significant improvement for Gurson-based models as the nucleation parameters can be easily determined and contains a fracture criterion based on the physical mechanism of coalescence.

2.7.2.2 Plastic Limit-Load Coalescence Criterion

To account for the physical mechanism of void coalescence, Thomason (1985a, b, 1990) theorized coalescence as the competition between stable homogeneous and

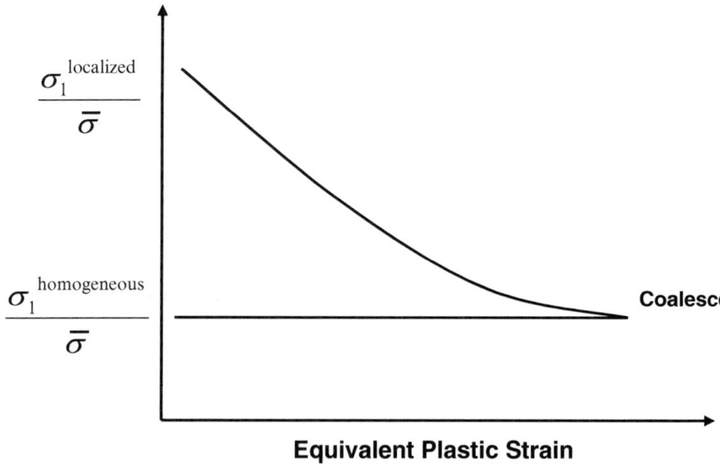

Fig. 2.9 Qualitative sketch of competing deformation modes as envisaged by Zhang (1998)

unstable localized deformation modes. Initially, the influence of voids is small and the deformation mode is homogenous. As voids nucleate and grow during further plastic deformation, the energy required to achieve an unstable localized deformation mode decreases. The point at which the homogenous and localized deformation modes become equal is taken as the onset of void coalescence and localized deformation as shown in Fig. 2.9. At this point, the plastic limit of the inter-void ligament has been reached and internal necking of the matrix begins, resulting in sudden localized fracture.

Zhang and Niemi (1994b) modified Thomason's plastic limit-load criterion to incorporate it into the Gurson-Tvergaard-Needleman (GTN) constitutive model to form the so-called complete Gurson model (CGM). More recently, Zhang et al. (2000) further extended his model to include the effect of hardening. In the modified plastic limit-load criterion, the voids are assumed to remain spherical and the constraint factor is expressed as

$$C_{f-Z} = (0.12 + 1.68n)\left(\chi^{-1} - 1\right)^2 + 1.24\chi^{-1/2} \qquad (2.25\text{a,b})$$

where n is the hardening exponent and χ is the void spacing ratio that is defined for spherical voids as

$$\chi = \left(\frac{f}{\gamma_{\text{cell}}} e^{\frac{3}{2}\left(\varepsilon_1 - \varepsilon_{\text{hyd}}\right)}\right)^{1/3} \qquad \gamma_{\text{cell}} = \begin{cases} 2/3 \text{ cylindrical unit cell} \\ \pi/6 \text{ cubic unit cell} \end{cases} \qquad (2.26\text{a,b})$$

where $\left(\varepsilon_1 - \varepsilon_{\text{hyd}}\right)_c$ is the principal deviatoric strain. The variants of the plastic limit-load models by Pardoen and Hutchinson (2000) and Benzerga (2002) found in Eqs. (1.15) and (1.16) could also be employed by setting the aspect ratio equal to unity.

This modified version of Thomason's criterion is best suited for materials with initial porosities less than 1 % and has been applied to steel and aluminum alloys (Chen and Lambert 2003; Zhang et al. 2000). Another successful variant of the plastic limit-load model was proposed by Ragab (2004a) and has been applied to many materials by Ragab (2004a). This variant was found to predict fracture strains in much better agreement with the experimental fracture strains than other variants of the Thomason (1990) model. For spherical voids, plastic constraint factor in the Ragab (2004b) coalescence model is:

$$C_{f-R} = \left(1 + \frac{2}{\chi^{-1} - 1}\right) \ln\left[1 + \frac{1}{2}(\chi^{-1} - 1)\right]\left(\frac{\varepsilon_{\text{lig}}}{\varepsilon_{\text{uc}}}\right)^{\text{n}} \tag{2.27}$$

where ε_{lig} is the ligament strain; ε_{uc} is the average effective strain in the unit cell; The strain in the ligament can be related to the geometry of unit cell for spherical voids and small void volume fractions as

$$\varepsilon_{\text{lig}} = 2\ln\left[\left(\frac{1-\chi^{-1}}{1-\chi_{\text{o}}^{-1}}\right)\left(\frac{f}{f_{\text{o}}}\right)^{\frac{1}{3}}\right] \tag{2.28}$$

and the average macroscopic strain for the unit cell is taken as $\varepsilon_{\text{uc}} = \varepsilon_1$ (Ragab 2004b). The adoption of the ligament strain hardening term in Eq. (2.27) improves the physical foundation of the model since the strain in the intervoid ligament increases faster than the bulk strain in a unit cell. However, most engineering materials do not infinitely harden and in the author's numerical experience, coalescence does not generally occur until later in deformation when the flow stress has all but saturated. Consequently, the ligament hardening term in Eq. (2.27) can be omitted in materials that possess a flow stress response that saturates at higher plastic strains. Additionally, complications arise if void nucleation is considered because the initial geometry used in Eq. (2.28) is no longer valid.

2.7.2.3 Post-coalescence Treatment When Using a Plastic Limit-Load Coalescence Model

The post-coalescence model of Tvergaard and Needleman (1984) in Eq. (2.24) is retained when using a plastic limit-load coalescence model by identifying the porosity when the coalescence condition is satisfied as the critical porosity, $f_c = f$. The fracture porosity can also be removed as a material parameter by re-working the physically-sound post-coalescence model of Scheyvaerts et al. (2010) as presented in Eq. (1.19). For the Gurson-Tvergaard model where the void is spherical at coalescence and the initial aspect ratio of the unit cell is unity, the fracture porosity in a specific unit cell geometry is

$$f_f = \begin{cases} 4f_c/\chi_c & \chi_c \geq \frac{1}{2} \\ \left(\chi_c + \dfrac{1-2\chi_c^2}{(9/2-\chi_c^2)}\right)\frac{f_c}{\chi_c^3} & \chi_c < \frac{1}{2} \end{cases} \qquad \chi_c = \left(\frac{f_c}{\gamma_{\text{cell}}} e^{\frac{3}{2}(\varepsilon_1-\varepsilon_{\text{hyd}})_c}\right)^{1/3} \tag{2.29a, b}$$

where $(\varepsilon_1 - \varepsilon_{\text{hyd}})_c$ is the principal deviatoric strain when the critical porosity is reached. Equation (2.29) can also be used to estimate the failure porosity when using the critical porosity coalescence model. Note that Eq. (2.29) is approximate since the void shape and the initial aspect ratio are not considered, introducing error into the computation of the spacing ratio, χ, which tends to be overestimated. For example, from unit cell simulations of an axisymmetric unit cell containing an initially spherical void at a stress triaxiality of unity and a hardening exponent of 0.10, $f_c = 6.17\,\%$ at a strain of 0.46. The void aspect ratio at coalescence is 1.56 and the actual spacing ratio is $\chi \sim 0.49$. By assuming the void remains spherical, the spacing ratio is is $\chi \sim 0.57$ and this will result in a larger predicted porosity at failure in Eq. (2.29). This overestimation generally increases with decreasing stress triaxiality since the void shape will not be spherical at coalescence. Fortunately, the porosity at failure is not a critical parameter and this will not have an overly deleterious influence on the fracture strains in a finite-element simulation of a metal forming operation.

The above modifications to the GTN framework dramatically improve the predictive capabilities of the model by removing two material parameters, f_c and f_f, and computing them using physically-sound models that predict coalescence and fracture as a consequence of the evolution of the stress state and the microstructure. Only the parameters related to the nucleation model remain to be identified, paving the way for the identification of unique-nucleation parameters that are transferrable to different stress states and this will be addressed in Chap. 4.

2.8 Lower Bound Solution of Sun and Wang

Following the success of the Gurson-Tvergaard constitutive model, Sun and Wang (1989) derived the analogous lower bound solution using the spherical unit cell geometry of Gurson to obtain a conservative estimate for the yield stress and formability. While Gurson applied a velocity field to the surface of the spherical unit cell, Sun and Wang (SW) applied a prescribed traction to obtain the lower bound solution. Similar to the original Gurson formulation, Sun and Wang's yield criterion does not account for void coalescence in a meaningful way and requires large porosities before the load carrying capacity is lost. Fortunately, the same treatments of void coalescence described previously for the Gurson model can also be applied to the SW model. The Sun and Wang yield criterion is expressed as

$$\Phi = \frac{\Sigma_{\text{eq}}^2}{\bar{\sigma}^2} + \frac{b_1 f \cosh\left(\frac{3}{2}\frac{\Sigma_{\text{hyd}}}{\bar{\sigma}}\right)}{\sqrt{1 + b_3 f \sinh^2\left(\frac{3}{2}\frac{\Sigma_{\text{hyd}}}{\bar{\sigma}}\right)}} - b_2 = 0 \tag{2.30}$$

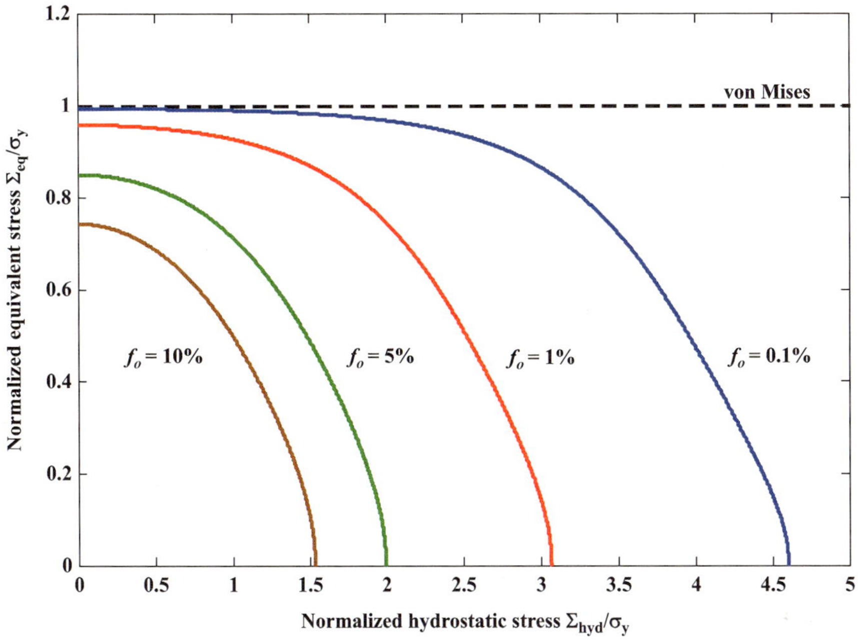

Fig. 2.10 Variation of the Sun and Wang (1989) yield surface with the hydrostatic stress and porosity

where

$$b_1 = 2 - \frac{1}{2}\ln f \qquad b_2 = 1 + f(1 + \ln f)$$

$$b_3 = \left(\left(\frac{b_1}{b_2}\right)^2 \coth^2\left(\frac{3\Sigma_{\mathrm{mt}}^0}{2\bar{\sigma}}\right) - \left(f^2 \sinh^2\left(\frac{3\Sigma_{\mathrm{mt}}^0}{2\bar{\sigma}}\right)\right)\right)^{-1}$$

$$\Sigma_{\mathrm{mt}}^0 = -0.65\ln(f)\bar{\sigma}$$

Sun and Wang's model converts to Gurson's upper bound solution for spherical voids when

$$b_1 = 2 \quad b_2 = 1 + f^2 \quad b_3 = 0$$

and to the von Mises yield criterion when $f = 0$. The variation of the SW yield surface with the hydrostatic stress and porosity is presented in Fig. 2.10.

2.8.1 Void Growth, Nucleation and Coalescence

The same void growth, nucleation and coalescence rules discussed previously for the Gurson model are also applicable for the Sun-Wang model. It is important to

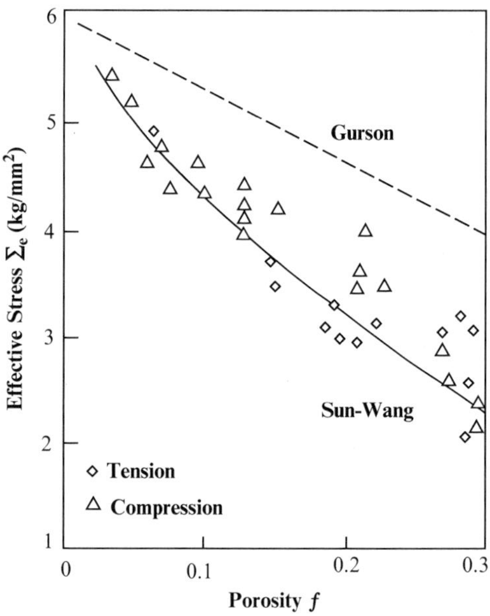

Fig. 2.11 Comparison of the *upper* and *lower* bound criteria with experimental results of Shima and Oyane (1976) for powder metallurgical materials (Reprinted with permission from Sun and Wang 1989. Copyright 1989 Springer)

note that the resulting void growth and coalescence predictions will be different than in the Gurson model due to the softer nature of the SW model which will promote higher plastic strains and thus larger void growth rates and additional material softening.

In a comparison of the upper and lower bound yield criteria with the experimental results of Shima and Oyane (1976), Sun and Wang (1989) observed yielding closer to the lower bound solution as shown in Fig. 2.11. Shima and Oyane (1976) used powder metallurgical samples to achieve porosities as high as 30 %. For a typical ductile metal, initial porosity ranges between 0.01 and 1 % (Pardoen and Hutchinson 2000). An additional work by Sun and Wang (1995) observed that the experimental porosities in sintered iron and titanium alloys were in very good agreement with the SW solution and well defined between the bounds of the SW and Gurson models as shown in Fig. 2.12. The good experimental agreement with the SW solution may be due to void distribution effects since voids that are arranged in heterogeneous clusters experience faster void growth and yielding at lower stresses and the SW model better represents that effect. Francescato et al. (2004) performed a numerical limit analysis of plasticity to compare the models of Gurson (1977), Tvergaard (1981), Richmond and Smelser (1985) and a modified form of Sun and Wang (1989) for cylindrical voids. This study observed that Sun and Wang's (1989) lower bound model closely approximates the actual solution for low porosities ($f < 5$ %) while the Richmond and Smelser (1985) model worked better for high porosities ($f > 5$ %).

The principal take-away from these figures and results is not that the SW model is superior to the Gurson model since there are surely cases where the experiment is

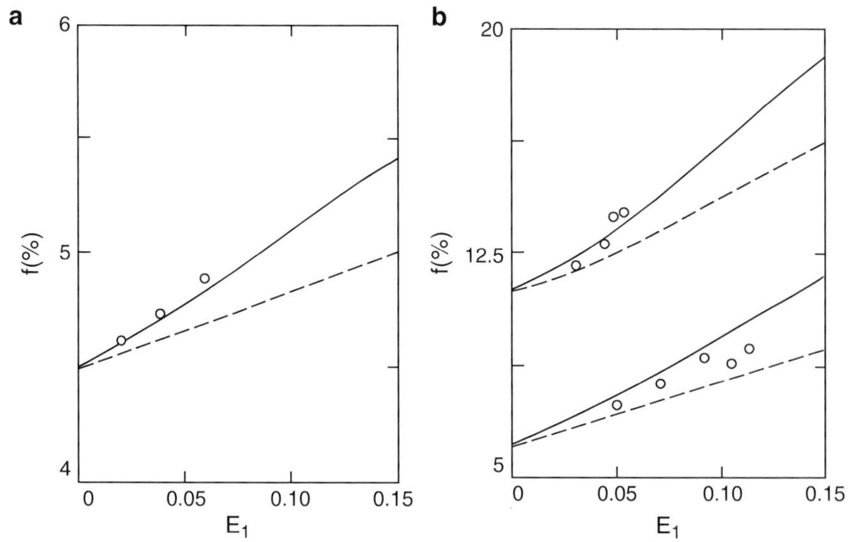

Fig. 2.12 Increase of the void volume fraction, f, as a function of the uniaxial strain: (**a**) experimental data for a Sintered CP Ti alloy from Marciniak and Kuczynski (1967). (**b**) Experimental data for Sintered iron by Bourcier et al. (1986). The solid line denotes the SW model and the dotted-line denotes the prediction of the Gurson (1977) model. Note that the SW model provides the upper limit for the porosity since it is softer than the Gurson model (Reprinted with permission from Sun and Wang (1995). Copyright 1995 Springer)

better described by the Gurson model, but rather that neither of these models is overly accurate. Only by utilizing both upper and lower bound solutions can the material behaviour be well represented in a meaningful way since these are approximate yield criterion with highly idealized void distributions.

2.9 Upper and Lower Bound Approach to Ductile Fracture of Porous Materials

The Gurson model has received significant attention since its introduction while the lower bound Sun-Wang model has gone largely unnoticed. An exhaustive number of Gurson-based models have since been developed to account for many different effects such as void shape (Pardoen and Hutchinson 2000; Ragab 2004a; Wen et al. 2005; Lassance et al. 2006), non-local damage development (Tvergaard and Needleman 1995; Leblond et al. 1994; Reusch et al. 2003), and anisotropic materials (Liao et al. 1997; Chein et al. 2001; Wang et al. 2004; Kelavelarma and

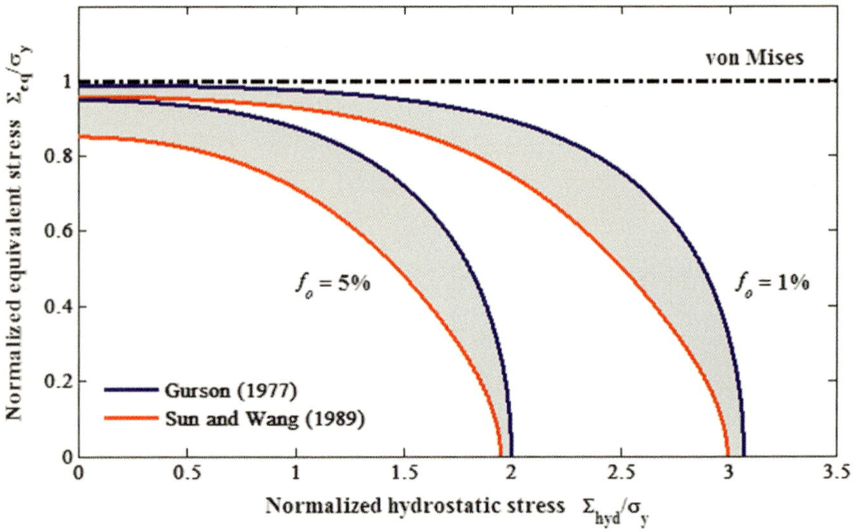

Fig. 2.13 Yield surfaces of the von Mises, Gurson (1977) and Sun and Wang (1989) models. The experimental yielding behaviour of a material should lie on or between the two bounds within the shaded band. The macroscopic equivalent and hydrostatic stress are normalized by the yield stress of the material (Reprinted with permission from Griffin et al. (2011). Copyright: Springer)

Benzerga 2009; Cazacu and Stewart 2009). Nowadays, the GLD (Gologanu, Devaux and Leblond) model (Gologanu et al. 1997) has largely replaced the GT model and its variants in the academic community due to its natural ease in describing the void shape effects that are critical to the accurate modeling of void coalescence. Unfortunatley, Sun and Wang's lower bound solution has only been modified to include kinematic hardening (Yan 1992), shear localization (Sun and Wang 1995; Sun 1995a) and a dual population of large and small voids (Sun 1995b). A lower bound solution to the ellipsoidal unit cell geometry of Golaganu et al. (1997) would provide an excellent counterpoint to the GLD model and enable the prediction of upper and lower forming limits using advanced void coalescence models.

The use of both upper and lower bound damage models can provide a novel and straightforward method to obtain estimates of the formability of ductile materials by acknowledging the inherent limitations of the models and exploiting the difference in the yield surfaces as shown in Fig. 2.14. The shaded band in Fig. 2.13 can be interpreted as the formability band for the actual yield stress of a material and it is expected to fall within this range. By employing the same void nucleation and coalescence models in both the upper and lower bound yield criteria, a range for the limiting strains or formability band for the material can be defined.

An important distinction in this approach is that although one of the yield criteria may be more accurate in an academic sense in that it matches the the numerical bound obtained from FE simulations of a spherical unit cell for a given stress state, this is of secondary importance because real materials do not adhere to such a rigid

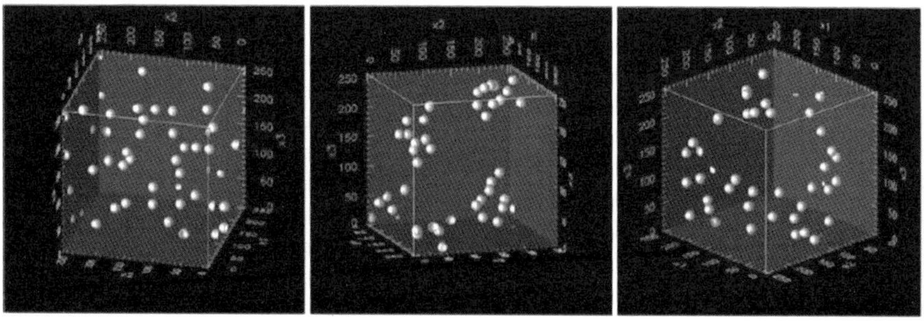

Fig. 2.14 Three-dimensional clustered microstructures of Bilger et al. (2005): random (*left*), disconnected clusters (*center*) and connected clusters (*right*) (Reprinted with permission from Bilger et al. (2005). Copyright 2005 Elsevier)

definition of the microstructure. The actual material behaviour will generally fall within the two bounds due to void distribution effects and other factors that are not considered in the models such as shearing, particle-void interactions, etc. It is important to not interpret the predicted upper and lower forming limits defined by the GT and SW models in an absolute mathematical sense because they are approximations to the microstructure and spherical unit cells do not strictly adhere to the definition of a periodic microstructure. Consequently, it is possible for the experiment behaviour to lie outside of these bounds but together they generally provide very good estimates. The upper and lower bound approach accepts that these models are inherently approximate and that there is value in estimating the material behaviour within a range. The following sections will discuss the evolution of the dual-bound approach to ductile fracture developed by the authors and its application to a variety of metal forming operations.

2.9.1 Application of the Dual Bound Approach to Porous Materials with Void Clusters

Inherent in the upper and lower bound formulations of Gurson (1977) and Sun and Wang (1989) is the assumption that the voids remain spherical and the material can be composed of a periodic assembly of spherical unit cells. However, real materials often contain a dilute concentration of voids that are heterogeneously distributed in clusters and the periodic assumption becomes a necessary yet questionable assumption in order to obtain a tractable geometry to derive the yield criteria. In this instance, it is important to evaluate the performance of the damage-based yield criteria as neither is expected to perform overly well in this situation. The work of Bilger et al. (2005) provides an interesting opportunity to evaluate this condition since they considered the overall and local responses of porous media composed of a perfectly plastic matrix with spherical voids in various arrangements. Bilger et al. (2005) employed the Fast Fourier transform (FFT) method to numerically

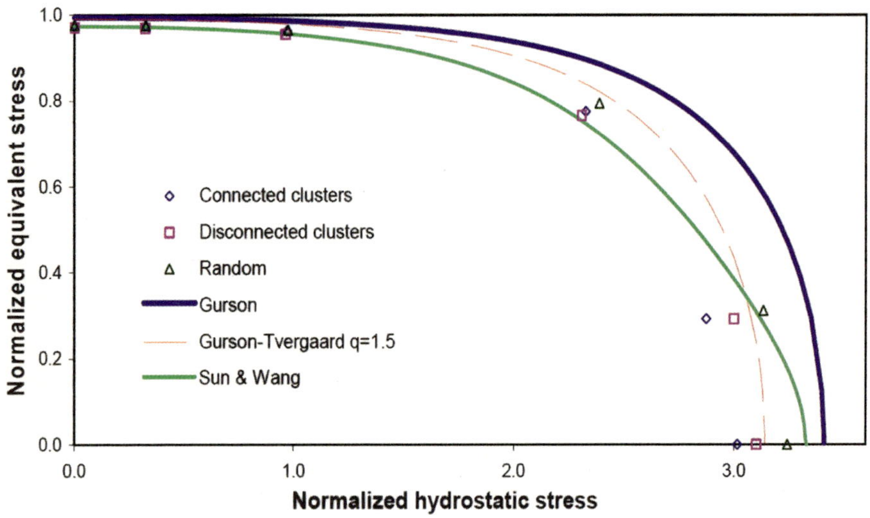

Fig. 2.15 The Gurson and SW yield surfaces have been superimposed onto the numerical results of Bilger et al. (2005) for type-A loading (pure shear with hydrostatic tension) (Reprinted with permission from Griffin et al. (2011). Copyright: Springer)

determine the onset of yielding in materials having microstructures characterized by one of three void spatial distributions:

i. a random void distribution without void clustering,
ii. connected clusters of voids, or
iii. disconnected clusters of voids.

The study considered two-dimensional and three-dimensional arrangements, each under two loading types (type-A, that combined pure shear with a superimposed hydrostatic tension, and type-B which did not have a shear component) over a large range of stress triaxiality. Only the three-dimensional arrangements are considered here because there is no analogous 2-D lower bound solution of the Sun-Wang model. The total void volume fraction in the microstructures is 0.6 % and the various cluster arrangements are presented in Fig. 2.14.

Bilger et al. (2005) compared their results with the Gurson-Tvergaard (GT) model which proved to be overly stiff and overestimated yielding in the different microstructures. Griffin, Butcher and Chen (2011) noticed the overestimation of the GT model and evaluated the lower bound criterion of Sun and Wang (1989) using the results of Bilger et al. (2005). The SW model is inherently more sensitive to the porosity than the Gurson model and experiences earlier yielding and additional material softening. The lower bound solution of SW was superimposed on the results of Bilger et al. (2005) as shown in Figs. 2.15, 2.16 and 2.17 to evaluate the performance of the dual bound approach first applied in Butcher et al. (2006). For clarity, Fig. 2.18 shows a close-up view of the results at low stress triaxialities. To provide a contrast with the traditional Gurson model, the more-commonly employed GT solution with fitting parameter $q = 1.5$ is shown for comparison with the original Gurson and SW models.

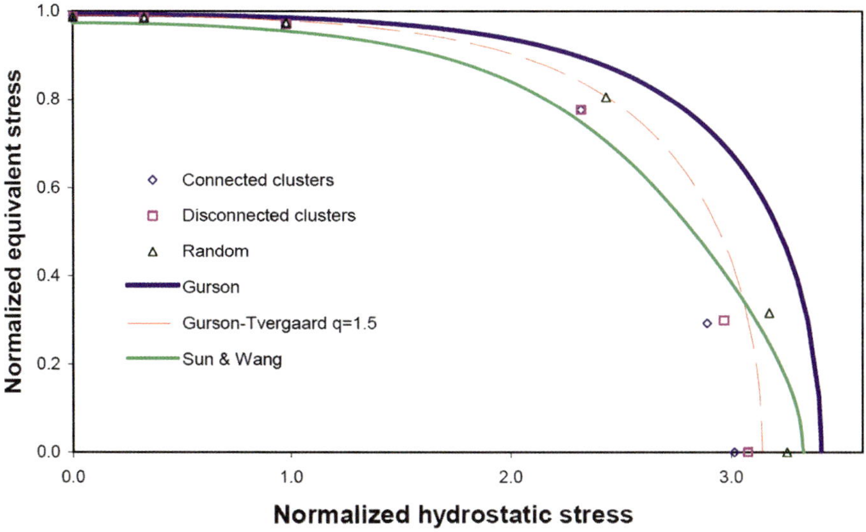

Fig. 2.16 The Gurson and SW yield surfaces have been superimposed on the numerical results of Bilger et al. (2005) for type-B loading (hydrostatic tension, no shear) (Reprinted with permission from Griffin et al. (2011). Copyright: Springer)

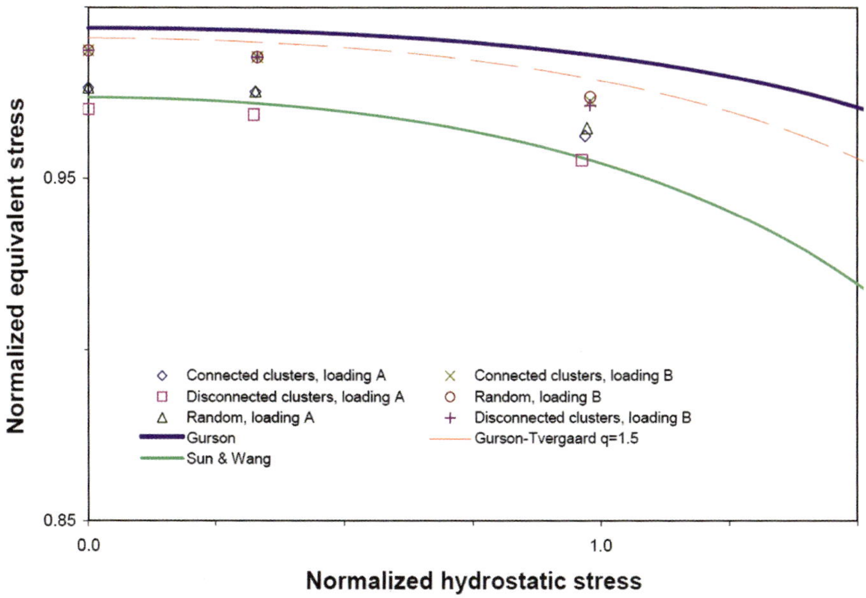

Fig. 2.17 Comparison of the three-dimensional macroscopic yield surfaces with the numerical results of Bilger et al. (2005) for two different loading conditions at low triaxialities (Reprinted with permission from Griffin et al. (2011). Copyright: Springer)

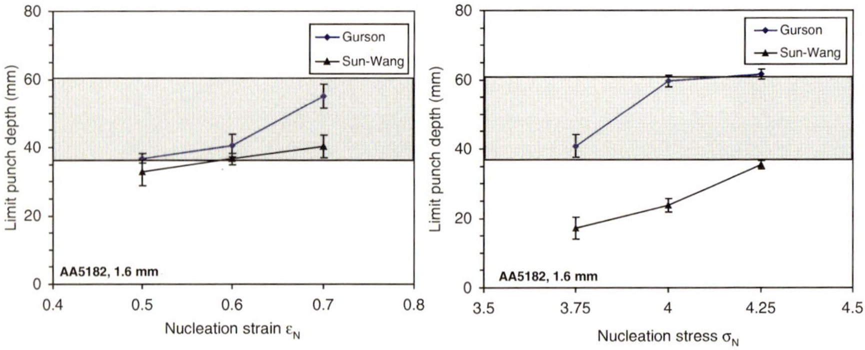

Fig. 2.18 Limit punch depth versus void nucleation strain (*left*) and stress (*right*) for AA5182 1.6 mm sheet; the *shaded band* represents the experimental results for failure due to circumferential cracking (Reprinted with permission from Butcher et al. (2006). Copyright: Springer)

For low stress triaxialities the results are well captured within the upper and lower bound predictions which is fortuitous since the practical range of stress triaxiality is less than unity for a sheet metal forming operation. Notice that none of the yield criteria were able to give great results for all of the stress states and microstructures considered. Despite the different loading conditions, stress triaxialities and microstructures, the dual bound approach is able to provide very good upper and lower estimates for the yield behaviour.

This is a significant advantage of the dual bound approach because although neither model was derived for void clusters or shear loading, a good representation of the material behaviour can still be captured between the two bounds. By obtaining upper and lower estimates for the material behaviour, the variation in the material behaviour due to deviations from the assumption of a periodic void distribution can be better captured than if using a single model. From a practical perspective, it is reasonable to expect the material response to usually fall within the upper and lower limits as predicted by the approximate models of Gurson and Sun and Wang. This result is very attractive to industry because the original Gurson and Sun and Wang models can be quickly implemented in a commercial finite-element code and employed to obtain a first-order prediction of the material behaviour in a forming process of interest.

2.10 Application of the Dual Bound Approach to a Stretch Flange Forming Process

The first application of the dual bound concept was performed by Butcher et al. (2006) for a stretch flange forming operation of AA5182 sheet. One example of a stretch flanging operation is to expand a cutout in a blank to create the openings for

Table 2.1 Controlling parameters in GTN-based and Sun and Wang material models

Model	q_1	q_2	q_3	f_o	f_c	f_f	f_u	f_n	s_n (% ε_N, σ_N)
GTN	1.25	0.95	1.5625	0	0.01	0.02	0.80	0.00768	20
Sun-Wang	–	–	–	0	0.01	0.02	0.80	0.00768	20

windows in car door panels. Details of the stretch flange forming process, experiments and finite-element modeling will be discussed in detail in Chap. 6. The lower bound SW model was implemented along with the GT model into LS-DYNA (Hallquist 2006), a commerical finite-element solver, using user-defined subroutines. Void nucleation was assumed to be either stress- or strain-controlled using the nucleation models of Chu and Needlemen (1980) described in Eq. (1.5). The rather simplistic critical porosity coalescence model of Tvergaard and Needleman (1984) was used since a detailed metallographic and numerical analysis of the material was performed by Chen (2004) and an appropriate critical porosity was identified. The controlling parameters used in the simulations are presented in Table 2.1. A minor limitation in this study was that the q_i parameters were not set to unity so that the formal upper bound of Gurson (1977) was slightly artificially softened by using the suggested q_i values of Koplik and Needleman (1988).

2.10.1 Predicting the Limit Punch Depth

Cutout sizes ranging from 88 to 98 mm were considered and the effect of the cutout size on the limiting punch depth was small. No correlation was observed between the onset of a radial crack at the cutout edge or the formation of a circumferential crack. The FE model was only able to predict the onset of a circumferential crack because the model was axisymmetric. Consequently, the experimental limit punch depths for the circumferential cracks for all of the cutout sizes are presented as a single shaded band in Fig. 2.18 (Chen 2004). The nucleation stress and strain were then parametrically identified in the SW and GT models until the predicted formablity band matched the experimental band for the limit punch depth. It was observed that a nucleation strain of 0.70 and a nucleation stress of $4.25\sigma_y$ (~500 MPa) gave good results.

2.10.2 Damage Evolution During Forming

Damage evolution is expressed as a function of punch displacement in the element of interest (EOI) and the punch displacement is measured relative to the main punch. The backup punch moves upwards to close the drawbead and stretching begins as the main punch moves downward as described in Chap. 6. To clearly

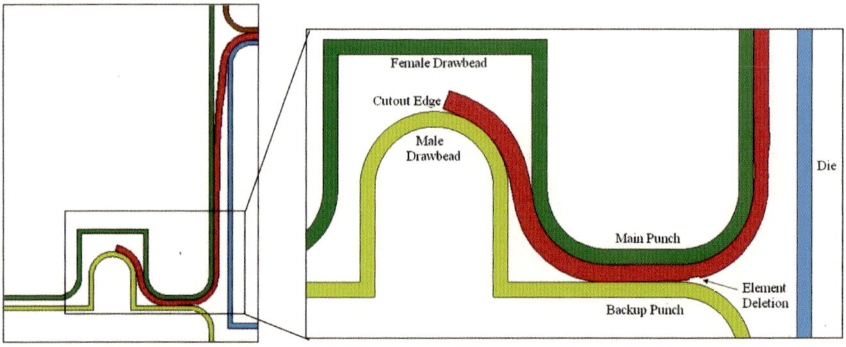

Fig. 2.19 Element deletion due to void coalescence at limit punch depth (Reprinted with permission from Butcher et al. (2006). Copyright: Springer)

present damage evolution as a function of punch displacement, the change in direction of the punches is neglected and drawbead closure corresponds to a displacement of 0 mm. Damage development is typically negligible until the blank leaves the drawbead and enters the punch profile. Upon entering the punch profile, void growth and nucleation occur resulting in the formation of a circumferential crack in the side-wall of the flange or at the punch nose as shown in Fig. 2.19. In the simulation, this coincides with the onset of element deletion, forcing the porosity measurement to zero and the termination of the simulation. The limit punch depth is then extracted. The element initially in contact with the male drawbead typically exhibits the highest damage rate during stretch-flange forming. Therefore, it is chosen as the element of interest (EOI) to characterize damage evolution.

2.10.3 Comparison of the Predicted and Measured Porosity

To validate the predicted damage evolution in the element of interest, the porosity history obtained using the strain and stress-based nucleation rules are compared with the damage measurements of Chen (2004) in Figs. 2.20 and 2.21, respectively. To experimentally measure damage evolution in the element of interest, Chen (2004) performed a series of interrupted tests to obtain damage measurements as a function of punch displacement. In each test, the area of interest which was sectioned and the through-thickness porosity measured using standard thresholding techniques.

It is important to note that the porosity measurements of Chen (2004) are not definitive due to the digital imaging process. To measure porosity, the sample is digitally scanned with a high resolution digital camera. From these images, the background can be eroded to leave the voids which will be darker than the background. This process can be very sensitive to the parameters used in the imaging process and small variations in the parameters may lead to different porosity measurements. This sensitivity is further compounded by the small damage level required to cause fracture in the 5xxx series alloys. With this in mind, the porosity

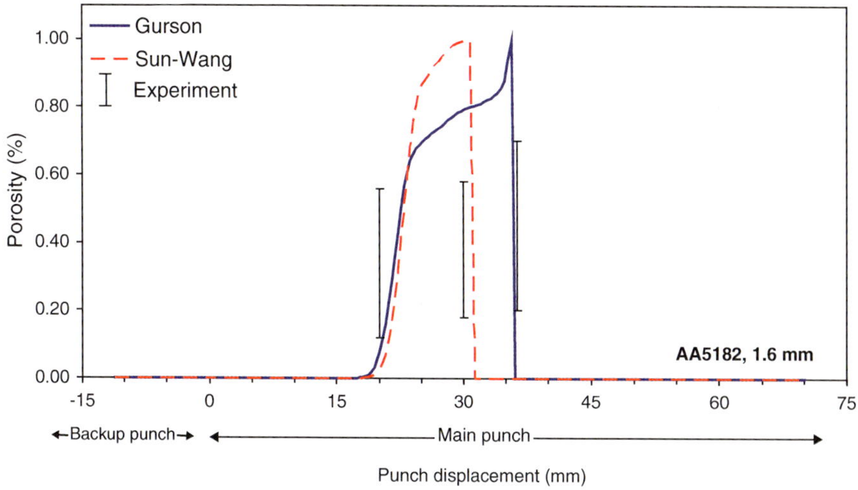

Fig. 2.20 Comparison of experimental and predicted damage evolution in the EOI for an 88 mm cutout and nucleation strain of 0.5

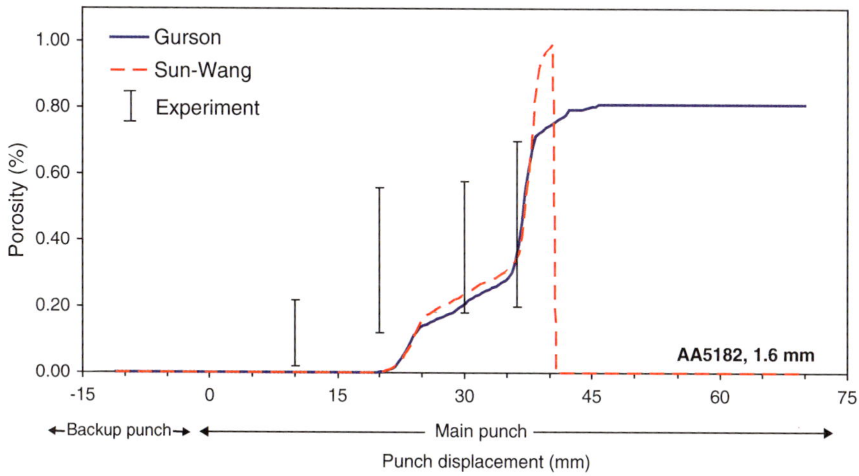

Fig. 2.21 Comparison of experimental and predicted damage evolution in the EOI for an 88 mm cutout and nucleation strain of 0.7

results are presented to provide a guideline of actual damage evolution in the stretch flange forming.

For strain-controlled nucleation, Figs. 2.20 and 2.21 demonstrates that porosity is not accurately predicted as both upper and lower bound models predict marginal damage development until a punch depth greater than 15 mm. However, the Gurson model gives a reasonable prediction of porosity for a nucleation strain of 0.5 while both models significantly underestimate damage for a strain of 0.7.

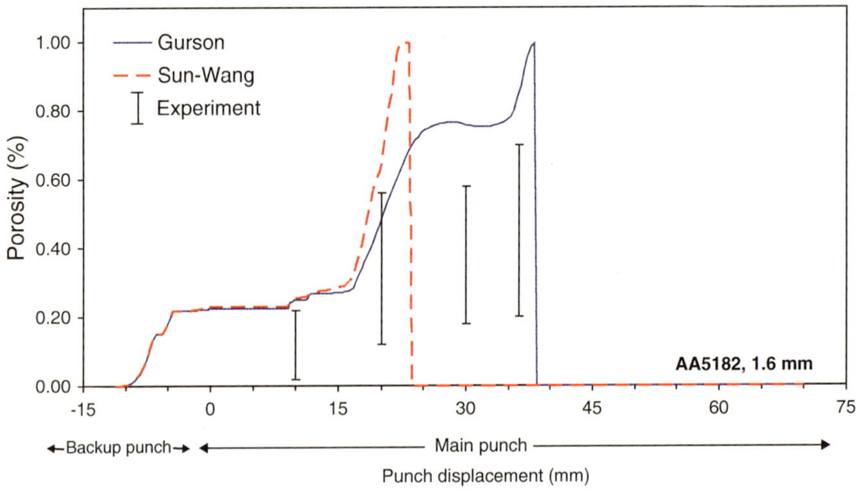

Fig. 2.22 Comparison of experimental and predicted damage evolution in the EOI for an 88 mm cutout and nucleation stress of $3.75\sigma_Y$

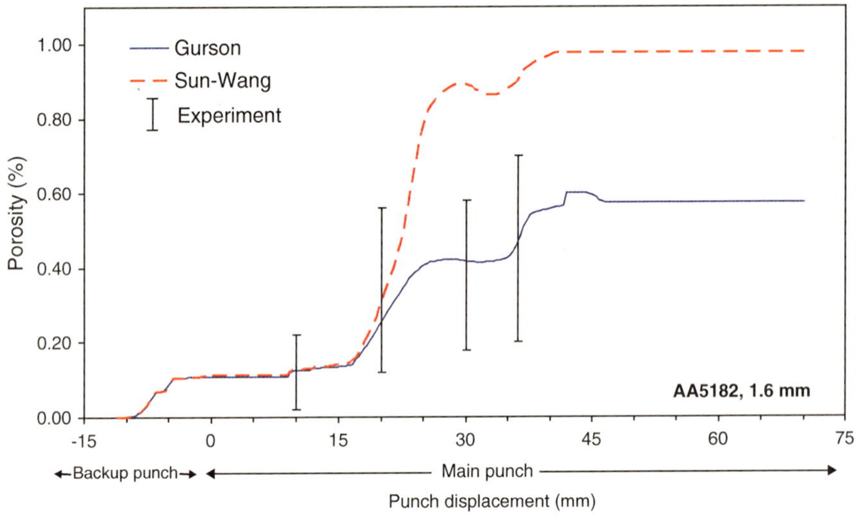

Fig. 2.23 Comparison of experimental and predicted damage evolution in the EOI for an 88 mm cutout and nucleation stress of $4.25\sigma_Y$

Damage evolution for the stress-based nucleation model clearly gives much better agreement with the experiment as demonstrated in Figs. 2.22 and 2.23. Both upper and lower bound models capture the experimentally measured porosity with the Gurson model giving a better prediction since the lower bound model fails prematurely at this nucleation strain. For a nucleation strain of $4.25\sigma_y$ (Fig. 2.24), both upper and lower bound models give very good agreement with the experiment.

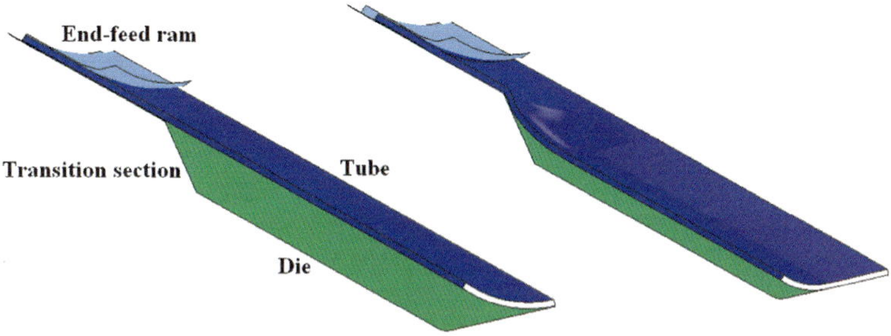

Fig. 2.24 One-eighth finite-element model of the initial tube (*left*) and formed tube (*right*) (Reprinted with permission from Butcher et al. (2009). Copyright: Springer)

The lower bound model provides an upper limit on porosity with the Gurson model giving a lower limit. Despite the scatter in the porosity measurements, stress-based nucleation gives realistic damage predictions in both models.

2.11 Application of the Dual Bound Approach to Ductile Fracture in Tube Hydroforming

Tube hydroforming is attractive to the automotive industry since it can produce low weight, high strength and uniform parts while eliminating traditional welding and stamping operations. A typical hydroforming operation involves pressurizing the inside of a tube to conform to the cross-section of a die. The hydroforming of advanced high strength steels (AHSS) such as DP600 offers benefits such as a more continuous yielding behaviour, higher work hardening limit and total elongation-to-failure compared to traditional high-strength low alloy steels (HSLA). Despite the apparent advantages of hydroformed AHSS components, formability is limited due to the initiation of microvoids (damage) that grow and coalesce leading to sudden fracture.

While void damage has typically been neglected in hydroforming simulations, Baradari (2006), Varma et al. (2007) and Butcher et al. (2009) have shown that good predictions of formability can be obtained using Gurson-based constitutive models. A previous study by the authors (Butcher et al. 2009) applied an advanced Gurson-based constitutive model to the hydroforming of DP600 steel that accounted for void shape effects, stress- and strain-based nucleation and coalescence due to both internal necking and shearing. While this model has shown some success in predicting formability, a large number of material parameters were required to describe each stage of damage evolution. The subsequent identification of these parameters can be prohibitive to the adoption of these models by industry. The dual bound approach provides a straightforward solution to this problem by focusing on using simpler damage-evolution rules and exploiting the upper and lower bound nature of the models. Instead of expecting a complicated single-bound model to provide an absolute prediction of ductility, the focus is to capture the behaviour within a band. In this manner, it is not essential that a single-bound completely describe the material behaviour just that the formability lies within the limits defined by both models.

The dual bound approach was applied to the straight-tube hydroforming model of Butcher et al. (2009) for DP600 steel tubes to determine the burst pressure, formability and failure location. Compressive axial loads of 0 and 133 kN were applied to the tube during forming to evaluate performance of the models in different loading conditions. A simple strain-controlled void nucleation rule was adopted for both damage models and calibrated to provide forming limits that capture the experiment data. Finally, the performance of the calibrated dual-bound models is compared with the formability predictions of Butcher et al. (2009) who used an advanced variant of the Gurson (1977) model.

2.11.1 Constitutive Modeling

The dual constitutive models are the same as used in the previous section with the exception that the q_{i} parameters in the GT model are set to unity to recover the upper bound solution of Gurson (1977) and that void nucleation is strain-controlled. The continuous nucleation model of Gurland (1972) was adopted where the nucleation rate is proportional to the plastic strain rate

$$\dot{f}_{nucleation} = A_N \dot{\varepsilon}^{\mathrm{p}} \tag{2.31}$$

where A_N is the nucleation intensity and $\dot{\varepsilon}^{\mathrm{p}}$ is the plastic strain rate. The continuous nucleation model only requires one parameter, A_N, to be identified from experiment compared to three parameters in the well-known Chu and Needleman (1980) model. Zhang and Niemi (1994a) demonstrated that the simpler continuous nucleation model can perform equally as well as the Chu and Needleman model (1980) where the nucleation intensity is assumed to follow a normal distribution. The authors have also observed that the continuous strain nucleation model can provide nearly equivalent predictions to the more complicated Chu and Needleman (1980) in some unpublished numerical studies.

2.11.2 Material Characterization

2.11.2.1 Material Properties

Tensile samples were obtained from the tubes at orientations of 3, 6 and 9 o'clock relative to the weld seam and averaged to determine the flow stress-strain relation (Bardelcik 2006). Although the stress state in hydroforming is biaxial stretching,

Table 2.2 Mechanical properties of DP600 steel tubes reported by Bardelcik (2006)

Material	E (GPa)	v	σ_y (MPa)	K (MPa)	n
DP600	206	0.30	413.54	795.8	0.115

Table 2.3 Controlling parameters in the material models

f_o	f_c (%)	f_f (%)	A_N
0	0.7	2	Identified parametrically

using the uniaxial flow stress relation is reasonable for DP600 since the forming limit curve (FLC) determined using sheet specimens gives very good agreement with the FLC determined using hydroformed tubes (Asnafi and Skogsgardh 2000). The averaged flow stress curve was converted to a plastic strain-true stress curve in the form of $\bar{\sigma} = K\varepsilon^n$ and linearly extrapolated from approximately 0.15 strain (ultimate tensile strength point) to 0.60. The tube stock had an average thickness of 1.85 mm with an outer diameter of 76.2 mm. The mechanical properties of DP600 are presented in Table 2.2.

2.11.2.2 Selection of Damage-Based Material Parameters

The DP600 steel tubes are composed of 5.5 % martensite in a banded-type formation and is considered to be initially damage-free with an initial porosity, f_o, of zero (Winkler et al. 2008). Uniaxial tensile tests were conducted by Winkler et al. (2008) using the same tube material used in the hydroforming experiments. 2-D metallographic analysis revealed the porosity (area fraction) near the fracture surface to be about 0.7 % and this value is taken as the critical porosity. The porosity at fracture is assumed to be 2 % (Maire et al. 2008). The only parameter required to be identified is the nucleation intensity, A_N, which is determined parametrically by comparing the numerical formability predictions of both damage models to the hydroforming test data. The value of A_N is selected that yields the best agreement with the experimental forming limits for both end-feed loads. The controlling parameters used in the damage-based constitutive models are shown in Table 2.3. Note the few parameters that exist in the model with the major parameters being the critical porosity and the nucleation intensity. If the plastic-limit load coalescence model was used then the critical porosity term can be removed from the analysis (Butcher et al. 2009). The initial porosity can safely be set to zero or a small value for a clean material such as DP600 and the porosity at fracture has minimal impact on the solution.

2.11.3 Finite-Element Model

The tube was modeled with 80,000 eight-node constant stress brick elements and the surfaces of the steel die and end-feed rams are rigid and meshed using

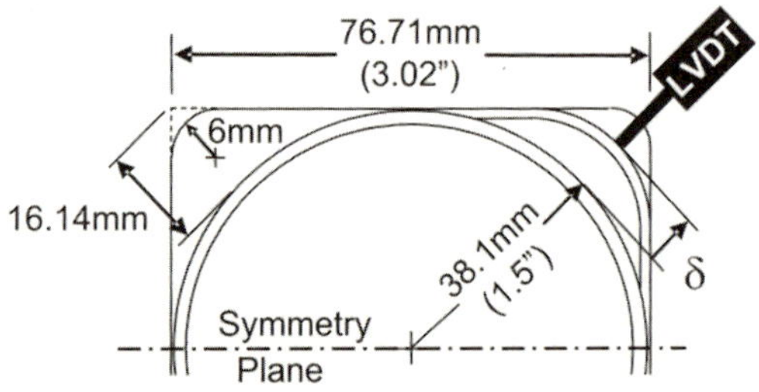

Fig. 2.25 Section of the hydroforming die showing corner-fill expansion (Reprinted with permission from Bardelcik (2006). Copyright 2006 Bardelcik)

quadrilateral shell elements. Contact between the tube, die and rams is modeled using a penalty-based contact algorithm available in LS-DYNA (Hallquist 2006). The coefficients of friction (COF) between the tube and die and tube and rams were experimentally determined using a twist-compression test as 0.035 and 0.08, respectively (Bardelcik 2006).

A null shell mesh coincident with the solid mesh of the tube is used to apply the pressure load. The tube is pre-pressurized to 35 MPa to avoid buckling before the end-feed load is applied and maintained using load control. The full EF load is applied from 35 to 38 MPa and then pressure is linearly increased to 152 MPa. For the zero end-feed case, the rams remain stationary and the end-feed load is adjusted to counteract the internal pressure on the face of the ram. The initial and formed tube models are shown in Fig. 2.24. The forming limit is detected at the onset of element deletion signifying localized necking and the onset of failure. The satisfaction of the post-coalescence criterion in Eq. (2.24) triggers element deletion in the finite-element model with widespread fracture of the tube occurring at the onset of element deletion.

2.11.4 Measuring Formability

Formability is defined using the corner-fill expansion (CFE) which measures the change in the cross-section of the tube from circular to rectangular with a CFE of 100 % corresponding to a perfect square. A section of the hydroforming die showing corner-fill expansion is presented in Fig. 2.25.

Corner-fill expansion is expressed as

$$\text{CFE }(\%) = \frac{100\delta}{\delta_{\max}} \tag{2.32}$$

The maximum corner distance between the tube and die is 16.14 mm. However, the maximum CFE achievable is 84.3 % of this distance due to the fillet corner radius as shown in Fig. 2.25. To present formability using an absolute scale, CFE in the present work is defined as

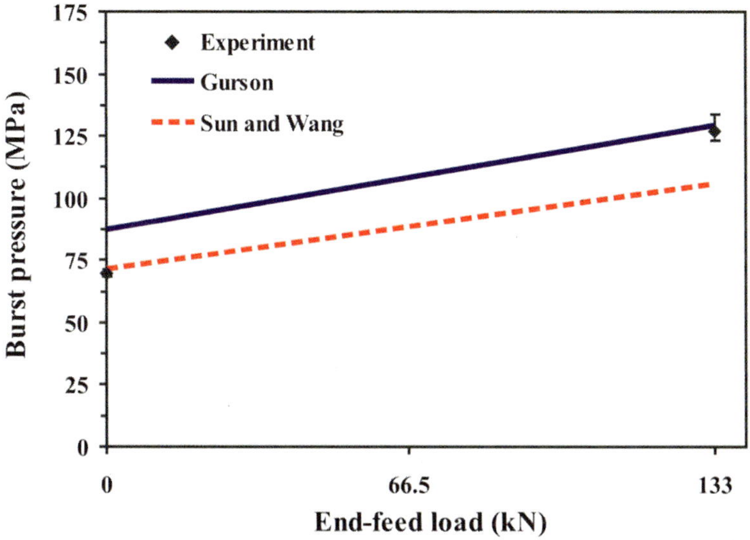

Fig. 2.26 Comparison of the experimental burst pressure with the *upper* and *lower* limits obtained using the Gurson and Sun and Wang material models for end-feed loads of zero and 133 kN. The nucleation intensity is 0.021

$$\text{CFE } (\%) = \frac{100}{0.843} \frac{\delta}{16.14 \text{ mm}} = 7.35\delta \tag{2.33}$$

Corner-fill expansion was measured at 90 % of the average burst pressure for each end-feed load and the results of Bardelcik (2006) have been restated according to Eq. (2.33). Further details related to the hydroforming experiments and development of the finite-element model can be found in Bardelcik (2006).

2.11.5 *Results*

A parametric study was conducted to determine the void nucleation intensity that provides good agreement with the experimental burst pressure, formability and failure location for each end-feed load and material model. A nucleation intensity of 0.021 successfully captures the experimental burst pressure and formability within the band defined by the upper and lower bound material models.

2.11.5.1 Tube Burst Pressure

The resulting band for burst pressure is compared with the experimental results in Fig. 2.26. The dual bound approach is able to capture the experimental burst

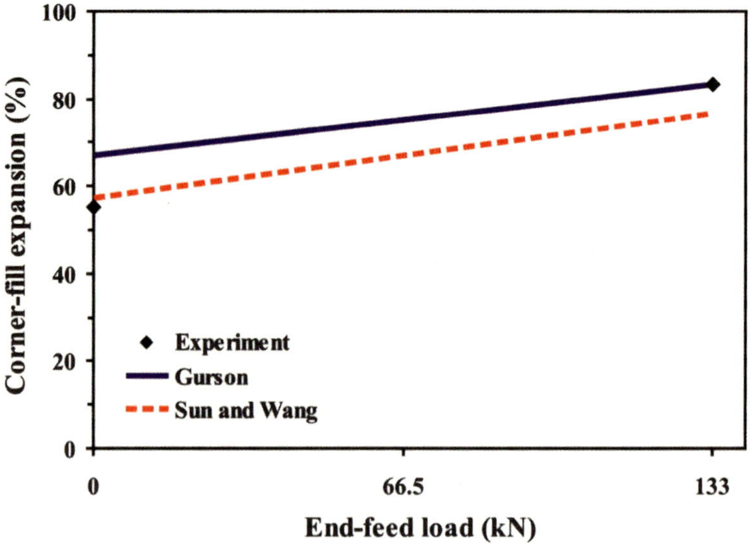

Fig. 2.27 Comparison of the experimental corner-fill expansion with the upper and lower limits obtained using the Gurson and Sun and Wang material models for end-feed loads of zero and 133 kN. The experimental and numerical corner-fill expansion is obtained at 90 % of the burst pressure. The nucleation intensity is 0.021

pressure for both end-feed loads. The lower bound model gives best agreement in the 0 kN EF case while the upper bound Gurson model performs best for the 133 kN EF load. Conversely, the Gurson and SW models over/underestimate the burst pressure for zero and 133 kN EF, respectively. It is likely that the upper bound Gurson model performs well in the high EF case as EF promotes a more uniform stress state, thus the material can be formed to higher strain levels, which better matches the upper bound approximation of the material. The addition of a compressive axial load during tube expansion reduces the stress triaxiality resulting in a lower void growth rate that is better described by the more rigid Gurson model. The dual bound approach can reliably predict fracture in both EF cases as well as define a range for the burst pressure.

2.11.5.2 Corner-Fill Expansion

The formability band for corner-fill expansion is compared with the experiment in Fig. 2.27. The experimental and numerical CFE were obtained at 90 % of the burst pressure. No standard deviation for CFE was reported as the tests showed little variation (Bardelcik 2006). Similar to the trend for burst pressure, the lower bound model is able to accurately predict the CFE for zero EF while the upper bound model obtains excellent agreement for the 133 kN EF case. Overall, the

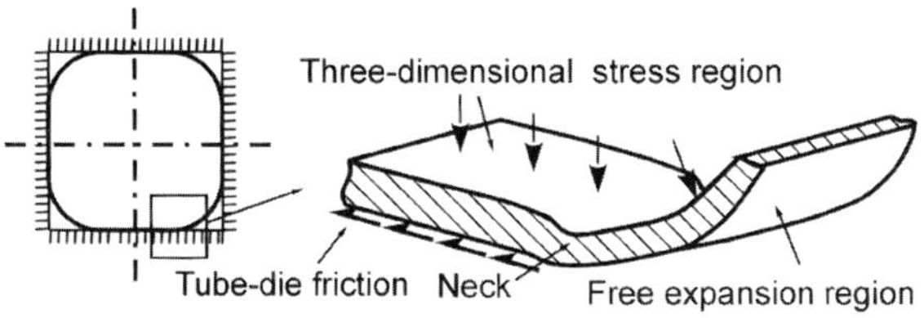

Fig. 2.28 Schematic of the stress state and localization during straight tube hydroforming (Reprinted with permission from Simha et al. (2007). Copyright 2007 ASME)

experimental results of CFE are captured within the formability band defined by the upper and lower bound models, which clearly demonstrates the advantage of the dual bound approach since CFE could not be described by a single bound model.

2.11.5.3 Failure Location

Regardless of the end-feed load, tube failure occurs in the transition region where the tube loses contact with the die and becomes free to expand (Bardelcik 2006). The material in this 'free-expansion zone' in Fig. 2.28 is under a state of plane-stress while the remaining material is under a three-dimensional state of stress. The friction between the tube and die and through-thickness compressive load due to internal pressure retard material flow into the plane-stress free-expansion zone, leading to the formation of a localized neck. The formation of a localized neck increases the local plastic strain and stress triaxiality, driving void nucleation and damage development leading to coalescence and fracture. As shown in Fig. 2.29, the application of a compressive end-feed load reduces the severity of the stress state and delays damage evolution enabling forming to higher strains.

The experimentally observed failure locations and porosity contours for 0 and 133 kN end-feed are compared with the failure location in the FE models in Figs. 2.30 and 2.31, respectively. No quantitative results for failure location were reported by Bardelcik (2006) so the comparison is qualitative. The upper and lower bound models both predict that damage becomes localized in the transition region bordering the free-expansion zone.

2.11.5.4 Void Damage

The void damage histories are presented in Fig. 2.32, which were obtained from the first element to fail for each end-feed case. The porosity histories for both upper and lower bound models exhibit a similar trend, with the softer SW model experiencing a faster rate of damage development compared to the more rigid Gurson model. End-feed effect suppresses nucleation because the compressive load forces more material into the die, reducing the effective strain and damage evolution.

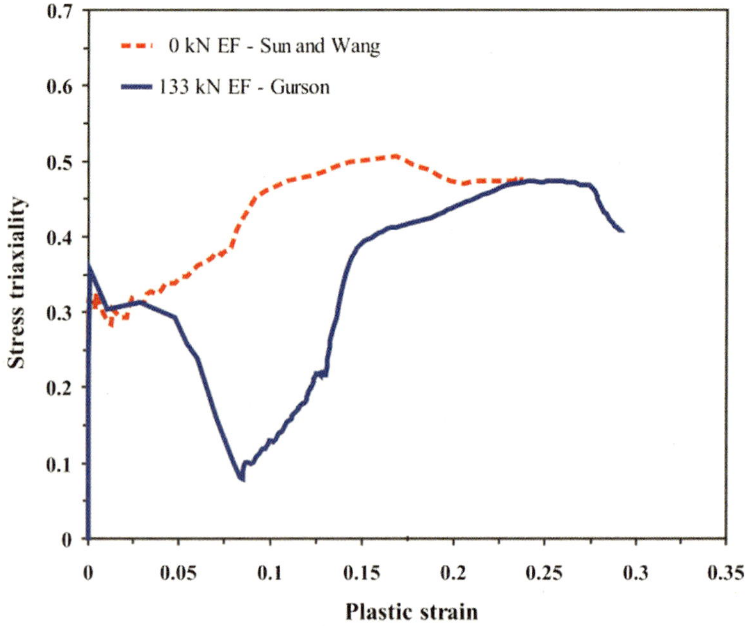

Fig. 2.29 History of stress triaxiality within the first element to fail for end-feed loads of zero and 133 kN. The stress state is obtained from the material model which gives the best performance for a particular end-feed load

Fig. 2.30 Experimental failure locations for end-feed loads of: (**a**) 0 kN and (**b**) 133 kN (Reprinted with permission from Butcher et al. (2009). Copyright: Springer)

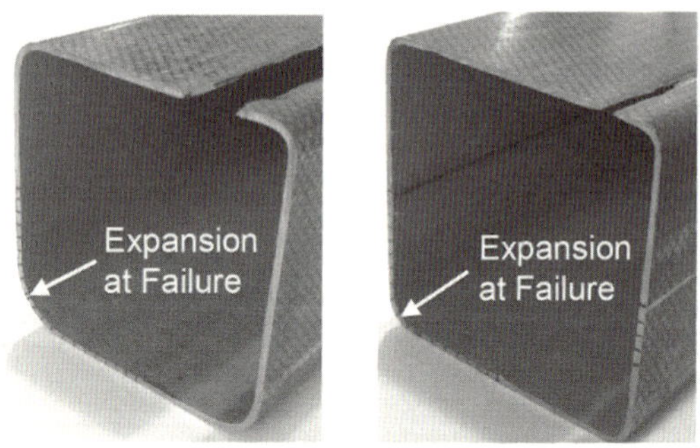

The dual bound approach enables a porosity band to be obtained for each loading condition which is valuable for comparison with experiment due to the variation present in porosity measurements. Experimental porosity data was not available for the fractured tubes and the porosity trends should be considered qualitative in nature. Overall, the porosity trends seem to be physically reasonable since the initial, critical and final porosities are based upon experimental observations.

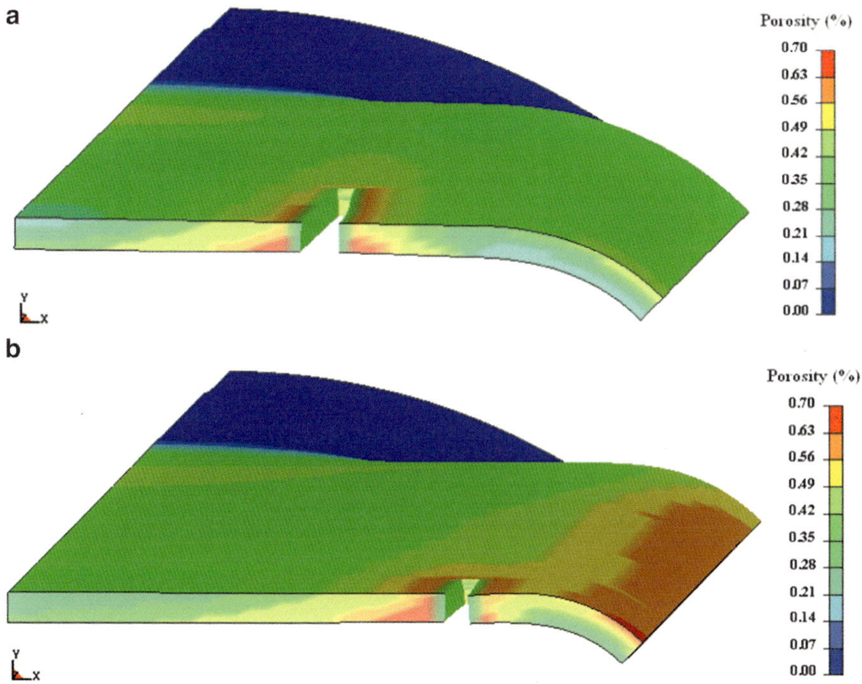

Fig. 2.31 Failure location and porosity contours for end-feed loads of (**a**) 0 kN using the Sun and Wang model and (**b**) 133 kN using the Gurson model. The failure location is obtained from the material model which gives the best performance for a particular end-feed load. One-eighth of the tube is modeled due to symmetry

2.11.6 Evaluation of the Dual Bound Approach for Tube Hydroforming

The advantages of the dual bound approach are clearly illustrated in Fig. 2.33 which compares the dual bound burst pressures to the results of Butcher et al. (2009) who employed a more sophisticated variant of the Gurson-Tvergaard-Needleman (GTN) model. This extended model accounted for the influence void shape evolution and coalescence due to internal necking and shearing of the inter-void ligaments. The stress- and strain-based nucleation models of Chu and Needleman (1980) were also considered. Despite the improved physical foundation of this model, it performed no better than the Gurson (1977) model. Only the dual bound approach could capture the burst pressure for each end-feed load.

It is important to mention that the small difference between the burst pressures obtained using the Gurson (1977) and the variant employed by Butcher et al. (2009) is coincidental because the two independently calibrated models possess different

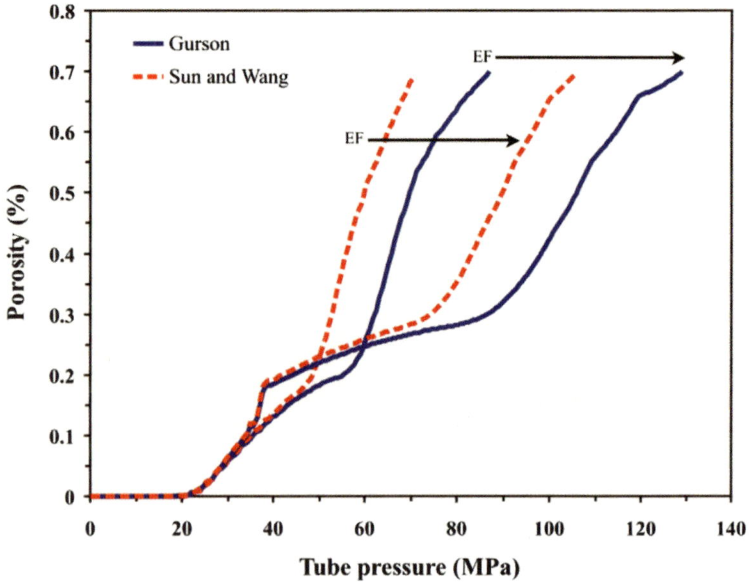

Fig. 2.32 Damage evolution within the first element to fail for end-feed loads of zero and 133 kN obtained using the Gurson and Sun and Wang constitutive models. The nucleation intensity is 0.021

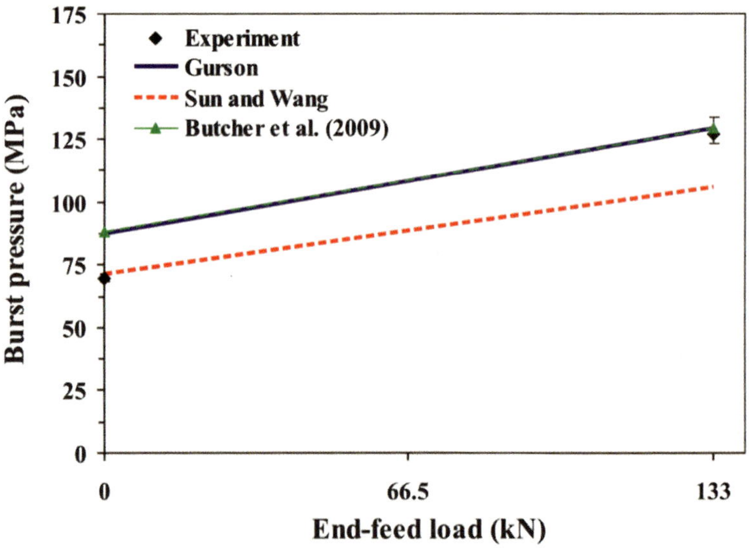

Fig. 2.33 Comparison of the experimental burst pressure with the forming limits obtained using the dual bound approach and the extended Gurson-based model of Butcher et al. (2009) for strain-controlled nucleation

models for damage evolution. Consequently, there are many combinations of models and parameters that can produce similar results. While a direct comparison is not possible, Fig. 2.33 reveals that none of the three models considered could capture the burst pressure for both end-feed loads. Only by combining the upper and lower bound models into a formability band could the burst pressure be properly described. This is a very attractive result because the dual bound approach employs simpler damage models with fewer parameters and may obtain burst pressures as good as, or better than, a single advanced model.

2.12 A Dual Bound Approach to Determining the Void Nucleation Parameters in Sheet Materials

The success of Gurson-based constitutive models in predicting ductile fracture depends on the identification of the material parameters governing void nucleation and coalescence. The void growth rule in the Gurson (1977) model does not require any material parameters since the voids are assumed to remain spherical. By adopting a coalescence rule based upon the stress state and microstructure geometry, the plastic limit-load criterion of Thomason (1990) has enabled void nucleation to be directly linked to fracture by removing the critical porosity as a material constant (Zhang 1996). The nucleation parameters in a Gurson-based material model can now be uniquely determined using notched tensile tests.

However, calibrating the nucleation parameters using a single-bound model like the GT yield criterion introduces an unavoidable bias because the parameters will reflect the formulation of the yield criterion. For example, void nucleation is used to effectively soften the Gurson (1977) model by increasing the porosity (lower nucleation stress/strain) to achieve coalescence at the experiment fracture strain. Conversely, the calibrated nucleation stress/strain would be higher in the Sun and Wang (1989) model to effectively stiffen the material by suppressing damage to avoid premature failure before the experimental strain is reached in the simulation. This bias can be mitigaged by using a dual bound approach since the nucleation parameter is identified using both bounds and thus captures the material behaviour within a band as shown in Fig. 2.34.

In this section, the dual bound approach to ductile fracture is used to identify the nucleation stress and strain parameters in AA5182 sheet using notched tensile specimens. Three specimen geometries were considered to identify the parameters over a range of stress states commonly found in a sheet metal forming process. The geometry of the microstructure is characterized from particle field measurements and implemented into the coalescence model to obtain physically reasonable porosities at fracture.

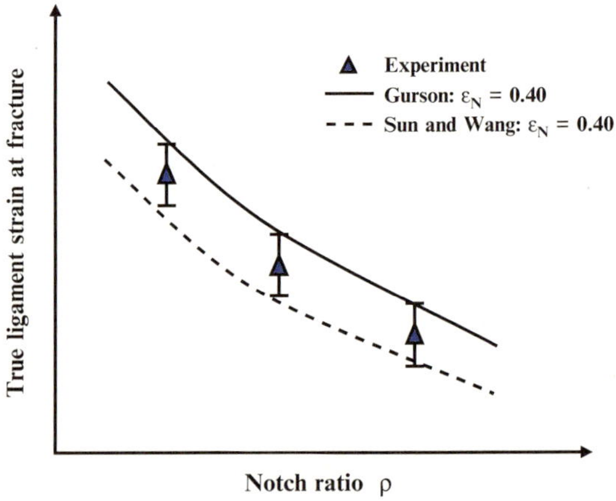

Fig. 2.34 Schematic representation of fitting the nucleation strain using the dual bound approach for notched tensile tests. In notched tensile specimens, a higher notch ratio corresponds to a higher stress triaxiality

2.12.1 Constitutive Modeling of Ductile Fracture

The dual bound constitutive models described in the previous sections were adopted with the exception of removing the critical porosity coalescence criterion and replacing it with a modified plastic-limit load criterion that eliminates the critical porosity as a material parameter. The same stress- and strain-controlled nucleation models of Chu and Needleman (1980) were again considered for this alloy, AA5182.

2.12.1.1 Void Coalescence

The PLL of Thomason (1990) was derived using an upper bound approximation to the material behaviour and should not be implemented into the lower bound SW model to ensure a consistent upper and lower bound analysis. This inconsistency is overcome by adopting the PLL model of Ragab (2004b) in Eq. (2.27) who replaced the plastic constraint factor derived by Thomason (1990) with the notch constraint factor of Bridgman (1952) for axisymmetric loading that was not derived using either upper or lower bound approximations. It is important to note that the use of a Thomason-based variant of the plastic constraint factor in the lower bound SW model is a minor issue and is only addressed in this work to provide a strict representation of the dual bound approach.

An additional consideration in selecting the Ragab variant of the plastic limit load in this application is because it has been shown to give much better fracture predictions than the variants of Pardoen and Hutchinson (2000) or Benzerga (2002) when applied to a large range of real materials (Ragab 2004b). While these variants of the plastic limit-load are accurate when compared to unit cell simulations, they tend to overpredict fracture in actual materials because the influence of secondary voids, shearing and other conditions are not considered. The Ragab model provides

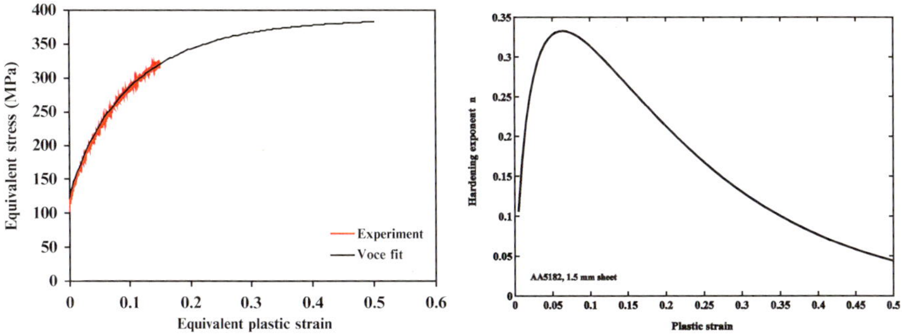

Fig. 2.35 Average experimental flow stress relation for AA5182 sheet with its Voce law representation (*left*) and the variation of the material hardening exponent obtained from the Voce law

a much lower plastic constraint condition and coalescence occurs sooner and is in better agreement with the experiments. From a physics perspective, the Pardoen and Hutchinson or Benzerga variants are the best but from an engineering perspective, the Ragab model is preferable. Coalescence occurs for spherical voids in the Ragab (2004b) PLL model when the condition in Eq. (2.27) is satisfied.

2.12.1.2 Material Properties

The flow stress relation for 1.5 mm thick AA5182-O sheet was determined using three standard tensile test specimens as shown in Fig. 2.35. The average flow stress response was expressed as a Voce hardening law (Voce 1948):

$$\bar{\sigma} = \sigma_s - \left(\sigma_s - \sigma_y\right) \exp\left[-\alpha(\bar{\varepsilon}^p)^\beta\right] \tag{2.34}$$

with $\sigma_y = 122.7$ MPa, $\sigma_s = 398.1$ MPa, $\alpha = 7.631$ and $\beta = 0.905$. AA5182 alloys exhibit different hardening regimes as shown in Fig. 2.35b where it initially resembles a high-strain hardening material with a peak hardening exponent of about 0.33 before the flow stress saturates and the matrix resembles that of a perfectly plastic material, $n \to 0$. The strain hardening parameter of AA5182 is computed at each stage during deformation using the relation

$$n = \frac{\bar{\varepsilon}^p}{\bar{\sigma}} \frac{d\bar{\sigma}}{d\bar{\varepsilon}^p} \tag{2.35}$$

The modulus of elasticity was estimated from the tensile tests to be 65.33 GPa and the Poisson's ratio was assumed to be the standard 0.33 for aluminum alloys.

Table 2.4 Controlling parameters in the Gurson and SW material models

f_o (%)	f_n (%)	f_f (%)	λ_{2o}	c	s_N (% ε_N, σ_N)	ε_N	σ_N
0.053	1.23	2.00	8.68	0.35	15	Identified parametrically	Identified parametrically

2.12.1.3 Selection of Damage-Based Material Parameters

The average area fraction of second-phase particles is 1.23 % (Chen 2004; Lievers et al. 2004; Orlov 2006) and is essentially damage-free with an initial porosity of 0.053 % (Orlov 2006) and failure porosity of 2 % (Chen 2004). The second-phase particles are essentially spherical with an average aspect ratio of 1.14 (Chen 2004) and exhibit significant clustering with an average particle spacing ratio, χ_p, of 0.588 (Butcher and Chen 2009b, c). The initial unit cell aspect ratio must be determined from the particle distribution to characterize the microstructure in the coalescence model. Since voids are nucleated at the particles, the initial unit cell geometry for a periodic arrangement of cubic unit cells can be obtained from the particle field data. Therefore, the aspect ratio of the cubic unit cell containing a spherical particle is determined to be $\lambda_{2o} = \frac{\pi}{6}\frac{\chi_p^3}{f_p} = 8.68$. It is assumed that area-based estimates of the microstructural variables are reasonable approximations to the volume-based measurements used in the damage-based constitutive model.

The only remaining parameters are related to void nucleation. Typically, nucleation in this alloy has been assumed to obey the strain-controlled model in Eq. (1.5) with nucleation strains reported in the range of 0.20–0.70 (Lievers et al. 2004; Butcher et al. 2006; Chen and Worswick 2008; Butcher and Chen 2009a, b, c) and nucleation stress of 440–500 MPa (Butcher et al. 2006). The standard deviation of the nucleation stress/strain in Eqs. (1.5) and (1.7) is assumed to be 15 % the nucleation stress or strain (Chen 2004; Orlov 2006). The nucleation stress and strain will be determined parametrically through calibration with the tensile test data (Table 2.4).

2.12.2 *Notch Tensile Test Experiment*

Notched tensile specimens of 1.5 mm thick AA5182 sheet with a notch radius, R, of 3 mm and gage length, L, of 12.5 mm were tested to failure in an Instron 1,332 testing machine with a crosshead velocity of 0.03 mm/s. The notch ligament length was varied to achieve various stress states and was characterized using the notch ratio defined as

$$\rho = \frac{2R}{w} \tag{2.36}$$

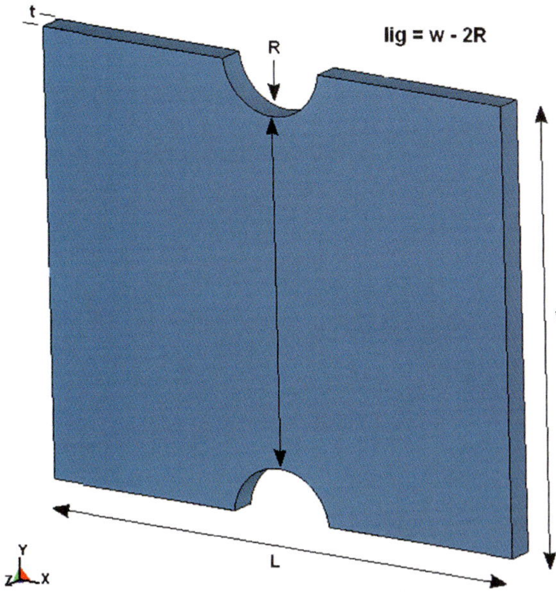

Fig. 2.36 Geometry of a typical notched tensile sheet specimen. The loading and rolling directions are collinear with the x-axis. The specimen is symmetric about the *x*, *y* and *z* axes

where w is the sample width. The notch ratios considered were 0.25, 0.33 and 0.50 and correspond to ligament lengths of 18, 12 and 6 mm. A total of six specimens for each notch ratio were tested to failure. All of the samples were fabricated with the loading direction aligned with the rolling direction of the sheet. The typical specimen geometry is presented in Fig. 2.36.

Notched tensile specimens are used to alter the stress state in the ligament and force localization to occur within the ligament. Two experimental fracture strains are used in the calibration of the nucleation models: (i) the ligament strain, $\varepsilon_{\text{lig}_f}$, and (ii) the axial strain ε_{a_f}. The ligament strain is adopted because it is representative of deformation where the specimen fractures. The axial strain at failure is used to provide a metric that is independent of the fractured region and representative of bulk deformation. The axial and ligament strains at fracture are defined as

$$\varepsilon_{\text{a}_f} = \ln\left(\frac{L_f}{L_o}\right) \qquad \varepsilon_{\text{lig}_f} = \ln\left(\frac{\text{lig}_f}{\text{lig}_o}\right) \tag{2.37–2.38}$$

where the initial ligament length is $lig_o = w - 2R$. The axial strain at failure is recorded at the appearance of a macro-crack at the notch root and not final failure since the objective of the finite-element models is to predict the formation of a macro-crack and not the subsequent tearing process of the ligament. Finite-element modeling of this process would require additional modelling considerations such as the use of cohesive elements.

Tensile specimens with notch ratios of ~1/3 or lower will exhibit visible cracking at the notch root prior to fracture as shown in Fig. 2.37. The onset of cracking at the

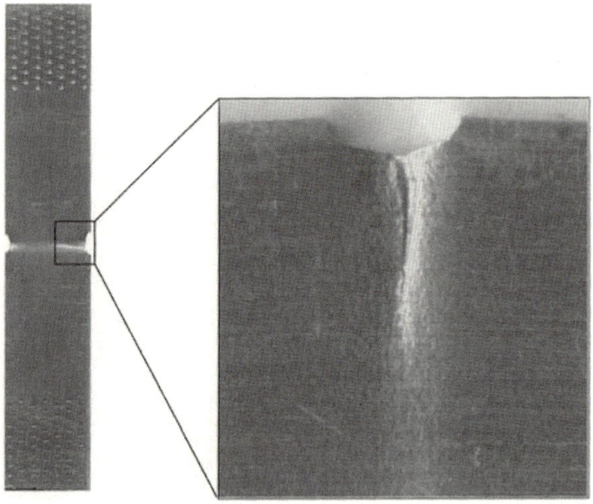

Fig. 2.37 Crack initiation at the notch root in a 1.5 mm thick AA5182 tensile specimen with a notch ratio of 0.125 and notch radius of 1 mm

notch root is readily identified from the experimental load-elongation curve as the appearance of the crack corresponds to an abrupt load drop as shown in Fig. 2.38.

2.12.3 Finite-Element Model

Due to symmetry, only one-eighth of the geometry is represented in the finite-element model. A velocity boundary condition is applied to the free-end of the specimen. A sinusoidal velocity profile is used to limit inertial effects with a peak velocity corresponding to a strain rate of 100 s^{-1}. As shown in Fig. 2.39, the tensile specimens were modeled with eight-node constant stress brick elements with 24,000, 18,000 and 12,000 elements for notch ratios of 0.25, 0.33 and 0.50, respectively. The mesh sensitivity for each notch ratio was negligible.

2.12.4 Identification of the Fracture Strains

The finite-element simulation is terminated at the onset of element deletion signifying localized necking and the onset of failure. The satisfaction of the post-coalescence criterion in Eq. (2.24) triggers element deletion in the finite-element model with widespread fracture of the specimen occurring at the onset of element deletion. The resulting ligament strain and elongation-to-failure are then obtained from the finite-element model and compared with the experiment values. The optimal nucleation parameter is identified when the formability band best captures the material behaviour over the range of notch ratios. Since the objective is to

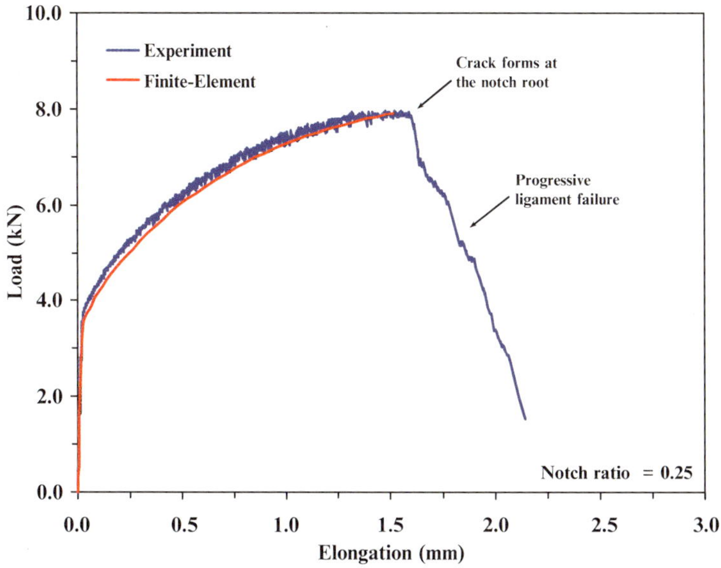

Fig. 2.38 Comparison of the average experimental and predicted load-curve for an AA5182 notched tensile specimen using the GT model. The predicted load-curve using the SW model is not presented because there was no visible difference in the response. The insensitivity of the load curve to the damage model is due to the small initial porosity and the fact that significant damage evolution occurs in a very small region of the notch. Tensile samples with a smaller notch ligament and a higher porosity at fracture would result in a larger difference in the predicted load-displacment responses

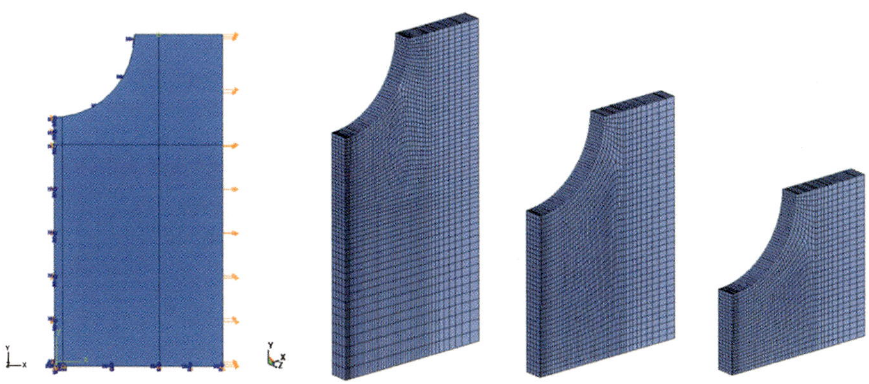

Fig. 2.39 Typical boundary conditions of the one-eighth FE model of the tensile specimens and the respective meshes for notch ratios of 0.25–0.50

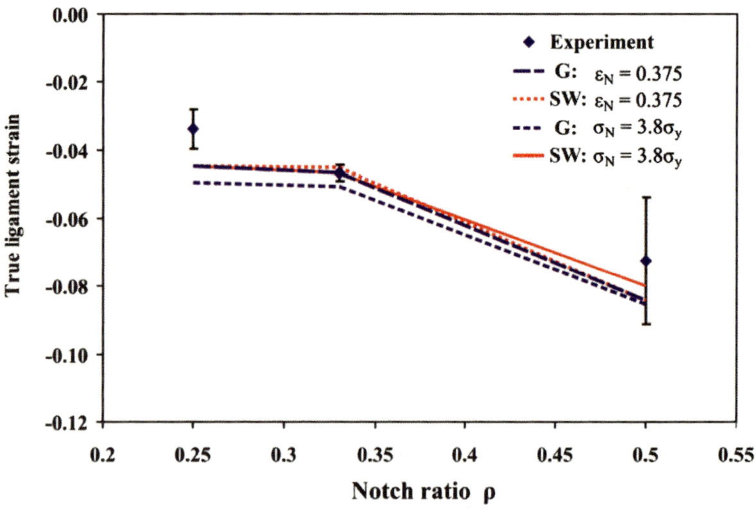

Fig. 2.40 Comparison of the experimental and numerical true ligament strains determined by the Gurson and SW models using stress- and strain-controlled nucleation for various notch ratios. The experimental true ligament strain is presented as a 95 % confidence interval

capture the fracture behaviour within a band, the experimental fracture strains are expressed using two-tailed 95 % confidence intervals.

2.12.5 *Results and Discussion*

2.12.5.1 Ligament Strain and Elongation-to-Failure

A parametric study was conducted to calibrate the void nucleation models in the Gurson and SW material models using notched tensile test data. Stress- and strain-controlled nucleation models with values of $\sigma_N = 3.8\sigma_y$ (466.26 MPa) and $\varepsilon_N = 0.375$ can provide good agreement with the experimental ligament strain and elongation-to-failure as shown in Figs. 2.40 and 2.41, respectively. The nucleation stress is in good agreement with the range of 440–500 MPa suggested by Butcher et al. (2006). The nucleation strain is also reasonable as Butcher and Chen (2009a, b, c) and Orlov (2006) suggested values of 0.42 and 0.30 in uniaxial tension. The performance of the dual bound approach is sensitive to the assumed nucleation model. A single bound model is sufficient if strain-controlled nucleation is adopted since the difference in the fracture predictions of the Gurson and SW models is marginal. The opposite behaviour is observed using stress-based nucleation which yields a meaningful formability band that captures the material behaviour and highlights the benefits of the dual bound approach.

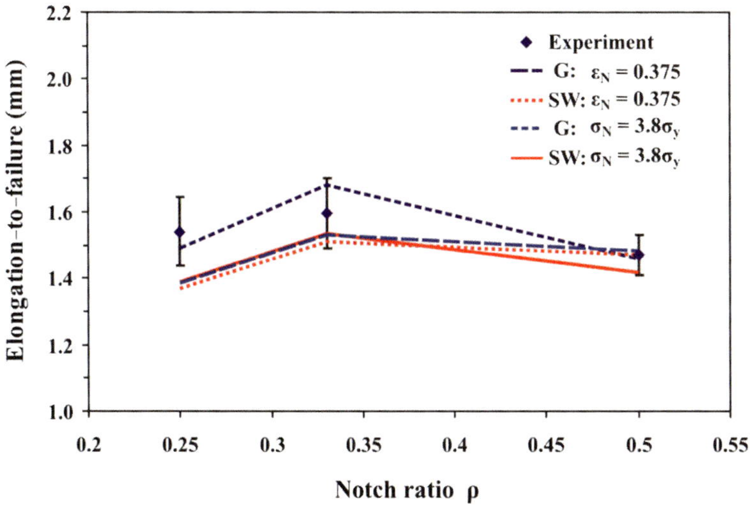

Fig. 2.41 Comparison of the experimental and numerical elongation-to-failure determined by the Gurson and SW models using stress- and strain-controlled nucleation for various notch ratios. The experimental elongation is presented as a 95 % confidence interval

2.12.5.2 Load-Elongation

In addition to providing good agreement with the ligament strain and elongation-to-failure, both the Gurson and SW models provide excellent agreement with the experimental load-elongation curves for each notch ratio as shown in Fig. 2.42. A crack originated at the notch root for notch ratios of 0.25 and 0.33 which then propagated throughout the ligament leading to fracture. The onset of cracking in these specimens is accompanied by a sudden load drop as seen in Fig. 2.42. Fracture occurred abruptly for a notch ratio of 0.50 with no visible cracks at the notch root.

2.12.5.3 Fracture Location

The predicted fracture locations and porosity contours obtained using both stress and strain-controlled nucleation are presented in Figs. 2.43 and 2.44, respectively. Damage development in the stress-controlled nucleation model mimics the stress triaxiality distribution and damage occurs over a larger area compared to the strain-controlled nucleation model. The fracture locations for both nucleation models are in good qualitative agreement with the experiment as fracture originates at the notch root for $\rho = 0.25$ and 0.33. The failure location for $\rho = 0.50$ is unknown since the ligament abruptly fails with no visible cracking at the notch root.

Void damage is highly localized in the strain-controlled nucleation model as the plastic strain is highest at the notch root for each notch ratio. Void nucleation occurs

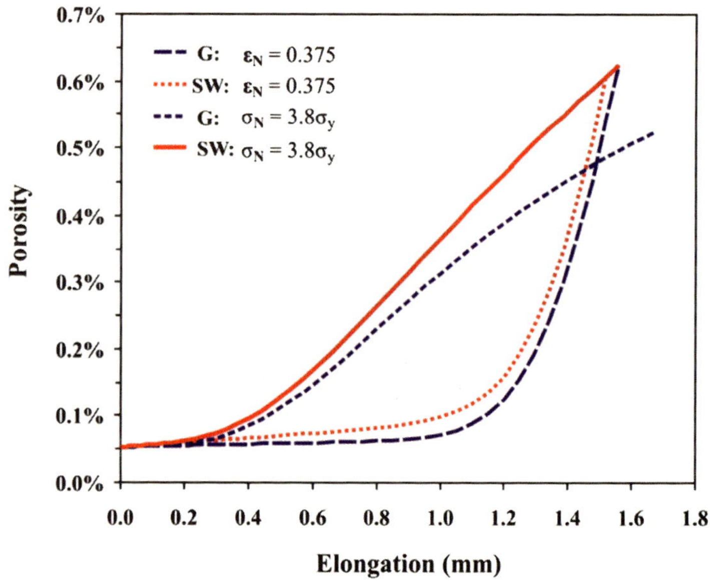

Fig. 2.45 Damage evolution within the first element-to-fail for a notch ratio of 0.33 using both *upper* and *lower* bound models and stress and strain-controllednucleation rules. The trends are similar for the other notch ratios

SW model and slower in the more rigid Gurson model. Consequently, the fracture strains predicted by the Gurson and SW models diverge and define a formability band.

In the strain-based nucleation model, void nucleation is negligible until the latter stages of deformation where rapid nucleation occurs leading to sudden fracture. This burst of nucleation overshadows any difference in the porosity between the Gurson and SW models due to void growth and fracture occurs at similar strains. In this case, coalescence is nucleation driven with marginal void growth.

The porosities at the onset of coalescence are physically reasonable which suggests the microstructure characterized using the particle distribution is representative of the material in the coalescence model. The porosity at coalescence ranges from about 0.5 to 0.7 % which is in general agreement with the experimental results of Smerd et al. (2005) who reported porosities near the fracture surface of tensile specimens of 0.5–1 % for various strain rates. A detailed analysis of the AA5182 microstructure in the notch tensile tests is discussed in Chap. 7 in regards to damage percolation modelling. Overall, both stress and strain-controlled nucleation models can be adopted for AA5182 sheet. It is important to mention that the Gurson and SW models are approximate yield criteria that are based on many simplifying assumptions and the calibration of the nucleation models will artificially correct for some of their deficiencies. The present results should only be considered valid for these specific nucleation rules and are dependent upon the coalescence model and its parameters.

Chapter 3
Anisotropy

Cold worked and extruded materials always exhibit a measure of anisotropy, or "texture" where the mechanical properties exhibit directional properties. The rolling process used to create sheet metals orients the material grains and precipitates/inclusions in the rolling direction and thus induces anisotropy. Typically, bulk materials that exhibit texture effects are treated as orthotropic while sheet metals are commonly assumed to possess planar isotropy and normal anisotropy. To characterize the anisotropy of sheet materials, uniaxial tensile tests are performed with the samples fabricated from the material in different directions relative to the rolling direction. The R-value quantifies the measure of anisotropy and is defined as the ratio of the transverse strain to the through-thickness strain as shown in Fig. 3.1 and Eq. (3.1)

$$R = \frac{\varepsilon_w}{\varepsilon_t} \tag{3.1}$$

An isotropic material will have an R-value of unity while a higher R-value indicates that the material has a higher resistence to thinning with higher through-thickness strength. An average value of the anisotropy parameter should be computed from the weighted average of samples obtained from the material at orientations of 0°, 45° and 90° to the rolling direction as

$$R = \frac{R_0 + 2R_{45} + R_{90}}{4} \tag{3.2}$$

The R-value is generally taken as a material constant evaluated at a typical strain of 20 % in the tension test although some anisotropic yield criterion are capable of modeling a dynamic R-value that evolves with deformation.

Z. Chen and C. Butcher, *Micromechanics Modelling of Ductile Fracture*,
Solid Mechanics and Its Applications 195, DOI 10.1007/978-94-007-6098-1_3,
© Springer Science+Business Media Dordrecht 2013

Fig. 3.1 Definition and measurement of normal anisotropy from a uniaxial tensile test

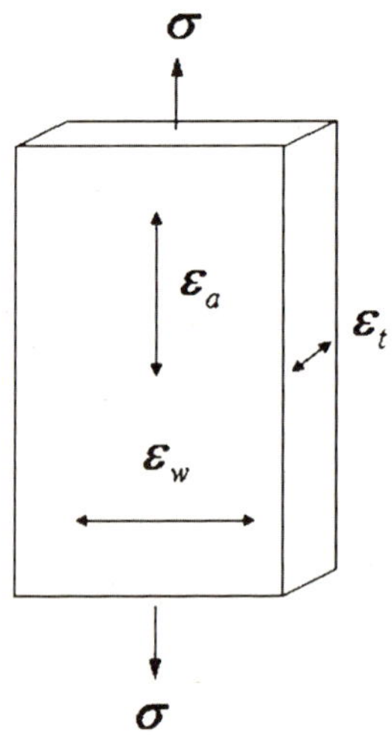

3.1 The Hill-48 Anisotropic Yield Criterion

The yield criterion of Hill (1948) has been widely used to characterize the anisotropy of sheet metals and can be considered the default or standard anisotropic model like the von Mises criterion is for isotropic materials. The Hill-48 criterion is not the ideal anisotropic yield criterion for many materials but it is a straightforward model that is readily implemented into numerical codes and well suited for analytical modeling. Additionally, the criterion requires a small number of physically-based parameters that can be directly identified from a series of tensile tests. For sheet metals where the stress state is approximately plane stress, only four parameters are required. More advanced non-quadratic anisotropic yield criterion such as the Barlat yield functions (Barlat 1987; Barlat and Lian 1989; Barlat et al. 1991, 1997; Cazacu and Barlat 2003; Cazacu et al. 2006) are widely used in industrial applications but are not well suited for analytical study due to the large number of phemenological calibration parameters and complex flow rules. The Hill-48 quadratic yield criterion for orthotropic materials can be expressed as follows

$$2\Phi(\sigma_{ij}) = F(\sigma_{22} - \sigma_{33})^2 + G(\sigma_{33} - \sigma_{11})^2 + H(\sigma_{11} - \sigma_{22})^2 + 2L\sigma_{23}^2 + 2M\sigma_{31}^2 + 2N\sigma_{12}^2 = 1 \quad (3.3)$$

where the six material constants, F, G, H, L, M, N, define the anisotropic properties of the yield surface. If the tensile yield stresses in the principal anisotropic directions are denoted as σ_0, σ_{90} and σ_t, that correspond to the rolling, transverse and thickness directions of sheet materials, the anisotropic constants are expressed as

$$\frac{1}{\sigma_0^2} = G + H \quad \frac{1}{\sigma_{90}^2} = H + F \quad \frac{1}{\sigma_t^2} = F + G$$
$$2F = \frac{1}{\sigma_{90}^2} + \frac{1}{\sigma_t^2} - \frac{1}{\sigma_0^2} \quad 2G = \frac{1}{\sigma_t^2} + \frac{1}{\sigma_0^2} - \frac{1}{\sigma_{90}^2} \quad 2H = \frac{1}{\sigma_0^2} + \frac{1}{\sigma_{90}^2} - \frac{1}{\sigma_t^2} \tag{3.4a–f}$$

The remaining parameters can be obtained from the shear yield stresses as

$$2L = \frac{1}{\tau_0^2} \quad 2M = \frac{1}{\tau_{90}^2} \quad 2N = \frac{1}{\tau_t^2} \tag{3.5a–c}$$

For plane stress, the yield criterion reduces to

$$2\Phi(\sigma_{ij}) = (G + H)\sigma_{11}^2 - 2H\sigma_{11}\sigma_{22} + (H + F)\sigma_{22}^2 + 2N\sigma_{12}^2 = 1 \tag{3.6}$$

and the anisotropy coefficients can be related to the R-values using the associated flow rule to obtain

$$R_0 = \frac{H}{G} \quad R_{90} = \frac{H}{F} \quad R_{45} = \frac{N}{F + G} - \frac{1}{2} \tag{3.7a–c}$$

The relationship between the yield stresses and R-values in the plane of the sheet can be defined as

$$\frac{\sigma_0}{\sigma_{90}} = \sqrt{\frac{R_0(1 + R_{90})}{R_{90}(1 + R_0)}} \tag{3.8}$$

From Eq. (3.8), the condition for the applicability of the Hill-48 criterion is that when $R_0 > R_{90}$, $\sigma_0 > \sigma_{90}$ and this is not the case for some materials such as certain aluminum alloys. Assuming planar isotropy, $R_0 = R_{45} = R_{90} = R$, the yield criterion reduces to

$$\frac{1}{1 + R}\left[\sigma_{11}^2 + \sigma_{22}^2 - R(\sigma_{11} - \sigma_{22})^2 + 2(2R + 1)\sigma_{12}^2\right] - \sigma_0^2 = 0 \tag{3.9}$$

3.2 Material Anisotropy in Porous Ductile Materials

The majority of research in developing damage-based constitutive models has focused on assuming a void with a constant shape (typically spherical or cylindrical) embedded within an incompressible isotropic matrix. In the past decade, a concerted effort has been made to account for void-induced anisotropy effects by

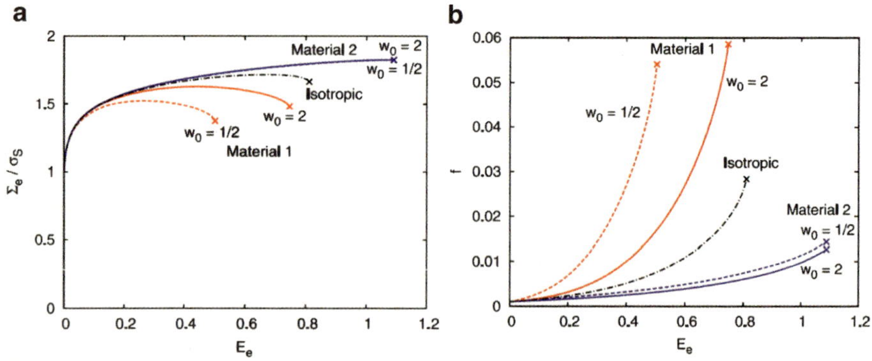

Fig. 3.2 Results of unit-cell calculations for two transversely isotropic matrix materials containing either oblate (w0 = 1/2) or prolate (w0 = 2) voids. (**a**) Normalized effective stress with the effective strain and (**b**) evolution of the void volume fraction. The initial porosity is 0.1 % with a matrix hardening exponent of 0.1 with a constant stress triaxiality ratio of unity. For comparison, the results for an initially spherical void in an isotropic matrix are shown. (Reprinted with permission from Keralavarma and Benzerga 2010. Copyright 2010 Elsevier)

relaxing the constraint that the void shape remains constant and instead modelling the void as a spheroid (ellipsoid of revolution) that evolves with deformation. It has been shown in many works that void shape-induced anisotropy can have a considerable influence on the material response (Budiansky et al. 1981; Gologanu et al. 1997; Kailasam and Ponte Castenada 1998; Pardoen and Hutchinson 2000; Aravas and Ponte Casteneda 2004; Danas and Ponte Casteneda 2009). However, relatively few studies have addressed the issue of anisotropic behaviour of the matrix material. Liao et al. (1997) extended Gurson's (1977) model for cylindrical voids to account for anisotropy in an approximate manner that is suitable for materials with normal anisotropy. Similar extensions were performed by Chein et al. (2001) and Wang and Pan (2004). The assumption of a constant spherical void in these anisotropic models is justified since considering void shape evolution significantly increases the complexity of the model because the void orientation vectors must be considered along with the directions of anisotropy. The influence of mechanical anisotropy on the response of a porous material is presented in Fig. 3.2.

Benzerga and Besson (2001) first accounted for the influence of both void shape and orthotropy and recently Moncheit et al. (2008) performed a limit analysis to obtain an analytical solution for a matrix material with elliptical voids that obeys the Hill (1948) model for anisotropy. A few attempts have been made in this area in recent years through the continual development of advanced anisotropic damage-based yield criteria that also account for anisotropy effects due to void shape (Benzerga et al. 2004) and orientation (Danas and Aravas 2012; Keralavarma and Benzerga 2010).

The lack of attention to material anisotropy in the development of damage-based material models can perhaps be attributed to the inherent difficulty in developing a general anisotropic model. Researchers in this area rely upon unit cell simulations

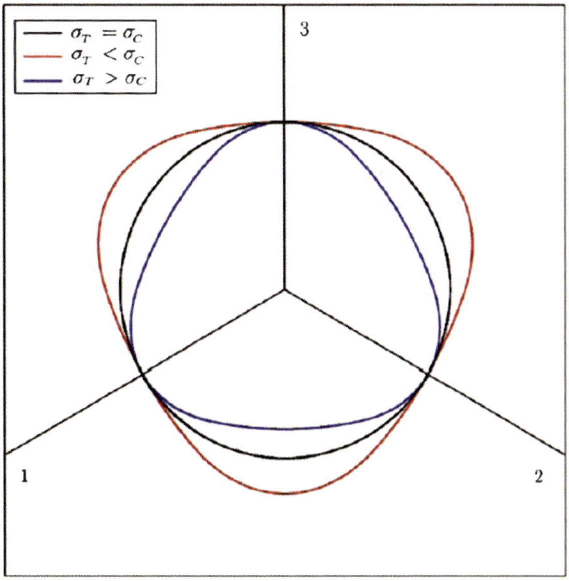

Fig. 3.3 Representation of the octahedric plane of Cazacu et al.'s (2006) isotropic yield surface corresponding to a ratio between the yield stress in tension and compression: $\sigma_T/\sigma_c = 0.82$, $\sigma_T/\sigma_c = 1$ (von Mises) and $\sigma_T/\sigma_c = 1.21$ (Reprinted with permission from Cazacu and Stewart 2009. Copyright 2006 Elsevier)

to model void growth, shape evolution and coalescence under a variety of stress states and use this data to develop their constitutive models. From a practical perspective, the assumption of an isotropic matrix enables the results of these studies to be readily generalized to the whole range of isotropic materials through the matrix hardening exponent. Conversely, anisotropy cannot be characterized in such a general manner as the anisotropy coefficients are specific to the material of interest and the yield criterion used to describe the matrix. The choice of the appropriate yield criterion is related to the material crystal structure (FCC, BCC, HCP) and may use non-quadratic yield surfaces that contain many calibration parameters that are determined from a variety of experiments. As a result, it is difficult to generalize unit cell results for void growth and shape evolution to another material. This problem becomes particularly acute when considering materials with an HCP crystal structure.

Fortunately, there has been a renewed interest in accounting for anisotropy effects on ductile fracture as industry has moved to using lighter weight, higher strength components out of more exotic and advanced aluminum and magnesium alloys that display significant anisotropy and damage sensitivity. The aforementioned anisotropic models are suitable for materials with cubic crystal structures such as most steels and aluminums because they assume the same yield stress in both tension and compression. For magnesium and titatium alloys that possess an HCP crystal structure, a tension-compression asymmetry in yielding is observed because the deformation mechanisms are related to twinning in compression and non-Schmid slip in tension (Cazacu and Stewart 2009). The variation of an undamaged yield surface with varying degrees of tension-compression asymmetry is shown in Fig. 3.3.

Magnesium alloys are of great interest in particular to the automotive industry as they offer significant weight reduction and gains in fuel efficiency, but are notoriously anisotropic and prone to void-induced cracking. Cazacu and Stewart (2009), Yoon et al. (2011) and Stewart and Cazacu (2011) have made progress in modelling void damage in HCP materials by extending the Gurson (1977) model for spherical voids into the Barlat-type yield criterion of Cazacu et al. (2006). It should be noted that unlike the traditional Gurson-based models that predict no void growth or material softening in shear-dominated stress states, the HCP damage models contain an explicit dependence upon the third invariant of the stress deviator due to their tension-compression asymmetry.

The proceeding sections will review the unit cell concept and its application to a porous sheet metal to first establish a lower bound solution for an isotropic, rigid-plastic matrix. In the sequel, the matrix material is considered as anisotropic and a quasi-exact yield criterion will be developed using fundamental unit cell theory. Finally, the models will be compared with existing models in the literature and with experimental results for the yielding of porous materials.

3.3 An Approximate Unit Cell for Porous Sheet Metals

Following the approach of Gurson (1977), the microstructure of the porous material is idealized as a periodic distribution of cylindrical unit cells with an interior cylindrical void. For sheet metals, the geometry can be reduced from a cylinder to a disk from the assumption of plane stress and subjecting the cell to a general biaxial loading that mimics a sheet metal forming operation as shown in Fig. 3.4.

The isolated unit cell is best analyzed using a polar coordinate system, $r - \theta$, as shown in Fig. 3.4b. The cell model has an outer radius, b, an inner radius, a, and an infinitesimial thickness, t. The porosity of the unit cell is thus defined as

$$f = \frac{a^2}{b^2} \tag{3.10}$$

3.3.1 Stress and Strain Rate Fields Inside the Unit Cell

From an analytical perspective, the two extreme conditions for the stress field within the unit cell are either: completely elastic or completely plastic. In reality, the stress fields will lie between these two cases where plastic deformation occurs in the vicinity of the void with the remaining material in an elastic state. This situation can be classified as "partially plastic with rigid sections". The influence of the void distorts the stress and strain fields within the unit cell at distances up to ten times the

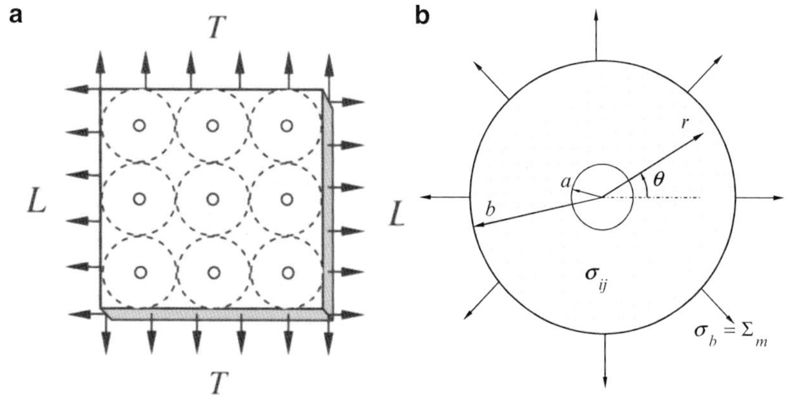

Fig. 3.4 (**a**) Idealized periodic microstructure of a porous sheet material (*left*) and (**b**) single unit cell approximated as a disc due to the assumption of plane stress (infinitesimal sheet thickness). The isolated unit cell experiences uniform radial stresses on its surface and the voids are assumed to remain cylindrical while subjected to in-plane tractions in the longitudinal direction, L, and transverse direction, T (Reprinted with permission from Landry and Chen (2011). Copyright: Elsevier)

Fig. 3.5 Finite-element description of the distortional effect of the void on the stress field. The *darkest area* neighbouring the void indicates the high stress level in the vicinity of the void and describes the microscopic stresses. The remaining region of the unit cell describes the uniform macroscopic stress distribution. The influence of the void on the strain distribution is similar

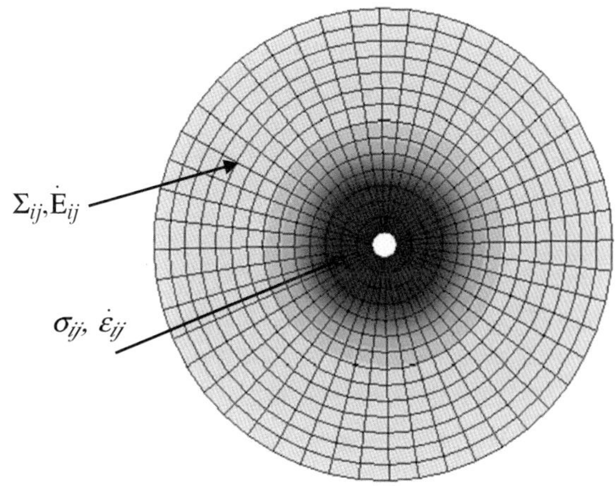

void radius and the stress concentration effect of the void is illustrated in Fig. 3.5. The analytical description of the two extreme stress conditions in the unit cell will be discussed in the subsequent sections.

3.3.2 Elastic Stress State in the Unit Cell

Since the stresses and strains are linearly related in the elastic regime, the principle of superposition can be used to obtain the resultant stress or strain in a system from the algebraic sum of their effects (Bayoumi 1999). The principle of superposition

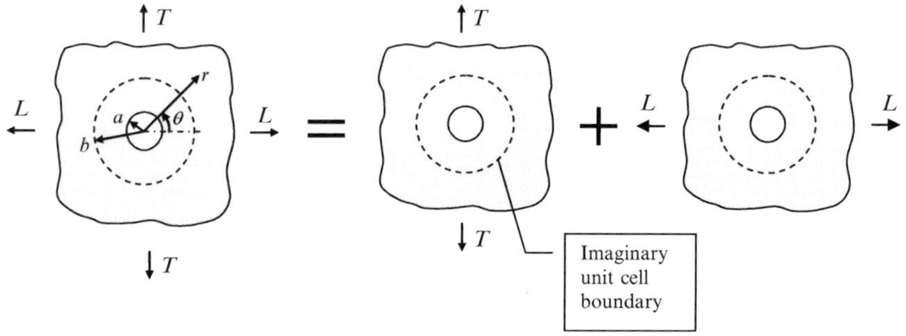

Fig. 3.6 Schematic representation of the linear decomposition of a biaxial stress state into a series of separate, isolated loadings using the principle of superposition

can be invoked to decompose the general biaxial loading on the unit cell into two separate uniaxial loading conditions as demonstrated in Fig. 3.6.

By applying the governing stress equations, biharmonic equations and equilibrium equations, we can obtain a solution for the microscopic stress fields as

$$\begin{aligned}
\sigma_r^L &= \frac{L}{2}\left[\left(1-x^{-2}\right)+\left(1-4x^{-2}+3x^{-4}\right)\cos(2\theta)\right] \quad x=r/a \\
\sigma_\theta^L &= \frac{L}{2}\left[\left(1+x^{-2}\right)-\left(1+3x^{-4}\right)\cos(2\theta)\right] \\
\sigma_{r\theta}^L &= -\frac{L}{2}\left[1+2x^{-2}-3x^{-4}\right]\sin(2\theta)
\end{aligned} \tag{3.11–3.13}$$

where σ_{ij}^L is the microscopic stress field due to the loading, L, and x is the normalized radial distance away from the void with $x = 1$ corresponding to the surface of the void. Similarly, the stress solutions for the transverse uniaxial case, σ_{ij}^T, are readily obtained by subtracting an angle of $\pi/2$ in Eqs. (3.11–3.13). The complete solution for the elastic microscopic stress fields is

$$\begin{aligned}
\sigma_{ij} &= \sigma_{ij}^L + \sigma_{ij}^T \\
\sigma_{r\theta} &= -\frac{L}{2}\left(1+2x^{-2}-3x^{-4}\right)\sin(2\theta) + -\frac{T}{2}\left(1+2x^{-2}-3x^{-4}\right)\sin(2\theta-\pi) \\
\sigma_{\theta\theta} &= \frac{L}{2}\left[\left(1+x^{-2}\right)-\left(1+3x^{-4}\right)\cos(2\theta)\right] + \frac{T}{2}\left[\left(1+x^{-2}\right)-\left(1+3x^{-4}\right)\cos(2\theta-\pi)\right] \\
\sigma_{rr} &= \frac{L}{2}\left[\left(1-x^{-2}\right)+\left(1-4x^{-2}+3x^{-4}\right)\cos(2\theta)\right] \\
&\quad + \cdots \frac{T}{2}\left[\left(1-x^{-2}\right)+\left(1-4x^{-2}+3x^{-4}\right)\cos(2\theta-\pi)\right]
\end{aligned} \tag{3.14–3.17}$$

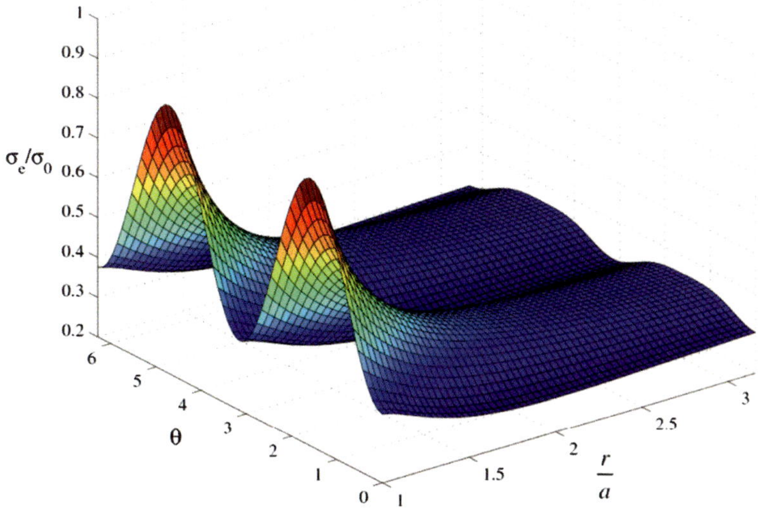

Fig. 3.7 Distribution of the microscopic effective stress in a disk-shaped unit cell containing an isotropic matrix material with a circular void with a porosity of 10 % (x = r/a, T = $0.25\sigma_0$, L/T = 1.5). Note that the stress response is periodic with respect to theta since the matrix is isotropic (Reprinted with permission from Landry and Chen (2011). Copyright: Elsevier)

A schematic plot of the microscopic effective stress is presented in Fig. 3.7 where a very large stress concentration exists on the boundary of the void ($x = r/a = 1$). Farther away from the void, the effective stress field becomes uniform and approaches a constant value.

3.3.3 Plastic Stress State in the Unit Cell

An analytical solution exists for the fully plastic state in the case of axisymmetric loading (Kachanov 1974). In an axially symmetric loading condition, the stress components, σ_r and σ_θ become the principal stresses and the yield surface on the (σ_r, σ_θ) plane is an ellipse as shown in Fig. 3.8 for an isotropic material.

The principal stress components can be readily expressed from the parametric equations for an ellipse as

$$\sigma_r = 2\sigma_0 \cos\left(\omega + \frac{\pi}{6}\right) \tag{3.18}$$

$$\sigma_\theta = 2\sigma_0 \cos\left(\omega - \frac{\pi}{6}\right) \quad \sigma_\theta > \sigma_r \tag{3.19}$$

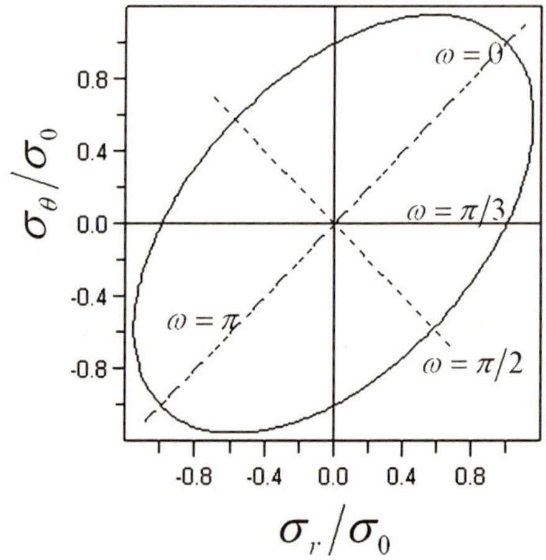

Fig. 3.8 Von Mises yield locus in the (σ_r, σ_θ) plane

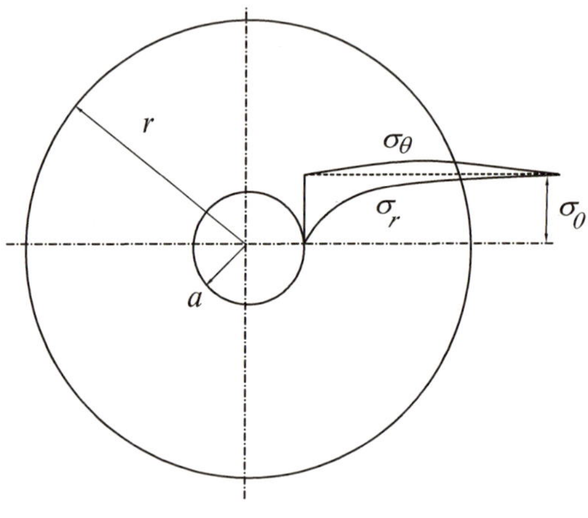

Fig. 3.9 Schematic of the stress distribution around the void in the fully plastic state under axisymmetric loading (Reprinted with permission from Xia and Chen (2007). Copyright: Springer)

where ω defines a position (stress state) on the yield locus in Fig. 3.8. For equilibrium in the radial direction

$$\frac{d\sigma_r}{dr} + \frac{\sigma_r - \sigma_\theta}{r} = 0 \tag{3.20}$$

we obtain the implicit solution to the simultaneous Eqs. (3.18), (3.19) and (3.20),

$$r^2 = \frac{C}{\sin\omega} e^{-\sqrt{3}\omega} \tag{3.21}$$

Where C is the integral constant that is determined by the boundary condition of the unit cell, i.e. $\sigma_r = 0$ on the surface of the void ($r = a$), see Fig. 3.9.

Table 3.1 Evaluation of the accuracy of the unit cell model for different unit cell sizes

Ratio (r/a)	Porosity $(a/b)^2$ (assuming $r = b$)	3.1 σ_r/σ_0 (unit-cell model)	Error percentage on σ_r/σ_0 (%)
1.5	0.444	0.360	63.98
1.7	0.346	0.452	54.81
2	0.250	0.557	44.32
5	0.040	0.894	10.57
8	0.017	0.955	4.49
10	0.010	0.971	2.94

It is straightforward to find that stress distribution decays to its uniform value in the radial direction according to

$$\left(\frac{r}{a}\right)^2 = \frac{\sqrt{3}}{2}\frac{1}{\sin\omega}e^{\sqrt{3}\left(\frac{\pi}{3}-\omega\right)} \tag{3.22}$$

A qualitative plot of the stress components σ_r and σ_θ is provided in Fig. 3.9. This method can be similarly applied to an anisotropic material through the anisotropic parameters.

A quantitative analysis of the problem shows that the distortion of the stress field distribution caused by the void tends to disappear when r/a is suffientiently large. According to the calculations in Eqs. (3.18–3.22), both σ_r and σ_θ reach σ_0 at infinity. Note that r is a measure of the distance from the void surface so that if we set a boundary in Fig. 3.6 to form an imaginary unit cell (namely set $r = b$, the outer radius of the imaginary unit cell), the porosity is $(a/b)^2$ for the planar case. However, as previously discussed, unit cell model assumes its outer boundary as the infinity. Therefore, errors are introduced in unit cell approximation with the error for various unit cell sizes presented in Table 3.1. If an error in the stress of 3 % is acceptable, the material located at a distance 10 times the void radius can be classified as being located at infinity.

3.4 Derivation of a Lower Bound Yield Criterion for Porous Sheet Metals

A lower bound solution for the isotropic matrix under a biaxial loading can be developed from the previous solutions for the microscopic stress fields in the elastic case in Eq. (3.14–3.17) since the material is elastic before the onset of yielding. As discussed in the previous section, the unit cell model creates artificial stress boundary conditions prescribed on the outer surface of the unit cell and consequently, the yield function is treated as a lower bound solution. The sheet metal matrix obeys von Mises yield criterion with the effective stress defined as

$$\sigma_{eq} = \frac{1}{\sqrt{2}}\left[\sigma_{rr}^2 + \sigma_{\theta\theta}^2 + (\sigma_{rr} - \sigma_{\theta\theta})^2 + 6\sigma_{r\theta}^2\right]^{\frac{1}{2}} \tag{3.23}$$

The matrix will begin to yield when the microscopic effective stress in a region of the cell reaches the yield limit

$$\max(\sigma_{eq}) = \sigma_0 \tag{3.24}$$

The macroscopic stresses on the outer surface of the cell are obtained by integrating over the cell volume as

$$\Sigma_{11} = \frac{1}{V}\int_V \sigma_{11} dV = (1-f)L \qquad \Sigma_{22} = \frac{1}{V}\int_V \sigma_{22} dV = (1-f)T$$
$$\Sigma_{12} = 0$$
$$(3.25\text{–}3.27)$$

3.4.1 Numerical Results

Figures 3.7 and 3.10 show the distribution of the effective microscopic stress over the unit cell for two specific loading ratios (*L/T*). For a given loading ratio, it is straightforward to obtain the solution for the macroscopic stresses at the onset of yielding. It is worth noting that an intense stress concentration occurs on the inner surface of the unit cell that results in considerably lower yield limits due to the retention of a predominantly elastic unit cell. Therefore, the outer surface is chosen to calculate the maximum effective stress. As shown in Fig. 3.10, the onset of yielding in the unit cell is defined when the outer surface of the cell has begun to yield

$$\max\left(\sigma_e\right)\Big|_{x=\sqrt{1/f}} = \sigma_0 \tag{3.28}$$

A similar technique was utilized by Gurson (1977) to account for the rigid sections in the unit cell and by Sun and Wang (1989) to construct the stress field causing a partly plastic state in the unit cell.

Figures 3.7 and 3.10 show the respective microscopic effective stresses over the entire unit cell for an entirely elastic loading scenario and an elastic–plastic loading. Note that the stresses in the elastic region of Fig. 3.10 have been scaled accordingly to generate an approximate stress surface that accommodates for load sharing between the elastic and plastic regions. The value of this scaling factor can be seen in Fig. 3.11. Approximating this value as unity imposes a maximum 5 % deviation under a void volume fraction of 0.05 for all loading paths.

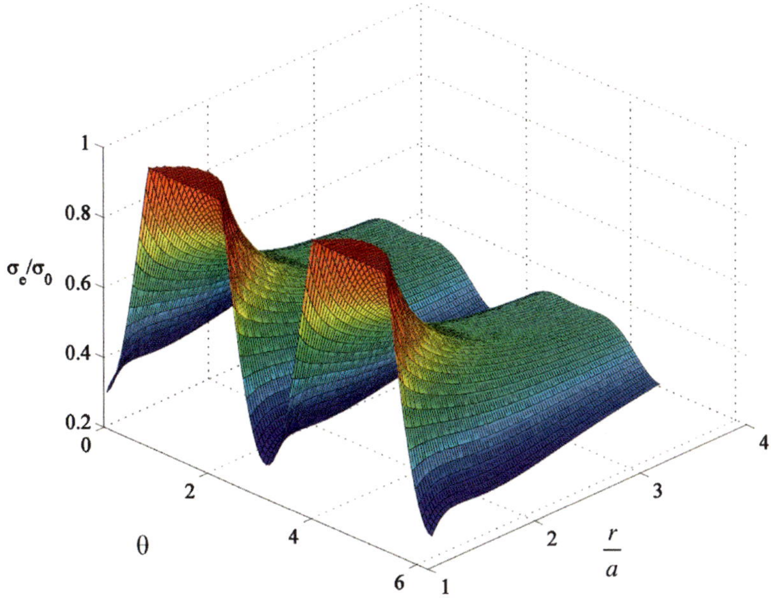

Fig. 3.10 Distribution of the microscopic effective stress in a disk-shaped unit cell containing an isotropic matrix material with a circular void with a porosity of 10 % (x = r/a, T = 0.25 σ_0, L/T = 2.0). The yield surface is truncated at unity where the material has yielded at the void surface (Reprinted with permission from Landry and Chen (2011). Copyright: Elsevier)

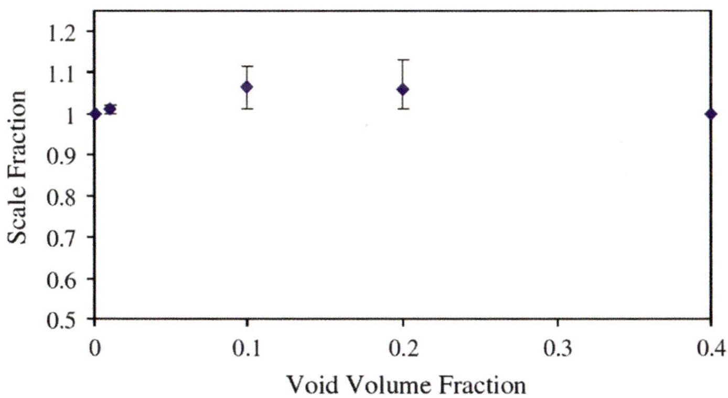

Fig. 3.11 The mean value and statistical dispersion of the load balancing scaling factor when evaluated under uniaxial, biaxial and shear loading for various void volume fractions (Reprinted with permission from Landry and Chen (2011). Copyright: Elsevier)

Following the calculation procedure discussed previously, the yield limits are numerically solved with a series of specified loading ratios (*L*/*T*). For simplicity, the sheet metals are assumed to have the same yield limits under tensile and compressive loadings. The numerical results are plotted with Gurson's (1977) upper bound criterion and von Mises yield function in Fig. 3.12. A comparison indicates that the

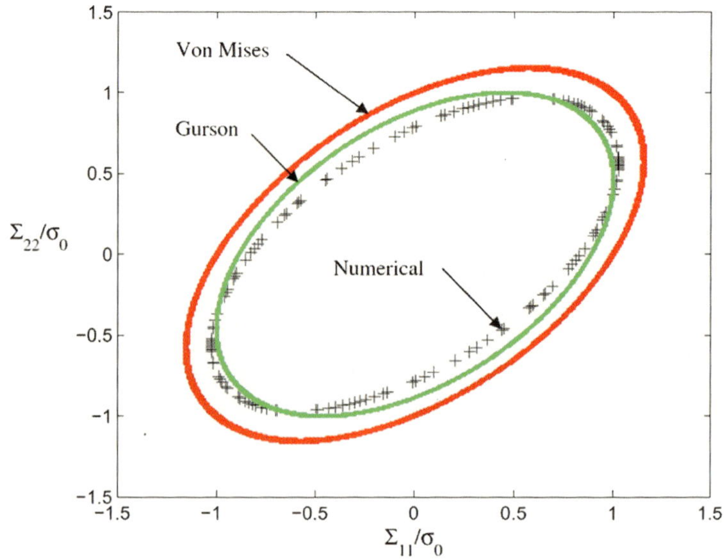

Fig. 3.12 Comparison of the numerical results for yielding in the unit cell with a porosity of 10 % compared with the yield surfaces of Gurson and von Mises (Reprinted with permission from Landry and Chen (2011). Copyright: Elsevier)

numerical results of the current model provide a more conservative yield point for uniaxial and shear loading. Due to the contrasting methodology used between the Gurson and current approach, the trend for biaxial loading shows that the Gurson yield criterion is in fact conservative in this instance. It is important to note that the yield locus never exceeds the von Mises yield locus even during equal biaxial loading due to the presence of the voids.

A closed form of the yield function is required for its practical application and this can be accomplished in a phemenological manner by introducing three fitting parameters into the equivalent stress equation in principal stress space as

$$\Phi = \Sigma_{eq} - \bar{\sigma} = 0$$

$$\Sigma_{eq} = \sqrt{1/2\left(q_1(\Sigma_{11} - \Sigma_{22})^2 + q_2\Sigma_{11}^2 + q_3\Sigma_{22}^2\right)}$$

$$q_{1-3} = \begin{cases} 8.586\,f^2 + 7.1329\,f + 1 \\ 5.857\,f^2 + 0.5734\,f + 1 & f \leq 0.20 \\ 15.258\,f^2 + 2.6973\,f + 1 \end{cases} \qquad (3.29-3.31)$$

These parameters are valid for porosities up to 20 % and allows the function to revert to the von Mises yield criterion when $f = 0$. The numerical process employed here to obtain a closed form of the yield function is effective but admittedly, not mathematically elegant. The performance of Eq. (3.30) can be seen in Fig. 3.13 for a relatively large porosity of 10 %.

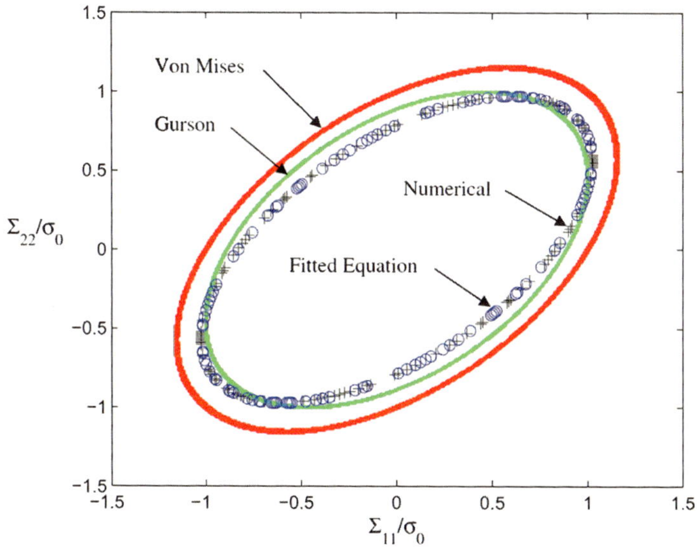

Fig. 3.13 Comparison of the numerical results for yielding in the unit cell with a porosity of 10 % compared with the fitted-yield surface in Eq. (3.25) along with the yield surfaces of Gurson and von Mises (Reprinted with permission from Landry and Chen (2011). Copyright: Elsevier)

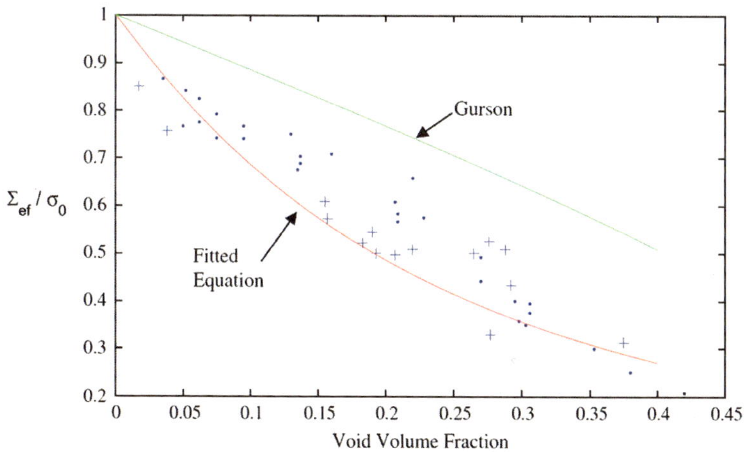

Fig. 3.14 Comparison of the fitted yield function in Eq. (3.25) and the Gurson yield criterion with the experimental results of Shima and Oyane (1976) for sintered copper specimens. The *closed dots* represent yielding in uniaxial compression while + indicates uniaxial tension (Reprinted with permission from Landry and Chen (2011). Copyright: Elsevier)

3.4.2 *Comparison of the Lower Bound Solution with Experiments*

The lower bound yield function in Eq. (3.29) is evaluated using the experimental results of Shima and Oyane (1976) for sintered iron and copper alloys in Figs. 3.14

Using Eqs. (3.53) and (3.54) in (3.52), σ_{rr} is solved as

$$\sigma_{rr} = \frac{\sqrt{2}}{2}\sigma_0 B\left[\sinh^{-1}x_a - \sinh^{-1}x\right] \tag{3.55}$$

where $x_a = \frac{\sqrt{2}A}{ca^2}$. Similarly, we define $x_b = \frac{\sqrt{2}A}{cb^2}$. According to the boundary condition (3.41), the boundary radial stress is

$$\Sigma_{rr} = \sigma_{rr}\big|_{r=b} = L = \frac{\sqrt{2}}{2}\sigma_0 B\left[\sinh^{-1}x_a - \sinh^{-1}x_b\right] \tag{3.56}$$

3.5.4 Solution for the Macroscopic Through-Thickness Stress

By definition, the macroscopic stress Σ_{zz} is

$$\Sigma_{zz} = \frac{2}{b^2}\int_a^b \sigma_{zz} r dr \tag{3.57}$$

The normal stress is manipulated as

$$\sigma_{zz} = \sigma_{rr} + (\sigma_{zz} - \sigma_{rr}) \tag{3.58}$$

Substituting Eqs. (3.42), (3.49) and (3.50) into (3.53c) gives

$$\begin{aligned} \sigma_{zz} - \sigma_{rr} &= \frac{1+R}{1+2R}\frac{\sigma_0}{d\varepsilon_{eq}}(d\varepsilon_{zz}R - \dot{\varepsilon}_{rr}) = \sigma_0 B\frac{\sqrt{2}x/2 + m}{\sqrt{x^2+1}} \\ m &= \sqrt{\frac{1}{2} + R} \end{aligned} \tag{3.58a, b}$$

Substituting Eqs. (3.55), (3.58) into (3.57) results in

$$\begin{aligned} \Sigma_{zz} = T = -B\sigma_0 x_b \Bigg\{ &\frac{\sqrt{2}}{2}\left(\sinh^{-1}x_a\right)\int_{x_a}^{x_b} x^{-2}dx - \frac{\sqrt{2}}{2}\int_{x_a}^{x_b} x^{-2}\left(\sinh^{-1}x\right)dx \\ &+ \frac{\sqrt{2}}{2}\int_{x_a}^{x_b}\frac{1}{x\sqrt{x^2+1}}dx + m\int_{x_a}^{x_b}\frac{1}{x^2\sqrt{x^2+1}}dx \Bigg\} \end{aligned}$$

Note the second term is

$$\frac{\sqrt{2}}{2}\int_{x_a}^{x_b} x^{-2}\left(\sinh^{-1}x\right)dx = -\frac{\sqrt{2}}{2}\left[\frac{1}{x}\sinh^{-1}x\right]_{x_a}^{x_b} + \frac{\sqrt{2}}{2}\int_{x_a}^{x_b}\frac{1}{x\sqrt{x^2+1}}dx$$

Therefore,

$$T = L + \sigma_0 B x_b m \left[\sqrt{\frac{1}{{x_a}^2} + 1} - \sqrt{\frac{1}{{x_b}^2} + 1} \right] \tag{3.59}$$

3.5.5 Solution for the Yield Function

Noting that $f = x_b/x_a$, Eq. (3.59) can be rewritten as

$$\left(\frac{T-L}{\sigma_0}\right)^2 = \frac{1+R}{2}\left[1 + f^2 - 2f\left(\sqrt{(1+{x_b}^2)(1+{x_a}^2)} - x_a x_b\right)\right] \tag{3.60}$$

Rearranging Eq. (3.56), we can obtain

$$\frac{\sqrt{2}L}{\sigma_0 B} = \frac{\sqrt{2}L}{\sigma_0}\sqrt{\frac{1+2R}{1+R}} = \left(\sinh^{-1}x_a - \sinh^{-1}x_b\right) \tag{3.61}$$

Using the following identity

$$\cosh\left(\sinh^{-1}x_a - \sinh^{-1}x_b\right) = \sqrt{(1+{x_b}^2)(1+{x_a}^2)} - x_a x_b \tag{3.62}$$

and combining Eqs. (3.60), (3.61) and (3.62) give

$$\left(\frac{T-L}{\sigma_0}\right)^2 = \frac{1+R}{2}\left[1 + f^2 - 2f\cosh\left(\sqrt{\frac{1+2R}{1+R}}\frac{\sqrt{2}L}{\sigma_0}\right)\right] \tag{3.63}$$

The macroscopic effective stress in the Hill-48 yield criterion is

$$\Sigma_{\text{eq}}^2 = \frac{1}{1+R} \times \left[(\Sigma_{22} - \Sigma_{33})^2 + (\Sigma_{33} - \Sigma_{11})^2 + R(\Sigma_{11} - \Sigma_{22})^2 + 2(1+2R)\Sigma_{12}\right] \tag{3.64}$$

Note $L = \frac{1}{2}(\Sigma_{11} + \Sigma_{22})$ and $\Sigma_{\text{eq}}^2 = \frac{2}{1+R}(T-L)^2$ thus, an equivalent form of Eq. (3.63) is obtained. Finally, the yield function is expressed as (Xia and Chen 2007)

$$\left(\frac{\Sigma_{\text{eq}}}{\sigma_0}\right)^2 + 2f\cosh\left(\sqrt{\frac{1+2R}{2(1+R)}}\frac{3\Sigma_{\text{hyd}}}{\sigma_0}\right) - 1 - f^2 = 0 \tag{3.65}$$

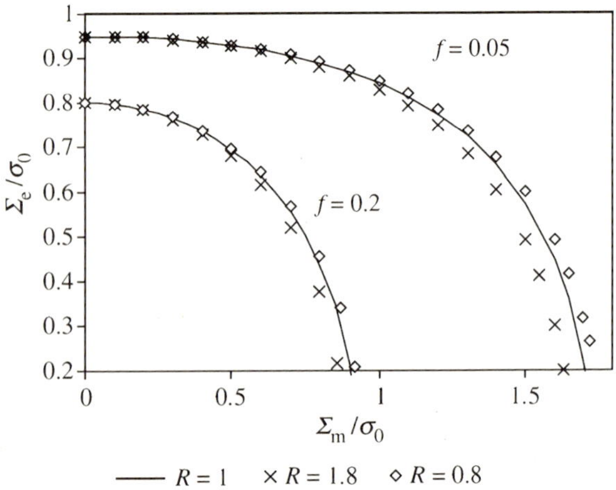

Fig. 3.16 Comparison of the yield loci of the current yield function (3.61) for various values of R and f. The *solid curves* indicate the yield loci for isotropic materials (R = 1) (Reprinted with permission from Xia and Chen (2007). Copyright: Springer)

When $R = 1$, the yield function (3.65) is reduced to the derivation of Gurson (1977) for a cylindrical unit cell subjected to axisymmetric loading

$$\left(\frac{\Sigma_{eq}}{\sigma_0}\right)^2 + 2f\cosh\left(\sqrt{3}\frac{3}{2}\frac{\Sigma_{hyd}}{\sigma_0}\right) - 1 - f^2 = 0 \tag{3.66}$$

3.5.6 Effect of Mechanical Anisotropy in a Porous Ductile Material

Two different R values, 0.8 and 1.8, are selected to quantify the influence of mechanical anisotropy on the yield surface in Fig. 3.16 for FCC and BCC materials, respectively. The material does not possess any void-induced anisotropy since the void is assumed to remain circular during deformation. The yield surface contracts slightly with an increase in the R-value for a constant porosity, indicating that anisotropy increases pressure sensitivity. The effect of anisotropy becomes more pronounced in plane stress loading as demonstrated in Fig. 3.17.

It is worth noting that for a given porosity value, the yield points on the vertical axis for various R values (Fig. 3.17) coincide with each other. This implies that for pure shear, the yield behavior of porous sheet metal is independent of normal anisotropy. It should be cautioned however that the influence of shear may be different if a cubic unit cell geometry was assumed. The cylindrical unit cell geometry is restricted to axisymmetric stress states resulting in a yield criterion that is independent of the third invariant of the stress deviator that is used to characterize shear loading via the lode parameter. The variational model of Danas and Ponte Casteneda (2009) shows evidence of the third stress invariant on

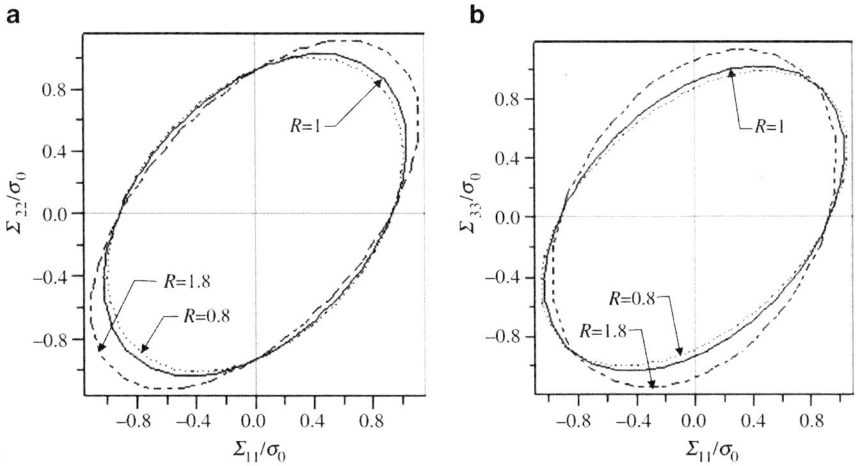

Fig. 3.17 Yield loci for various R values for a porosity of 5 % in the (**a**) Σ11–Σ22 plane and (**b**) Σ11–Σ33 plane (Reprinted with permission from Xia and Chen (2007). Copyright: Springer)

yielding of a porous ductile material. As expected, the influence of the *R*-value becomes more pronounced at higher stress triaxialities since the *R*-value acts as a scaling factor for the hydrostatic stress dependence of the yield surface.

3.5.7 Assessment of the Uniqueness of the Current Yield Function

A limitation of the unit-cell model is that the geometry of the unit-cell must be constructed so that the specified porosity is satisified even though the cell may not be large enough for the outer surface to be regarded as infinity (Gurson 1977). In this sense, the stress field inside the unit cell could be considered a statically admissible stress field and the yield function in Eq. (3.65) would be classified as an analytical lower bound solution rather than an exact solution. If one takes this view, the current yield function can provide a benchmark to assess other lower bound yield criteria. Conversely, if we ignore the limitation of the unit cell model, the only remaining approximation is that the normal strain rate is independent of the unit cell radius. This limitation can be reasonably ignored for a comparison between unit cell models and since the unit cell is assumed to be in a fully plastic state, the yield stress is likely overestimated since the actual unit cell would likely contain rigid regions. From this perspective, the current solution could be interpreted as an analytical solution and used to evaluate upper bound yield criteria. In other words, the uniqueness of the current model may make it a valuable reference for a variety of situations.

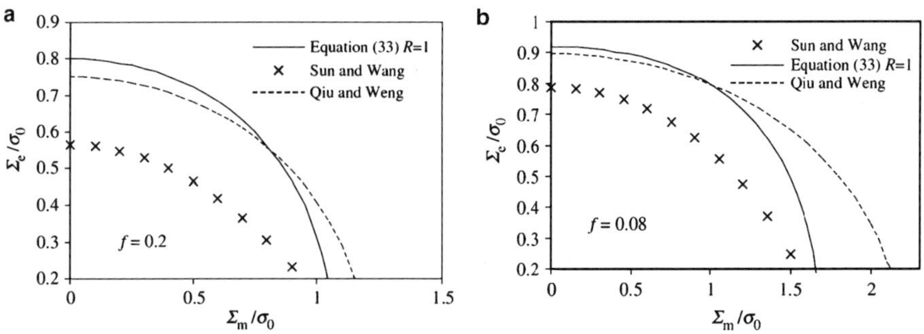

Fig. 3.18 Comparison between the current yield function (3.64) with Sun and Wang's lower bound yield function (Eq. 2.31), and Qiu and Weng's energy based yield function, Eq. (2.31) (Reprinted with permission from Xia and Chen (2007). Copyright: Springer)

As an example, the lower bound solution of Sun and Wang (1989) and the energy-based solution of Qiu and Weng (1993) are compared with the current yield function. In Sun and Wang's derivation, the stress field in a spherical unit-cell was decomposed into an elastic and plastic part. Qiu and Weng (1993) eschewed the unit cell approach and derived a yield criterion for randomly- oriented voids using an energy-based approach to obtain

$$\left(\frac{\Sigma_{eq}}{\sigma_0}\right)^2+\frac{f}{4\left(1+\frac{2}{3}f\right)}\left(3\frac{\Sigma_{hyd}}{\sigma_0}\right)^2-\frac{(1-f)^2}{1+\frac{2}{3}f}=0 \tag{3.67}$$

It is worth noting that the Sun and Wang and Qiu and Weng models were derived for spherical voids while the current yield function assumes circular (cylindrical) voids. To provide an equal comparison, a scale factor of $1/\sqrt{3}$ is applied to the the hydrostatic stress component in Eq. (3.65) with reference to the Gurson's (1977) conversion from the cylindrical model to the spherical void yield criterion as

$$\left(\frac{\Sigma_{eq}}{\sigma_0}\right)^2+2f\cosh\left(\sqrt{\frac{1+2R}{6(1+R)}}\frac{3\Sigma_{hyd}}{\sigma_0}\right)-1-f^2=0 \tag{3.68}$$

It is more reasonable to compare yield function (3.68) with Sun and Wang's yield function (Eq. 2.31) and Qiu and Weng's yield function (3.67).

Figure 3.18 shows the yield loci of Eq. (3.22) with $R = 1$ in comparison with Sun and Wang's (1989) lower bound yield function (Eq. 2.31) and Qiu and Weng's (1993) yield function in Eq. (3.67). Large discrepancies are observed between the current yield function and Sun and Wang's criterion, as shown in both figures. This is because the unit cell in Sun and Wang's lower bound approach is assumed to be partially plastic while the cell was assumed to be completely plastic in our derivation. In contrast, the current yield loci are in better agreement with Qiu and Weng's results who considered both the elastic and plastic energies in the material.

3.5.8 Evaluation of the Quasi-Exact Anisotropic Yield Criterion

If the prescribed stresses and normal strain rate are assumed to be actual quantities, the yield criterion should be considered an "exact" solution because the macroscopic stresses in the unit cell were direcly obtained from the equilibrium equation. However, it is better to describe the yield function as "quasi-exact" due to the assumed independence of the normal strain rate on the unit cell radius. The other principal source of error in the model can be attributed to the prescribed boundary condition on the unit cell that can lead to significant errors for high porosities as discussed in Sect. 3.3.3. Fortunately, this model has been developed for application to sheet materials where the porosities at failure are on the order of a few percent.

It is important to discuss the work of Liao et al. (1997) who obtained the same result as Eq. (3.65) using the upper bound theory of plasticity by decomposing the velocity field into volume and shape-changing fields and then constructing each field seperately. Although the resulting yield surfaces are equivalent, very different mathematical approaches were used to obtain the macroscopic yield stress. Consequently, the model of Liao et al. (1997) is approximate since Eq. (3.65) was derived from the analytical stress field solutions in the unit cell. From the perspective of extremum theory, the model of Liao et al. (1997) is an upper-bound solution whilst Eq. (3.65) is an "exact" solution.

The current result coincides with the yield surface of Liao et al. (1997) for the following reasons:

- The same disk-shaped unit cell geometry was used and subjected to axisymmetric loading
- Similar form of solutions to the velocity fields: Liao et al. (1997) utilized the flow rule and the equilibrium equation while the present method solves for the velocity fields directly from volume conservation.
- Both models assume that the matrix material in the unit cell has achieved a completely plastic state

It is also interesting to note that the anisotropic yield function of Benzerga and Besson (2001) for porous orthotropic materials also reduces to Eq. (3.68) when simplified to normal anisotropy. In the model of Benzerga and Besson (2001), the velocity field was envisaged as a linear combination of Rice and Tracey's (1969) solution for volume change and the other is for a uniform shape change. Strictly speaking, the present result should not be considered to be a special case of Benzerga and Besson (2001) since their result is also an approximate upper bound solution like the model of Liao et al. (1997).

Despite that the closed form yield function is derived under an axisymmetric loading condition, it still remains valid and provides acceptable predictions of plastic deformation under universal loading conditions as investigated by Liao et al. (1997) for rigid plastic materials, and Chien et al. (2001) for a three-dimensional unit cell of hardening materials. Liao et al. generated all possible planar deformation modes by specifying the macroscopic strain rate ratio

dE_{11}/dE_{22} and obtained the corresponding solutions for the macroscopic stresses Σ_{11} and Σ_{22}. Their numerical results indicate that the closed form yield criterion matches well with the numerical results. Also, as Benzerga and Besson (2001) discussed about Gurson's results (1977) although Gurson considered axisymmetric loading conditions, it was proved by Leblond et al. (1995) that the analysis and corresponding results remain valid for general loading conditions.

Chapter 4
Void Growth to Coalescence: Unit Cell and Analytical Modelling

4.1 Void Shape Evolution During Ductile Fracture

The voids in a ductile material subjected to plastic deformation change shape according to the local plastic flow of the material since the voids are not internally pressurized. As a result, the void growth rate and shape evolution are intrinsically linked because the void shape (and orientation) induce anisotropy, altering the stress state and the growth rate in a non-linear fashion. The standard Gurson-Tvergaard model maintains its isotropic formulation by enforcing the void to remain spherical. The influence of the void shape on the stress response of the material is shown in Fig. 4.1 as well as the variation in the growth rate in Fig. 4.2 for a practical stress triaxiality of 2/3 (equal-biaxial stretching).

To simplify the modelling procedure, it is often assumed that the voids can be approximated as spheroids (an ellipsoid of revolution) as shown in Fig. 4.3. In uniaxial tension, void shape evolution is significant as initially penny-shaped voids can rapidly elongate in the principal loading direction into prolate voids. Alternatively, at high stress triaxialities found ahead of a crack tip, the voids tend to grow laterally and become oblate regardless of the principal loading direction. Penny-shaped voids are the exception to this trend and only appreciably grow in the opening direction even at high stress triaxiality (Lassance et al. 2006; Butcher 2011). Furthermore, a specific stress triaxiality exists for each void shape that will enforce the void to grow in a self-similar manner and retain its shape. Consequently, the mechanics for modeling void shape evolution are complex and no analytical model has yet been developed that can account for the void shape evolution in different stress states. As such, the adoption of heuristic parameters in these models is currently unavoidable (Keralavarma and Benzerga 2010).

The evolution of the void shape is a function of the initial void size, f_i, initial aspect ratio, W_i, equivalent plastic strain, $\bar{\varepsilon}^p$, matrix hardening exponent, n, and stress state, σ_{ij}, in the general form

$$W = g(f_i, W_\mathrm{i}, T, n, \bar{\varepsilon}^p) \tag{4.1}$$

Z. Chen and C. Butcher, *Micromechanics Modelling of Ductile Fracture*, Solid Mechanics and Its Applications 195, DOI 10.1007/978-94-007-6098-1_4,

© Springer Science+Business Media Dordrecht 2013

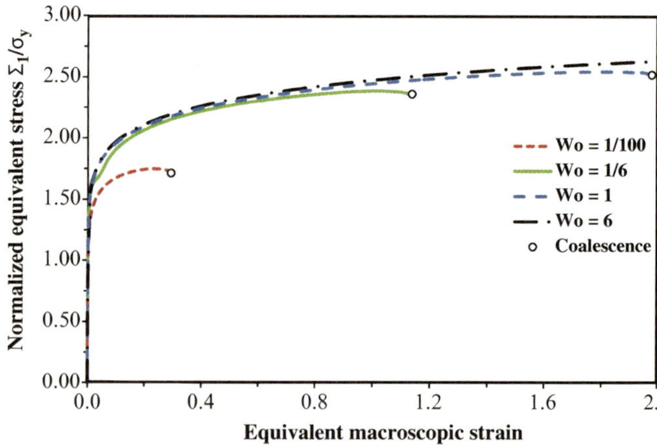

Fig. 4.1 Normalized equivalent stress response of an axisymmetric voided unit cell with various void shapes

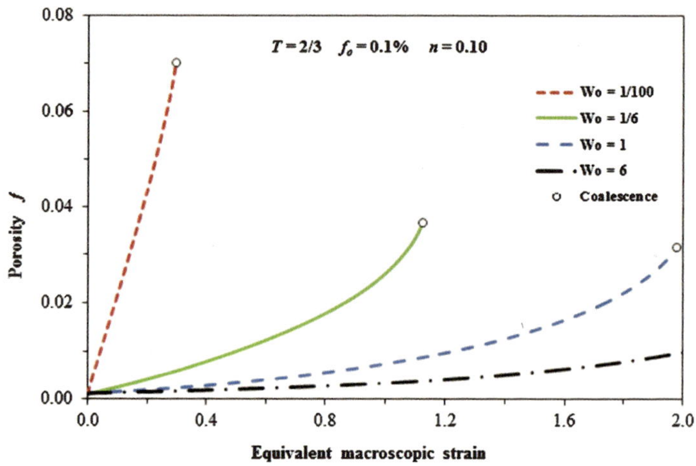

Fig. 4.2 Variation of void growth in an axisymmetric unit cell with the initial void shape. The assumption of a constant spherical void in the standard GT model significantly overestimates void growth for prolate voids and underestimate growth for oblate voids in the practical triaxiality regime ($T < 1$)

4.2 Damage-Based Material Models with Void Shape Effects

The physical foundation of a damage-based constitutive model rests upon the accuracy of the analytical sub-models that describe void nucleation, evolution, coalescence and material softening. While numerous micromechanical models have been proposed in the literature that can capture damage evolution, at least qualitatively, the veracity of these models must be evaluated over the large number of void shapes, sizes and stress states that will be found within a real microstructure.

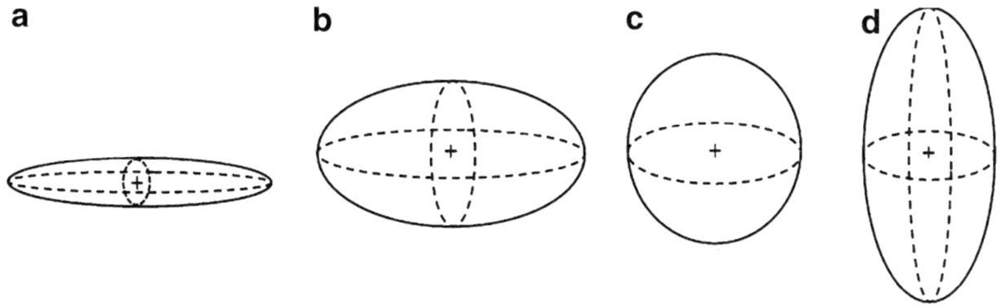

Fig. 4.3 Idealized spheroidal void shapes. (**a**) Penny-shaped spheroid: $W \to 0$, (**b**) oblate spheroid: $W < 1$, (**c**) sphere: $W = 1$, (**d**) prolate spheroid: $W > 1$

Finite-element simulations of axisymmetric unit cells containing different void geometries are required to provide the numerical benchmarks for these evolution models. An emphasis will be placed upon the evolution of penny-shaped voids because they are nucleated by particle cracking and have received scant attention in the literature. The organization of the section is as follows:

- Finite-element modeling and analysis of voided unit cells
- Discussion of finite-element simulations for a range of void aspect ratios and initial porosities
- Evaluation of current micromechanical models to predict damage evolution
- Development of unit cell correlations for damage evolution

4.3 Modeling Void Evolution Using a Unit Cell

The voids within a bulk material are assumed to be distributed throughout the material in such a manner that the material may be considered to be composed of identical unit cells, each containing a single void at its center. Typically, the unit cells are assumed to be cubic or axisymmetric. The axisymmetric unit cell is obtained by assuming the material is composed of interlocking hexagonal unit cells. These hexagonal unit cells can be approximated as cylindrical and then reduced to one-quarter, 2-D axisymmetric geometry for finite-element modelling as shown in Fig. 4.4. Similarly, the geometry of a 3-D cubic unit cell can be directly reduced to a one-eighth model due to symmetry.

The predicted trends for damage evolution are somewhat dependent upon the type of unit cell geometry (Kuna and Sun 1996). The evolution of porosity is lowest in a 3-D model and highest in the axisymmetric model due to the 3-D material containing more material to constrain the growth of the void and delay coalescence. As seen in Fig. 4.5, the cubic unit cell requires a significantly larger number of elements than the axisymmetric model and requires 8-node solid brick elements that are significantly more computationally expensive than the 4-node quadrilaterals used in the axisymmetric model. As a result, the run-times for cubic cells tend to be prohibitive unless a specific loading condition or geometry mandates their use. In the literature, most unit

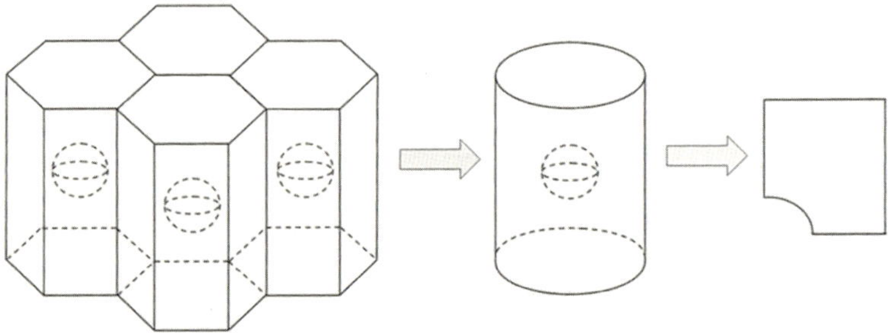

Fig. 4.4 Procedure for approximating a hexagonal distribution of unit cells into an axisymmetric cell model that can be reduced to a one-quarter, 2-D geometry for finite-element modeling

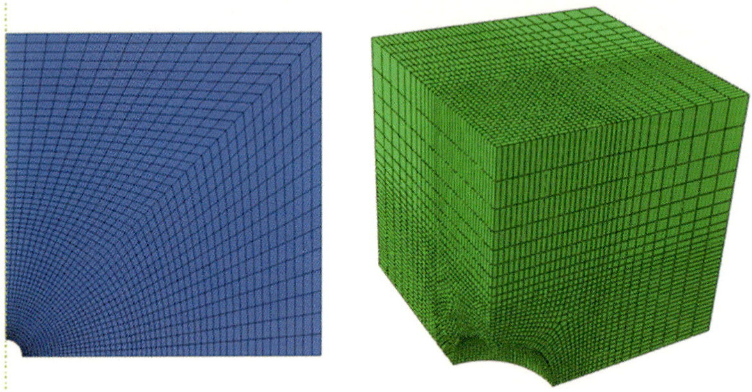

Fig. 4.5 Typical cell geometry and finite-element mesh for an initially spherical void in a one-quarter axisymmetric unit cell (*left*) and a one-eight cubic unit cell (*right*)

cell studies have focused on the use of axisymmetric geometries as they are efficient and provide a good estimate for damage evolution and fracture strains (Koplik and Needleman 1988; Kuna and Sun 1996; Brocks et al. 1996; Pardoen and Hutchinson 2000). It should be emphasized that axisymmetric models provide a lower, and therefore, more conservative estimate for the fracture strain (coalescence) than their 3-D counterparts. Axisymmetric unit cells are considered in the present work to generate the void evolution trends required for validation of the micromechanical models.

4.3.1 Analysis of an Axisymmetric Unit Cell

The axisymmetric unit cell is analyzed using a cylindrical coordinate system denoted by $(\mathrm{e}_r, \mathrm{e}_\theta, \mathrm{e}_z)$ along the r-radial, θ-ortho-radial and z-axial axes, respectively. As shown in Fig. 4.6, the geometry of the cylindrical unit cell is defined by a height

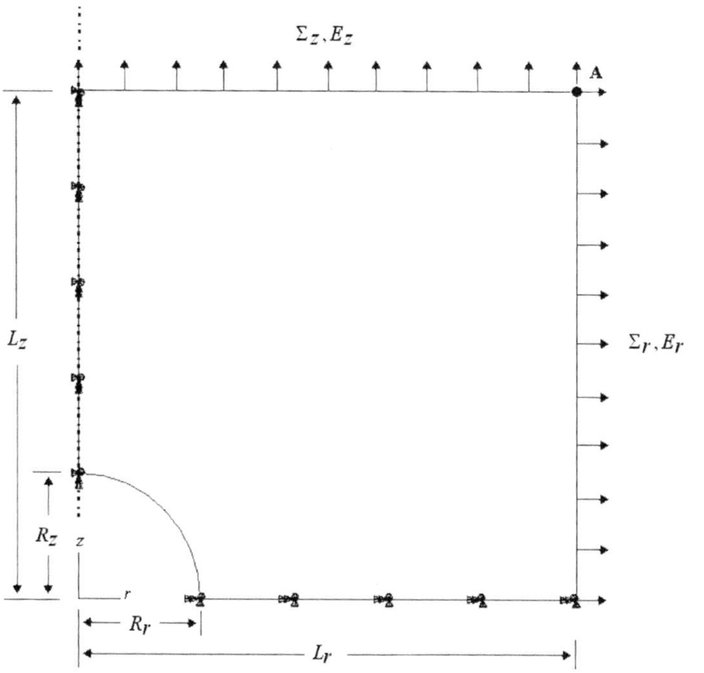

Fig. 4.6 General geometry and boundary conditions used to model the one-quarter axisymmetric unit cell

of $2L_z$ and radius, L_r and contains a spheroidal void with radii, R_z and R_r. The void shape is a prolate spheroid when $R_z > R_r$, an oblate spheroid when $R_z < R_r$, or a sphere when $R_z = R_r$. The geometry of the unit cell and void are defined using non-dimensional parameters. The void is defined by its volume fraction or porosity, *f*, and its aspect ratio, *W*, while the cell geometry is defined by its aspect ratio, λ. These parameters can also be combined to calculate the void spacing ratio (ligament size ratio), χ, that plays a critical role in void coalescence. These parameters all evolve during deformation of the cell and their initial values are defined as

$$f_0 = \frac{2R_{z0}}{3L_{z0}}\left(\frac{R_{r0}}{L_{r0}}\right)^2 \quad W_0 = \frac{R_{z0}}{R_{r0}} \quad \lambda_0 = \frac{L_{z0}}{L_{r0}}$$

$$\chi_0 = \frac{R_{r0}}{L_{r0}} = \left(\frac{3}{2}f_0\frac{\lambda_0}{W_0}\right)^{1/3} \qquad (4.2\text{–}4.5)$$

A number of important and useful relations can be obtained to describe the microstructure of a voided unit cell and its evolution. First, consider a general voided unit cell subjected to an arbitrary deformation process. The unit cell has the half-lengths L_1, L_2, L_3 that are coincident with the principal loading directions. The unit cell contains an arbitrarily oriented ellipsoidal void with semi-axes denoted as R_1, R_2 and R_3. The porosity (void volume fraction) of the cell is defined as

$$\frac{V_{\text{void}}}{V_{\text{cell}}} = \frac{f}{\gamma_{\text{cell}}} = \frac{R_1 R_2 R_3}{L_1 L_2 L_3} = \frac{W_1}{\lambda_1}\frac{W_2}{\lambda_2}\chi_2^3\chi = \chi_1\chi_2\chi_3 \tag{4.6}$$

where W_i are the void aspect ratios; λ_i are the cell aspect ratios; χ_i are the void spacing (or ligament size) ratios and γ_{cell} is a shape parameter specific to the assumed unit cell with $\gamma = 2/3$ for an axisymmetric unit cell and $\gamma = \pi/6$ for a cubic cell.

The void aspect ratios, spacing ratios, and cell aspect ratios can be defined as

$$W_1 = \frac{R_1}{R_2} \quad W_2 = \frac{R_3}{R_2} \quad \chi_i = \frac{R_i}{L_i} \quad \lambda_1 = \frac{L_1}{L_2} \quad \lambda_2 = \frac{L_3}{L_2} \tag{4.7–4.11}$$

Taking the derivative of Eqs. (4.6–4.11) with respect to time, it is straightforward to obtain the following relations as a function of the macroscopic principal strain rates as

$$\frac{\dot{f}}{f} = \frac{\dot{R}_1}{R_1} + \frac{\dot{R}_2}{R_2} + \frac{\dot{R}_3}{R_3} - \dot{E}_1 - \dot{E}_2 - \dot{E}_3 \tag{4.12–4.17}$$

$$\frac{\dot{W}_1}{W_1} = \frac{\dot{R}_1}{R_1} - \frac{\dot{R}_2}{R_2} \quad \frac{\dot{W}_2}{W_2} = \frac{\dot{R}_3}{R_3} - \frac{\dot{R}_2}{R_2} \quad \frac{\dot{\chi}_i}{\chi_i} = \frac{\dot{R}_i}{R_i} - \dot{E}_i \quad \frac{\dot{\lambda}_1}{\lambda_1} = \dot{E}_1 - \dot{E}_2 \quad \frac{\dot{\lambda}_2}{\lambda_2} = \dot{E}_3 - \dot{E}_2$$

From Eqs. (4.12–4.17), the growth rates of the void semi-axes can be expressed as

$$\frac{\dot{R}_1}{R_1} = \dot{E}_{\text{hyd}} + \frac{1}{3}\left(\frac{\dot{f}}{f} + 2\frac{\dot{W}_1}{W_1} - \frac{\dot{W}_2}{W_2}\right) \tag{4.18}$$

$$\frac{\dot{R}_2}{R_2} = \dot{E}_{\text{hyd}} + \frac{1}{3}\left(\frac{\dot{f}}{f} - \frac{\dot{W}_1}{W_1} - \frac{\dot{W}_2}{W_2}\right) \tag{4.19}$$

$$\frac{\dot{R}_3}{R_3} = \dot{E}_{\text{hyd}} + \frac{1}{3}\left(\frac{\dot{f}}{f} - \frac{\dot{W}_1}{W_1} + 2\frac{\dot{W}_2}{W_2}\right) \tag{4.20}$$

Alternatively, using only the void spacing ratios, the evolution rate of the porosity can be expressed as

$$\frac{\dot{f}}{f} = \frac{\dot{\chi}_1}{\chi_1} + \frac{\dot{\chi}_2}{\chi_2} + \frac{\dot{\chi}_3}{\chi_3} \tag{4.21}$$

The above Eqs. (4.6–4.21) are valid for any type of unit cell or ellipsoidal void geometry and reduce accordingly depending on the type of geometry considered. For example, an axisymmetric unit cell with the void semi-axes coincident with the principal loading directions: $\dot{W}_2/W_2 = 0$, $\dot{\lambda}_2/\lambda_2 = 0$ and $\dot{\chi}_2/\chi_2 = \dot{\chi}_3/\chi_3$.

4.3.2 Unit Cell Boundary Conditions

To enforce the assumption of a periodic array of unit cells, the faces of the unit cell must remain straight and move as rigid planes during the deformation process so that the cell remains cylindrical. The faces of the cell at $r = \mathrm{L}_r$ and $z = \mathrm{L}_z$ must have normal displacements and retain their mutual orientations. This condition is easily accomplished for one-quarter 2-D unit cell by designating the upper-right corner node as a master node denoted as '**A**' in Fig. 4.6 to control the displacements of the radial and axial planes by using multipoint constraints. These constraints enable the lengths of the unit cell to be defined at an arbitrary state as

$$L_r = L_{r0} + u_r^{\mathrm{A}}\; L_z = L_{z0} + u_z^{\mathrm{A}} \tag{4.22}$$

The formal displacement boundary conditions of the one-quarter unit-cell geometry are stated as

$$\begin{array}{llll} u_r = 0 & \text{along the axis} & r = 0, & 0 \le z \le L_z \\ u_z = 0 & \text{on the bottom} & 0 \le r \le L_r, & z = 0 \\ u_r = u_r^{\mathrm{A}} & \text{on the lateral surface} & r = L_r, & 0 \le z \le L_z \\ u_z = u_z^{\mathrm{A}} & \text{on the top} & 0 \le r \le L_r, & z = L_z \end{array} \tag{4.23}$$

The boundaries of the cell are free of shear tractions and the surfaces of the void are free of all tractions. Additionally, the matrix material is assumed to be a pure matrix containing no secondary voids or particles. A void nucleation rule governing the formation of secondary voids is not considered because the unit cell results would be dependent upon that specific nucleation model and its assumed parameters. The above conditions are applicable to all of the void geometries and stress states considered in this chapter.

4.3.3 Stress State and Microstructure Evolution

The microscopic stress and strain tensors describe the stress state within the unit cell and are denoted by σ_{ij} and ε_{ij}. The macroscopic stress and strain tensor are applied at the cell boundaries and are denoted as Σ_{ij} and E_{ij}, respectively. Since the cell surfaces are free of shear, the applied stress and strain directions coincide with the principal directions. The macroscopic principal and equivalent strains are given as

$$E_{\mathrm{r}} = \ln\left(1 + \frac{u_r^{\mathrm{A}}}{L_{r0}}\right) \quad E_{\mathrm{z}} = \ln\left(1 + \frac{u_z^{\mathrm{A}}}{L_{z0}}\right) \quad E_{\mathrm{eq}} = \frac{2}{3}|E_z - E_r| \tag{4.24–4.26}$$

The remote true macroscopic principal stresses are calculated at any instant as the average reaction force at the cell faces per current area through

$$\Sigma_r = \frac{1}{L_z}\int_0^{L_z} S_r|_{r=L_r}\,dz \quad \Sigma_z = \frac{2}{L_r^2}\int_0^{L_r} rS_z|_{z=L_z}\,dr \qquad (4.27, 4.28)$$

where S is the stress vector. The corresponding macroscopic hydrostatic stress, Σ_{hyd}, equivalent stress, Σ_{eq}, and stress triaxiality, T, are calculated as

$$\Sigma_{\text{hyd}} = \frac{\Sigma_1 + \Sigma_2 + \Sigma_3}{3} \quad \Sigma_{\text{eq}} = |\Sigma_z - \Sigma_r| \quad T = \frac{\Sigma_{\text{hyd}}}{\Sigma_{\text{eq}}} \qquad (4.29\text{–}4.31)$$

The analysis can be simplified by defining a loading parameter, α, that can be determined for a specified stress triaxiality as

$$\alpha = \frac{\Sigma_r}{\Sigma_z} = \frac{3T-1}{3T+2} \qquad (4.32)$$

where $\alpha = 0$ for uniaxial tension ($T = 1/3$) or -0.5 for pure shear ($T = 0$). The plastic strain increment, $\dot{\bar{\varepsilon}}^p$, and flow stress, $\bar{\sigma}$, within the unit cell can be determined from work equivalence as

$$\dot{\bar{\varepsilon}}^p = \frac{\Sigma_{ij}\dot{E}_{ij}^p}{\bar{\sigma}(1-f)} \qquad (4.33)$$

by neglecting the elastic strains and assuming an incompressible matrix so that $\dot{E}_{ij}^p \approx \dot{E}_{ij}$. The microscopic plastic strain and flow stress of the unit cell are related by the stress-strain relation of the matrix. In the present study, the ductile matrix is assumed to be an isotropic, rate-independent material that hardens according to the power-law relation

$$\frac{\bar{\sigma}}{\sigma_y} = \left(1 + \frac{E}{\sigma_y}\bar{\varepsilon}^p\right)^n \qquad (4.34)$$

where σ_y is the initial yield stress, E is the Young's modulus and n is the hardening exponent. A typical value of $E/\sigma_y = 500$ is valid for most ductile engineering alloys where n varies from 0 to 0.40.

The void volume fraction of the unit cell can be computed from the updated coordinates of the element nodes that define the surface of the void or from an approximate analytical formula proposed by Koplik and Needleman (1988) using the cell volume, V, as

$$f = 1 - (1 - f_0)\left(\frac{V_0}{V} - \frac{3(1-\nu)}{E}\Sigma_{\text{hyd}}\right) \quad \frac{V}{V_0} = \frac{(L_{r0} + u_r^A)^2(L_{z0} + u_z^A)}{L_{r0}^2 L_{z0}} \qquad (4.35)$$

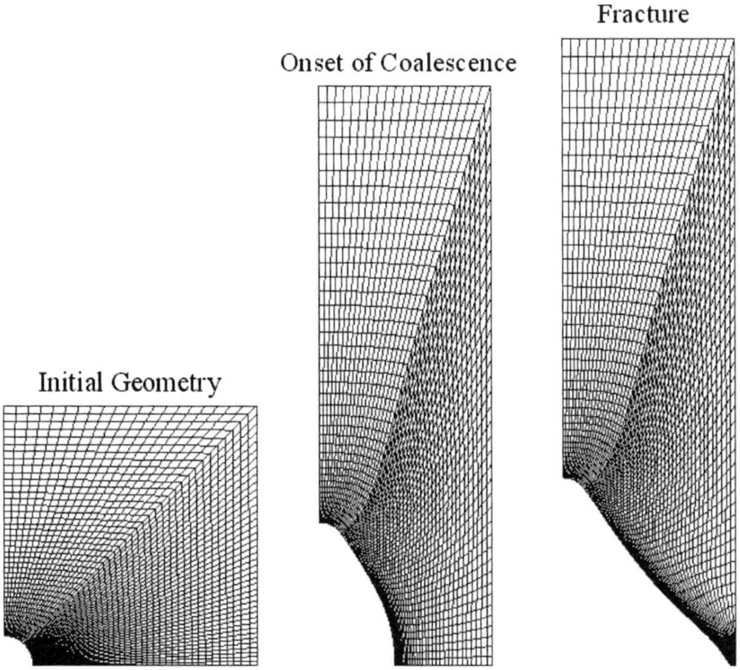

Fig. 4.7 Typical deformation history of a voided unit cell. Initially, the deformation mode is homogeneous and characterized by stable void growth until the onset of coalescence when the deformation mode becomes unstable and localized within the ligament. Note the abrupt change in the transverse growth of the void during coalescence compared to the vertical displacement of the cell. These results were obtained for an initially spherical void with an initial porosity of 0.1 % and subjected to a stress triaxiality of unity

This approximation provides very good agreement with numerical integration of the void surface as observed in previous studies (Koplik and Needleman 1988; Pardoen and Hutchinson 2000; Siad et al. 2008) and reduces the amount of post-processing. It should be noted that this approximation in Eq. (4.35) is not valid for a porous matrix material (secondary voids). The void aspect ratio, cell aspect ratio and the void spacing ratios can be readily determined from the geometry using Eqs. (4.2–4.5).

4.3.4 Identification of Void Coalescence

The termination of homogeneous deformation within the unit cell is marked by the onset of void coalescence when deformation becomes unstable and localized within the inter-void ligament while the material outside the ligament unloads elastically as shown in Fig. 4.7. The transition to void coalescence is identified by the radial strain rate approaching 0 as the unit cell deforms in a type of uniaxial stretching

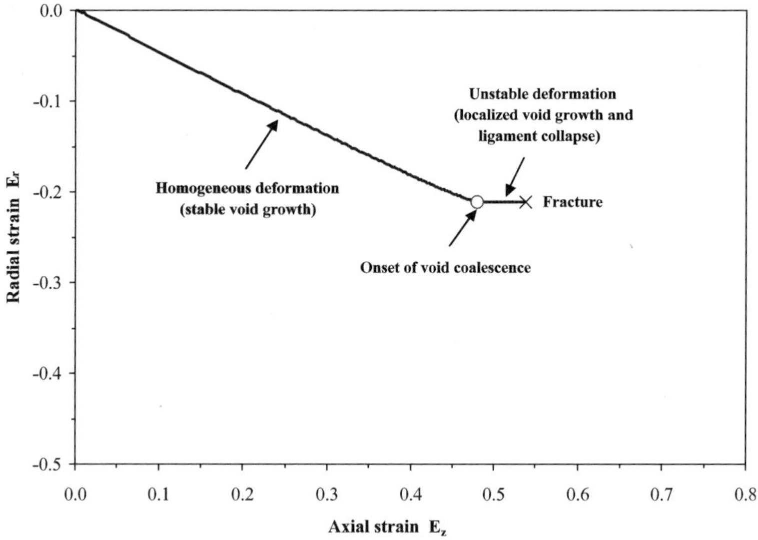

Fig. 4.8 Typical relationship of the macroscopic radial and axial strains for a voided unit cell. The onset of void coalescence is identified when the radial strain rate is 0

mode, regardless of the stress triaxiality. In this regime, the void expands rapidly in the transverse direction as the ligament length approaches 0 and the load-bearing capacity of the material is lost. Physically, this represents the collapse of the ligament as the void is beginning to link-up with a neighbouring void from an adjacent unit cell to form a larger void. However, in the finite-element simulation, the ligament will shrink and become highly deformed as the length approaches 0, ultimately requiring a re-meshing scheme to model the post-coalescence regime.

In the present study, the finite-element simulations are terminated after the onset of void coalescence as the results of the model in the post-coalescence regime are not of prime importance. For the practical range of stress triaxiality found in sheet metal forming operations $(T \leq 1)$, the collapse of the ligament occurs almost instantaneously and thus the coalescence strain can be used as an excellent approximation to the fracture strain. The procedure for identifying the onset of coalescence using the radial strain rates is demonstrated in Fig. 4.8 and the associated loss of load-bearing capacity with coalescence is illustrated in Fig. 4.9.

4.3.5 Numerical Solution Procedure

To obtain general trends for damage evolution and coalescence within a unit cell, the stress triaxiality must be kept constant throughout the cell during deformation. Since the compressibility of the cell varies during deformation, the stress triaxiality will steadily increase and fluctuate unless a control scheme is used to adjust the displacements of the cell faces. This problem becomes acute near the onset of void

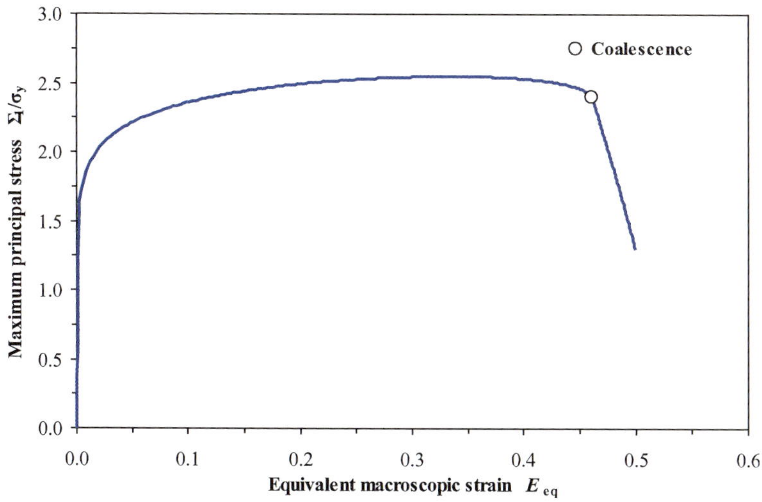

Fig. 4.9 Typical macroscopic stress and strain curve for a voided unit cell. Note that the loss of load-bearing capacity of the material is approximately linear in the post-coalescence regime

coalescence where deformation becomes unstable within the inter-void ligament. Fortunately, using ABAQUS finite-element software, the modified Riks-algorithm (Riks 1979) is ideally suited to incrementing the load on the unit cell to maintain equilibrium. The Riks method is an iterative scheme that allows equilibrium solutions to be found for nonlinear, unstable problems such as buckling or an abrupt change in the deformation mode such as a snap-through process or void coalescence.

The Riks-option within ABAQUS is employed to determine the displacements on the unit cell faces in such a way to enforce the loading parameter, α, in Eq. (4.32) to be a constant prescribed value based upon the specified stress triaxiality. The axial stress, Σ_z, is set to unity and the corresponding radial stress is $\Sigma_r = \alpha$. Consequently, the axial direction is the principal loading direction for all simulations in the present study since $\alpha < 1$ for the range of stress triaxialities considered $(1/3 \leq T \leq 3)$. Convergence studies were performed to test for mesh dependence and it was found to be minimal if the element shape within the ligament is rectangular and the mesh is sufficiently refined near the void surface as seen in Figs. 4.5 and 4.7. The accuracy of the unit cell results was verified through comparison with previous cell studies in the literature by Pardoen and Hutchinson (2000), Lassance et al. (2006) and Scheyvaerts et al. (2010).

4.4 Unit Cell Simulation Results

A large number of finite-element models have been developed with an emphasis placed on small void aspect ratios ($W < 1$) that best resemble the conditions observed in particle cracking and debonding. Typical void shape evolution models are only valid for an aspect ratio greater than 1/6 while voids nucleated by particle cracking have an aspect ratio close to 0. Knowledge of the deformation of the void at small aspect ratios is critical since many alloys nucleate voids by particle cracking. To best represent an isolated void within a real material the initial porosity of the voids is taken to be 0.01 and 0.1 %. Smaller porosities could be considered with the use of an adaptive remeshing algorithm since very large strains would be required to grow the void to coalescence. Finally, the simulations are terminated shortly after the onset of coalescence as mesh distortion effects become significant, especially for initially flat voids. Fortunately, the load drop, the void growth rate and shape evolution all become linear within the post-coalescence regime and their behaviour can be reliably extrapolated to estimate their value at fracture (Scheyvaerts et al. 2010).

The aspect ratios considered in the analysis are: 0.001, 0.01, 0.05, 1/6, 1 and 6 for porosities of 0.1 and 0.01 % for a general hardening exponent of 0.10. To enable the results to be generalized to any material, the parameters in the flow stress relation for the material in Eq. (4.34) are $E/\sigma_y = 500$, which is applicable to a wide range of engineering materials (Pardoen and Hutchinson 2000). The stress triaxialities range from 1/3 to 3 to capture void evolution in stress states ranging from uniaxial tension to the severe conditions found ahead of a crack tip. It is important to note that the stress states considered are purely triaxial and that no shear stress is present. The influence of shear on void growth and shape evolution is currently a very active area of research and complex boundary conditions are required to enforce a constant stress state. The interested reader is referred to Barsoum and Faleskog (2007a, b) and Scheyvaers et al. (2010) for further information.

4.4.1 Penny-Shaped Voids: $W_o = 1/100$

The evolution of the porosity and the aspect ratio of an initially penny-shaped void for a wide range of stress triaxialities are presented in Figs. 4.10 and 4.11, respectively. What is most remarkable about these figures is that the voids grow exceedingly fast and in an approximately linear fashion in each stress state. In uniaxial tension ($T = 1/3$), the void enlarges from 0.1 % to nearly 9 % while the aspect ratio changes by a factor of over 100 and is roughly spherical at fracture. The linear nature of the penny-shaped voids evolution is attributed to it opening in the load direction and experiencing negligible lateral growth. As most nucleated voids are initially penny-shaped, the proper modeling of this geometry is of prime importance for modelling damage in alloys where void nucleation is significant as in the 5xxx series Al-Mg alloys (Chen 2004; Orlov 2006; Butcher 2011).

The rapid expansion of initially penny-shaped voids is an interesting result since SEM micrographs commonly observe spherical dimples that are indicative of ductile fracture as shown in Fig. 4.12. During the final stage of ductile fracture, the high stress triaxiality surrounding the primary voids nucleates voids from the

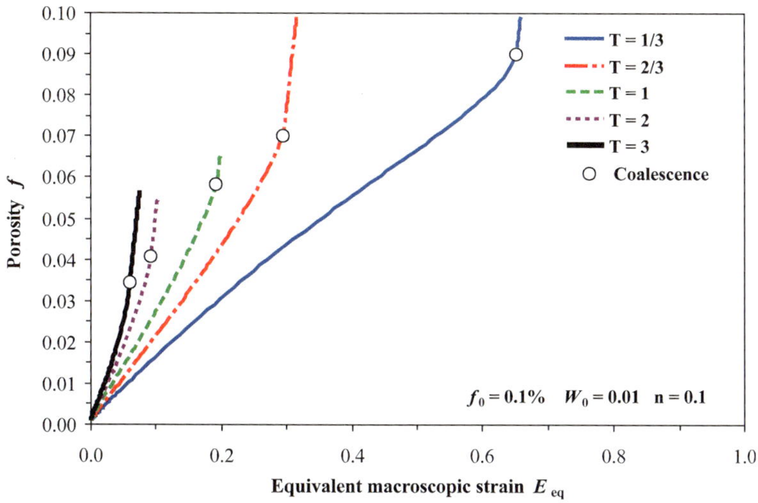

Fig. 4.10 Porosity history of a penny-shaped void subjected to various triaxial loadings

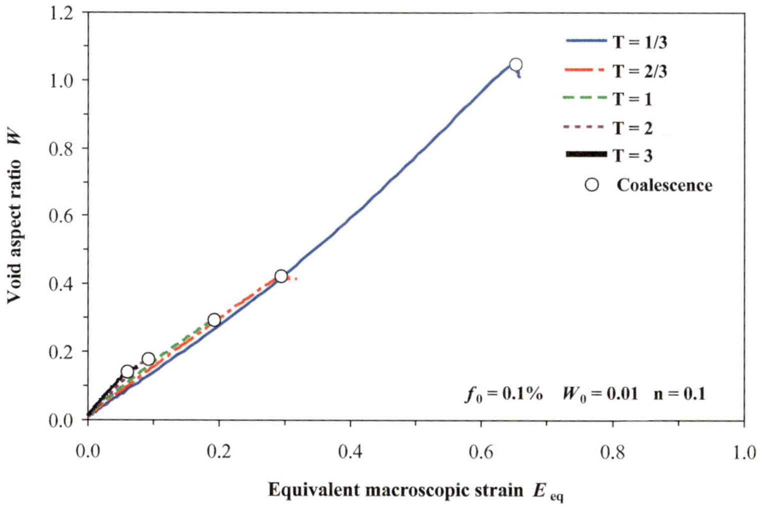

Fig. 4.11 Evolution of the void aspect ratio of an initially penny-shaped void subjected to various triaxial loadings

surrounding particles and the rapid growth of these penny-shaped voids can explain the spherical-shaped voids on the fracture surface. The high stress triaxiality will also encourage pre-existing voids to grow in a spherical manner and may even cause prolate voids to turn into oblate voids if the value is high enough.

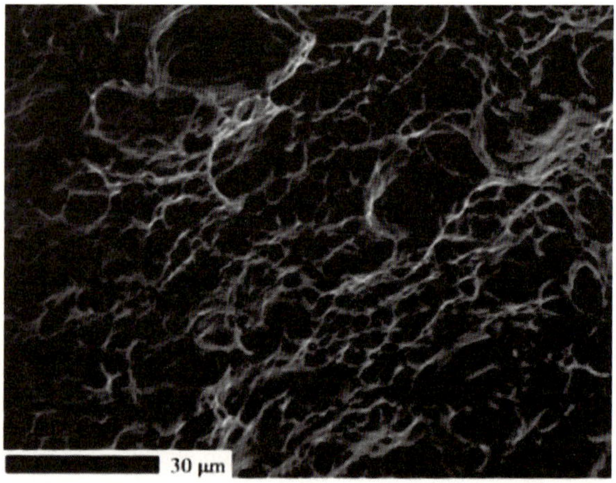

Fig. 4.12 SEM micrograph of the fracture surface in an AA5182 alloy. Note the presence of the smaller secondary dimples surrounding the dimple from a primary void in the top of the figure

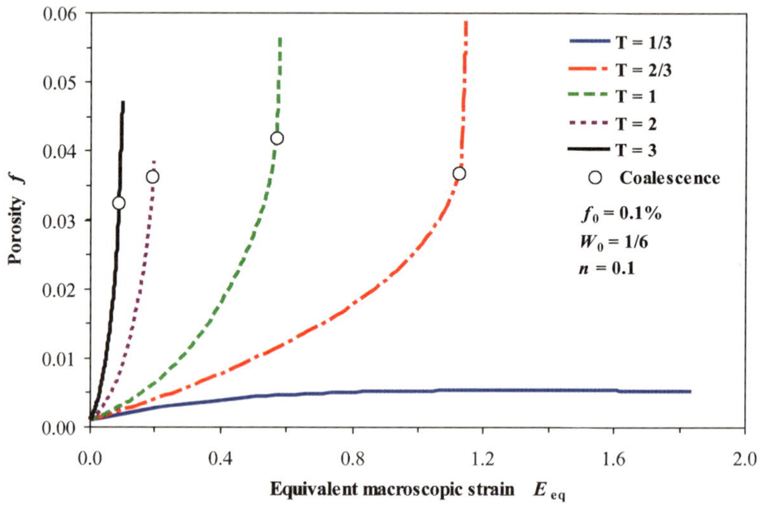

Fig. 4.13 Porosity history of an initially oblate void subjected to various triaxial loadings

4.4.2 *Oblate Voids:* $W_o = 1/6$

The growth and shape evolution of an initially oblate void is presented in Figs. 4.13 and 4.14, respectively. The expected non-linear response of the voids begins to emerge where void growth is slow at low triaxiality and rapid at high triaxiality. The high growth rate at high large stress triaxiality occurs because the void has sufficient height to be expanded by the lateral stress. This is in contrast with the

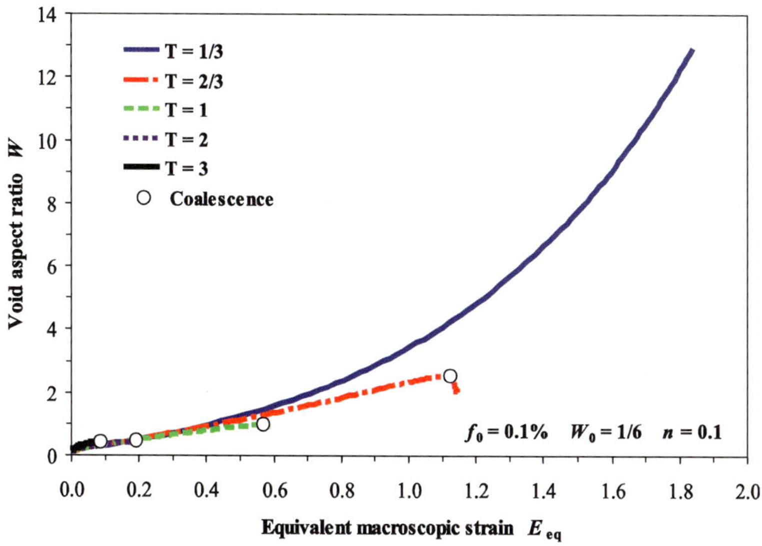

Fig. 4.14 Shape evolution of an initially oblate void subjected to various triaxial loadings

initially penny-shaped void whose growth rate is nearly linear and much less sensitive to the hydrostatic stress.

Conversely, in uniaxial tension, the void becomes prolate and the void growth rate is very slow compared to the penny-shaped case. Prolate voids do not grow significantly at low triaxiality because plastic flow of the material over the void surface becomes easier as the voids laterally contract and become needle-like. In the limit, the void essentially closes and the matrix can deform almost as if the void were not present.

4.4.3 Spherical Voids: $W_o = 1$

The well-known void evolution trends for spherical voids are presented in Figs. 4.15 and 4.16. The void growth trends for spherical voids are similar to that of the oblate void with the difference in the growth rates between stress states more exaggerated. The trend for the aspect ratio shows a new development compared to the previous cases in that the void becomes oblate at high stress triaxiality. The height of the void is sufficiently large for the hydrostatic stress at high triaxiality to cause the void to grow laterally and cause the aspect ratio to decrease despite the principal loading direction being vertical. This is a counterintuitive but well documented phenomenon for voids at high stress triaxiality and highlights the difficulties in developing a shape evolution model because the void grows laterally when one would expect it to grow axially.

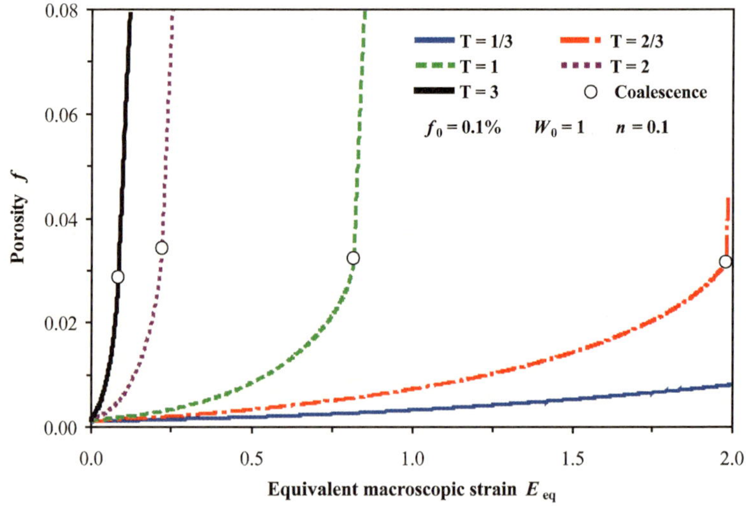

Fig. 4.15 Porosity history of an initially spherical void subjected to various triaxial loadings

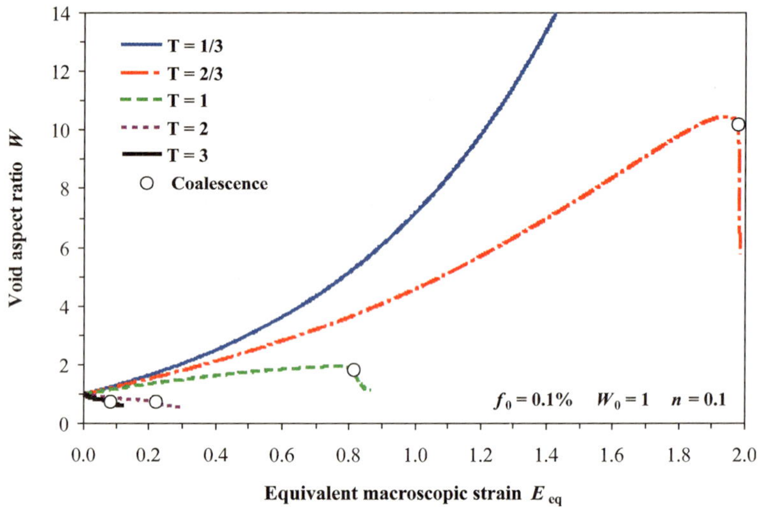

Fig. 4.16 Shape evolution (*right*) of an initially spherical void subjected to various triaxial loadings

4.4.4 Prolate Voids: $W_o = 6$

Finally, the void growth and shape evolution for initially prolate voids are presented in Figs. 4.17 and 4.18. The trends are again similar to the spherical case with the influence of the stress state more dramatic. What is most apparent is that the growth of the voids is completely negligible in uniaxial tension as the void becomes extremely

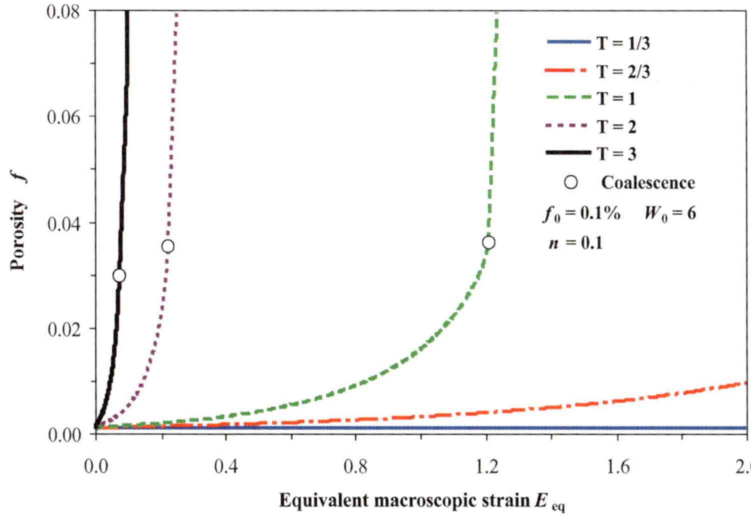

Fig. 4.17 Porosity history of an initially prolate void subjected to various triaxial loadings

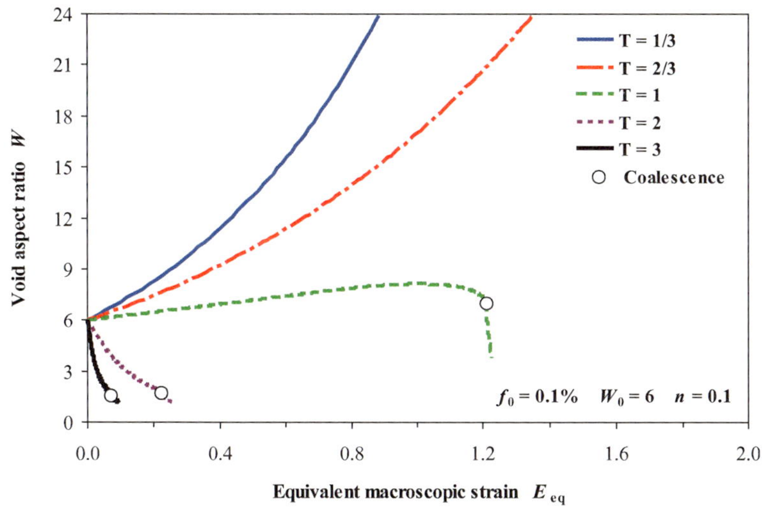

Fig. 4.18 Shape evolution of an initially prolate void subjected to various triaxial loadings

prolate but occupies nearly the same volume. Void growth remains very low at a triaxiality of 2/3 which corresponds to equal-biaxial tension. Since the void is initially prolate and much taller than it is wide, it is very susceptible to lateral growth at high triaxiality and the void shape becomes nearly spherical at fracture. From a modeling perspective, prolate voids are not of paramount importance since their growth rates are so small in the practical regime found in metal forming ($T < 1$).

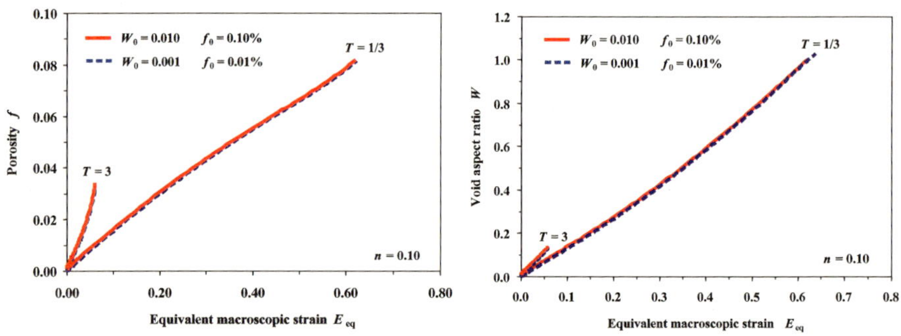

Fig. 4.19 Comparison of void growth and shape evolution for different penny-shaped voids for the extreme cases for the stress triaxiality

4.4.5 Selection of a Minimum Void Aspect Ratio

There are only so many finite-element simulations and unit cell geometries one can consider and the argument can always be made for why different stress states or geometries should have been included. Overall, a maximum initial aspect ratio of 6 seems reasonable for the modeling of prolate voids since most voids will be nucleated by particle cracking and debonding and best described as penny-shaped. If penny-shaped cracks are the most important void geometry then what is the minimum shape that can be considered? In the limit, the penny-shaped void has no initial height, the aspect ratio is 0, and it cannot be modeled using regular finite-element techniques. Mesh distortion issues also arise when modeling increasingly small voids so there is a practical limit to the geometry one can consider.

Fortunately, the void growth rate and shape evolution for extremely flat voids are essentially the same as shown in Fig. 4.19, which is caused by deformation primarily occurring in the opening direction of the void. The void opens in the loading direction at a high enough rate that the solutions for aspect ratios of 1/1,000 and 1/100 are quite similar. This is an extremely fortuitous result since any void that is nucleated by particle cracking can be assigned an arbitrary aspect ratio of 0.01, removing a parameter that otherwise would have had to been predetermined. A similar result has also been reported by Lassance et al. (2006).

4.5 Theoretical Models for Void Growth, Shape and Coalescence

With the benchmark trends for void growth and shape evolution established, the validity of the analytical evolution models can be evaluated. It is important to state that these models are all expected to perform at least reasonably well in some stress states, or they would be without merit in the first place. However, it is of great practical interest to evaluate the performance of these models in a variety of conditions to establish confidence in the model predictions for a wide range of stress states.

4.5.1 *Yield Criterion*

The yield criterion of Gurson and Tvergaard (GT) described in Eq. (1.23) is adopted to describe material softening and will be used to integrate the stress state in the analytical model. The extension of Ragab (2004a) is adopted to account for the influence of the void shape in the GT model. In this extension, q_1 and q_2 parameters in the yield function are no longer material constants but related to the stress state, void shape and material hardening.

4.5.2 *Void Growth, Shape Evolution and Coalescence*

Due to the large number of void shape growth and shape evolution models available in the literature, it is prohibitive to evaluate all of them since they are all valid for certain situations. Fortunately, the semi-empirical void growth and evolution models of Ragab (2004a) provide a logical and reasonable benchmark. Ragab (2004a) developed and validated a set of semi-empirical equations by performing a large-scale meta-analysis of the definitive analytical and numerical studies of void growth and shape evolution in the literature (McClintock 1968; Rice and Tracey 1969; Budiansky et al. 1982; Huang 1991; Lee and Mear 1992a, b; Yee and Mear 1996; Sovik and Thaulow 1997; Pardoen and Hutchinson 2000). In his analysis, Ragab (2004a, b) individually calibrated the q_i parameters in the GT yield criterion as functions of the void shape, porosity, stress state and hardening exponent. The calibrated (q_1, q_2) parameters not only enabled improved modelling of an isolated void but also improve the accuracy of the yield criterion by accounting for void shape effects while preserving the relatively simple Gurson framework. This is very advantageous since other damage-based models such as the GLD model and its variants (Gologanu et al. 1997; Benzerga 2002) that account for void shape effects are more cumbersome to implement compared to the GT model. An approximate method for estimating the influence of the void shape on the yield criterion is preferable to implementing these rigorous models.

It is important to note that the semi-empirical equations of Ragab (2004a) were not derived directly from finite-element simulations of a voided unit cell but from a large range of published analytical and numerical data for a large range of void geometries, stress states and material hardening exponents. As a result, it is of interest to directly compare the final equations with the finite-element simulations to best evaluate their performance. Butcher (2011) performed a large-scale finite-element study of cylindrical unit cells with void shapes ranging from penny-shaped to prolate to validate the Ragab (2004a, b) void growth and shape evolution models.

4.5.2.1 Void Growth

The associated flow rule of the GT model can be employed to obtain an alternate expression for the void growth rate using the first principal strain rate:

$$\dot{f}_{\text{growth}} = \frac{3f(1-f)q_1 q_2 \sinh\left(q_2 \frac{3}{2}\frac{\Sigma_{\text{hyd}}}{\bar{\sigma}}\right)}{3\left(\frac{\Sigma_{\text{I}}-\Sigma_{\text{hyd}}}{\bar{\sigma}}\right) + f q_1 q_2 \sinh\left(q_2 \frac{3}{2}\frac{\Sigma_{\text{hyd}}}{\bar{\sigma}}\right)} \dot{E}_1^{\text{p}} \tag{4.36}$$

and simplified by Ragab (2004a) by assuming small porosities and axisymmetric deformation to obtain

$$\frac{\dot{f}_{\text{growth}}}{f} = \frac{3}{2} q_1 q_2 \sinh\left(q_2 \frac{3}{2}\frac{\Sigma_{\text{hyd}}}{\bar{\sigma}}\right) \dot{E}_1^{\text{p}} \tag{4.37}$$

that was then calibrated to the void growth results available in the literature.

The semi-empirical equations of Ragab (2004a) for the variation of the q_{i} parameters in the GT model are valid for a large range of stress triaxiality from $1/3 \leq T \leq 8/3$, and for hardening exponents up to 0.40. The q_{i} relations are

$$\begin{aligned} q_1 &= A + BT + CT^2 + DT^3 \\ A &= 2.28 - 3.55n + 3.84n^2 \quad B = -0.92 + 1.32n - 0.32n^2 \\ C &= 0.53 - 2.31n + 2.35n^2 \quad D = -0.10 + 0.27n + 0.70n^2 - 1.78n^3 \end{aligned} \tag{4.38}$$

$$\begin{aligned} q_2 &= W^{\eta} \\ \eta(W < 1) &= 0.206\ln(T) - 0.266 - 0.02n \\ \eta(W \geq 1) &= -3.484 + 11.614T - 13.72T^2 + 6.54T^3 - 1.06T^4 + 0.2n \end{aligned} \tag{4.39}$$

4.5.2.2 Void Shape Evolution

For a constant strain-path, the evolution law of Ragab (2004a) for the void aspect ratio is

$$\ln(W/W_0) = 1.1(\omega + 2 - \Sigma_{\text{hyd}}/\bar{\sigma} + n)(1 - f_0)E_{\text{eq}}^{\text{p}} \tag{4.40a}$$

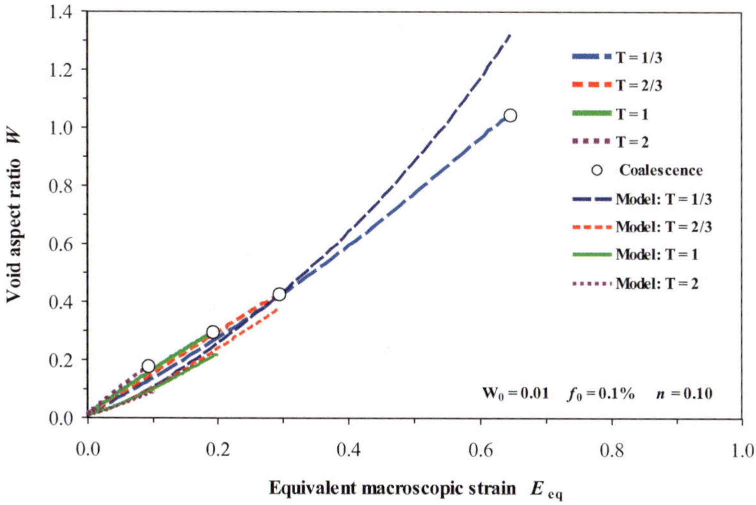

Fig. 4.20 Comparison of the analytical void shape evolution model with the unit cell data for an initially penny-shaped void

Where ω is defined for initially oblate $(1/6 \le W_0 \le 1)$ and prolate voids $(1 \le W_0 \le 6)$ as

$$\omega_{\text{oblate}} = -\ln(W_0)\Big/\left(0.109 + 1.224E^{\text{p}}_{\text{eq}}\right)$$

$$\omega_{\text{prolate}} = \left(-0.535 + 0.0235E^{\text{p}}_{\text{eq}}\right)\ln(W_0) \qquad (4.40\text{b, c})$$

4.5.3 Comparison with Unit Cell Results

4.5.3.1 Void Shape

The predictions of the semi-empirical model of Ragab (2004a) for the void aspect ratio for the range of initial shapes considered are presented in Figs. 4.20 through 4.23. Unlike the void growth model, the predictions for the initial penny-shaped void in Fig. 4.21 are in very good agreement with the unit cell data. This is a surprising result since the model was not calibrated for this regime. Overall, the model gives decent predictions for the void aspect ratio for oblate, spherical and prolate voids in most stress states. The predictions are the least accurate in the case of prolate voids but the accuracy is reasonable at low strains, say, of 0.30. This is not a significant limitation since initially prolate voids with an aspect ratio of 6 are not as common as initially penny-shaped or oblate voids since these shapes best resemble the void at nucleation by particle cracking and partial debonding. In

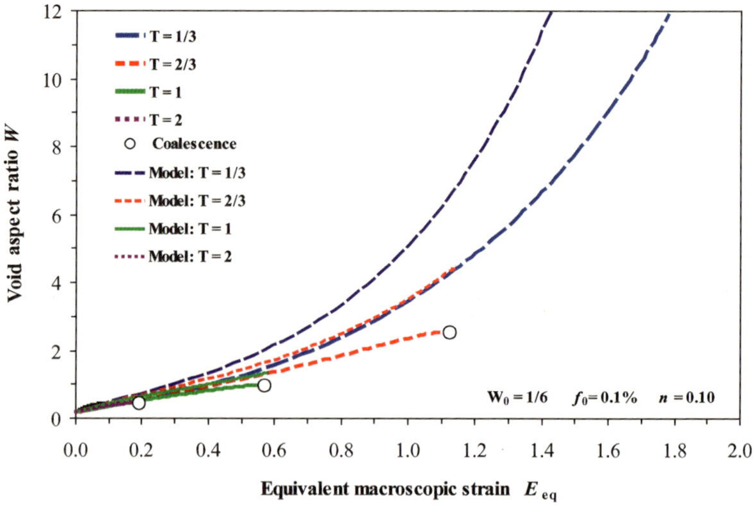

Fig. 4.21 Comparison of the analytical void shape evolution model with the unit cell data for an initially oblate void

general, the shape evolution model is quite reasonable and suitable for use in a typical damage-based constitutive model and is very attractive for its simplicity.

4.6 Calibration of the Void Evolution Models

4.6.1 Void Growth

The void growth model of Ragab (2004a) used a simplified version of the growth rule obtained for the Gurson model. In the present work, we will use the original form of the growth rule in Eq. (4.36) because unlike Ragab (2004a), the unit cell data has been obtained first-hand and not taken from the literature. The q_2 parameter cannot be solved for explicitly in Eq. (4.36) and a Newton-Raphson method is used to determine the q_2 value obtained at each time-step in the unit cell simulation. The calibrated q_2 values for each void geometry and stress state were determined to create a library of highly accurate void growth correlations in Table 4.1. The calibrated q_2 values for several of the void shapes considered are shown in Fig. 4.24. The calibrated q_2 value converges to a value of unity for a spherical void which is the value derived by Gurson (1975). The calibrated void growth model is compared with the unit cell data in Fig. 4.25 and the agreement is excellent.

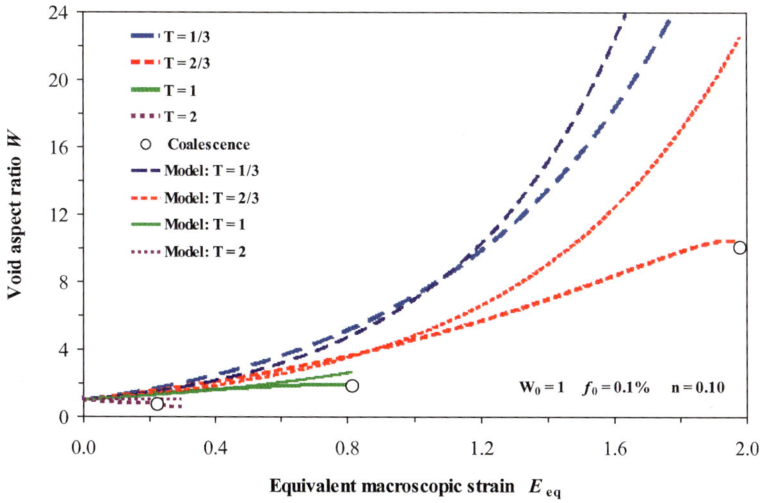

Fig. 4.22 Comparison of the analytical void shape evolution model with the unit cell data for an initially spherical void

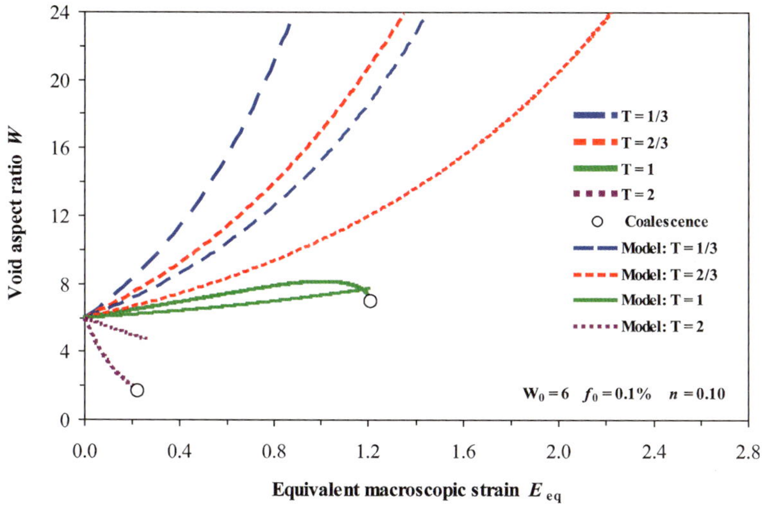

Fig. 4.23 Comparison of the analytical void shape evolution model with the unit cell data for an initially prolate void

4.6.2 Void Shape Evolution

Although the semi-empirical equations of Ragab (2004a) give pretty good results for the void aspect ratio, a calibration is required because void growth is tightly coupled to the void shape through $q_2 = W^{\eta}$. Due to this exponential dependence, the

Table 4.1 Void growth calibration parameters for various void volume fractions, void shapes and stress triaxiality ratios

				Range of the aspect ratio (W) that correlation is valid		Range of the equivalent strain (ε_{eq}) that correlation is valid	
f_o (%)	W_o	T	ζ	W_o	W_{limit}	ε_o	ε_{limit}
0.1	0.01	1/3	−0.4310	0.01	1.00	0	0.62
		2/3	−0.3150		0.42		0.29
		1	−0.2530		0.29		0.19
		2	−0.2081		0.18		0.09
		3	−0.2310		0.14		0.06
	0.05	1/3	−0.4785	0.05	1.00	0	0.50
			−0.7580	1.00	4.00	0.50	1.22
		2/3	−0.3300	0.05	1.00	0	0.51
			−0.1200	1.00	1.36	0.52	0.71
		1	−0.2440	0.05	0.69	0	0.40
		2	−0.1170		0.36	0	0.15
		3	−0.3210		0.31	0	0.08
	1/6	1/3	−0.5850	1/6	1.00	0	0.41
			−0.8879	1	6.00	0.41	1.33
		2/3	−0.3600	1/6	0.97	0	0.41
			−0.3000	1	2.39	0.43	1.01
		1	−0.2350	1/6	0.98	0	0.57
		2	−0.0850		0.47		0.19
		3	−0.1120		0.45		0.09
	1	1/3	−0.0700	1	33.25	0	2.00
		2/3	−0.1695		10.46		1.94
		1	−0.0780		1.94		0.76
		2	−0.0580		0.75		0.22
		3	−0.6290		0.59		0.12
	6	1/3	−0.6030	6	27.07	0	0.97
		2/3	−0.1900		17.97		1.06
		1	−0.0980		8.14		1.00
		3	0.0850		1.81		0.06
	0.001	1/3	−0.4045	0.001	1.00	0	0.62
		2/3	−0.3100		0.42		0.30
0.01		1	−0.2540		0.29		0.19
		2	−0.2180		0.17		0.09
		3	−0.2320		0.13		0.06
	1	1/3	−0.8881	1	35.43	0	2.03
		2/3	−0.7000		7.03		1.37
		1	−0.0890		2.14		1.06
		2	−0.2200		0.62		0.30
		3	−0.4140		0.71		0.11
	6	1/3	−0.1200	6	147.72	0	2.09
		2/3	−0.1240		39.03		1.84
		1	−0.0960		9.30		1.50
		2	0.0330		1.02		0.31
		3	0.1200		1.03		0.10

(continued)

Table 4.1 (continued)

				Range of the aspect ratio (W) that correlation is valid		Range of the equivalent strain (ε_{eq}) that correlation is valid	
f_o (%)	W_o	T	ζ	W_o	W_{limit}	ε_o	ε_{limit}
	1/6	1/3	−0.5180	1/6	1.02	0	0.47
			−0.5200	1	2.57	0.47	0.98
1		2/3	−0.3260	1/6	0.83	0	0.42
		1	−0.2210		0.53		0.25
		2	−0.1970		0.34		0.10
		3	−0.2920		0.28		0.05
	1	1/3	−0.7200	1	9.22	0	1.21
		2/3	−0.3100		3.52		0.85
		1	−0.1900		1.59		0.41
		2	−0.5000		0.88		0.13
		3	−5.8000		0.84		0.04
	6	1/3	−0.6750	6	137.63	0	2.05
		2/3	−0.1750		21.27		1.20
		1	−0.1050		7.30		0.64
		2	0.0050		2.95		0.14
		3	0.3410		3.29		0.04

A piece-wise correlation is used for certain void shapes at low triaxialities and the ranges that the calibration parameters are accurate are given for both the void shape and equivalent strain

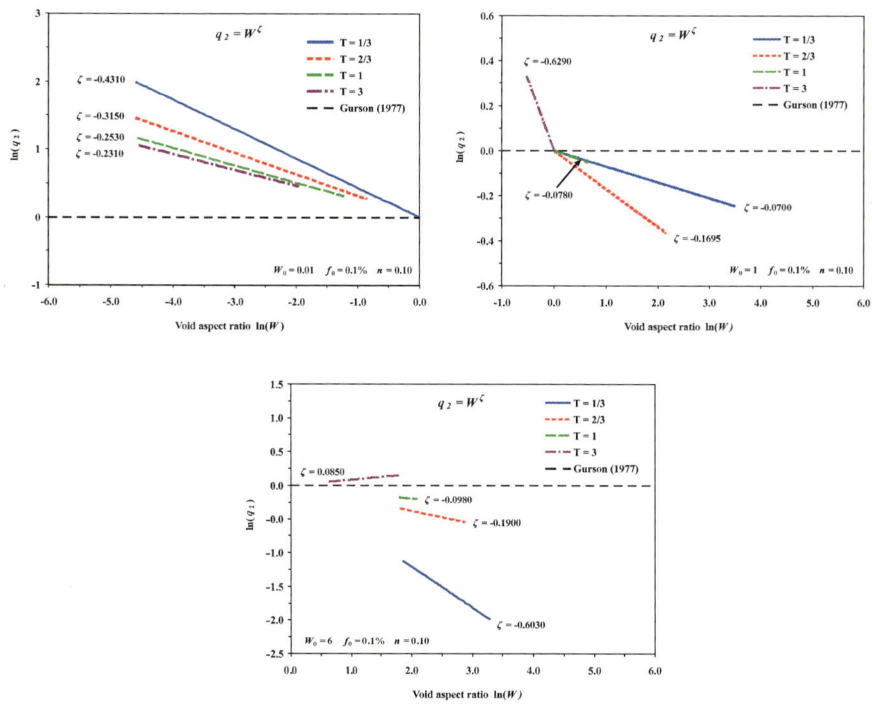

Fig. 4.24 Calibrated q2 parameter for an initially penny-shaped void (*top-left*), spherical void (*top-right*) and prolate void (*center*) subjected to various stress triaxialities

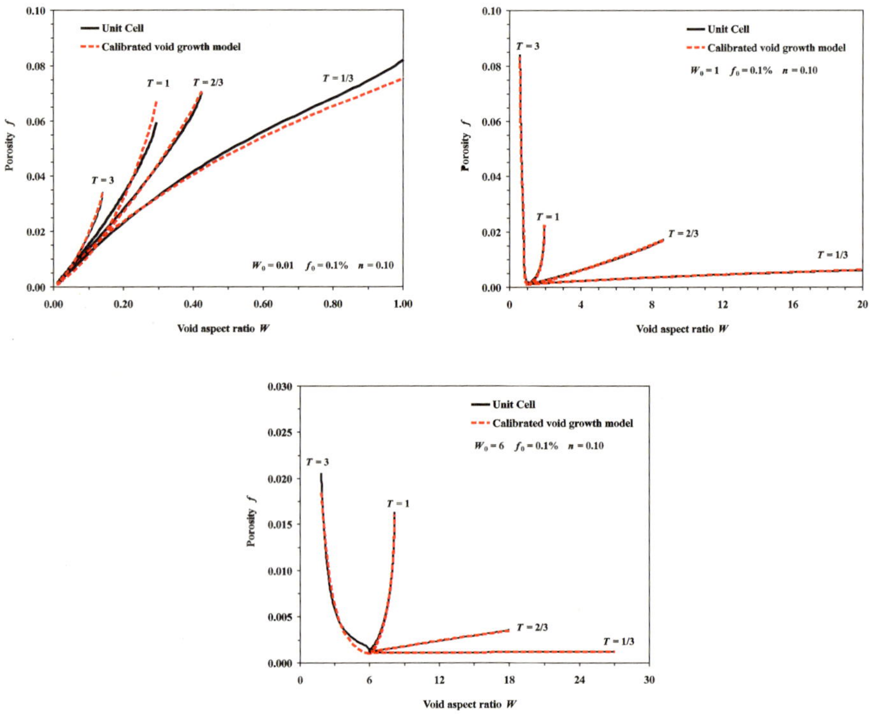

Fig. 4.25 Comparison of the calibrated void growth model with the unit cell data for an initially penny-shaped (*top-left*), spherical (*top-right*) and prolate void (*center*)

previous void growth calibration is in vain unless the aspect ratio is also highly accurate because errors in the aspect ratio will be magnified in the growth rule. Unlike the growth rule, the semi-empirical equation for the void shape is a pure correlation and its form is not readily amenable to calibration. To ensure that the void aspect ratio is modeled to a high degree of accuracy, a second-order polynomial was observed to be suitable to very large strains. If higher strains are required, the curve is linearly extrapolated with the slope selected to give good agreement to high strains. The correlation function for the aspect ratio is a function of the equivalent strain and initial aspect ratio as

$$\begin{aligned} W &= a_2 E_{\mathrm{eq}}^2 + a_1 E_{\mathrm{eq}} + W_{\mathrm{o}} \quad & E_{\mathrm{eq}} \leq E_l \\ W &= b_1 \left(E_{\mathrm{eq}} - E_l\right) + W_l \quad & E_{\mathrm{eq}} > E_l \end{aligned} \tag{4.41}$$

where E_l is the limit strain and a_1, a_2, and b_1, are the calibration coefficients. W_l is the aspect ratio at the limit strain. For nearly all geometries considered, b_1 was selected so that it can be computed as the derivative of quadratic correlation evaluated at the limit strain. The calibrated parameters are presented in Tables 4.2, 4.3 and 4.4.

Table 4.2 Parameters for the void aspect ratio for an initial porosity of 0.01 % and a hardening exponent of 0.10

Aspect ratio	T	a_2	a_l	W_0	Limit strain ε_l	R^2	Aspect ratio at strain limit W_l	Slope for extrapolation b_l
0.001	1/3	0.68	1.19	0.00	0.635	1.000	1.031	2.114
	2/3	−0.25	1.49	0.00	−0.275	1.000	0.393	1.357
	1	−1.25	1.73	0.00	0.190	1.000	0.300	1.256
	2	−3.37	2.16	0.00	0.090	1.000	0.168	1.553
	3	−5.76	2.57	0.00	0.059	1.000	0.133	1.894
1/6	1/3	2.69	1.84	0.17	0.900	0.999	4.008	12.000
	2/3	1.14	1.34	0.17	1.800	1.000	6.278	5.450
	1	−0.51	1.94	0.17	0.850	1.000	1.448	1.072
	2	−7.61	2.95	0.17	0.140	1.000	0.431	0.821
	3	−66.36	9.68	0.17	0.063	0.998	0.513	1.316
1	1/3	3.51	2.34	1.00	0.800	1.000	5.115	11.000
	2/3	1.86	1.83	1.00	1.500	1.000	7.921	7.400
	1	−0.73	1.87	1.00	0.950	1.000	2.118	0.482
	2	2.67	−2.00	1.00	0.250	1.000	0.667	−0.664
	3	49.38	−7.04	1.00	0.063	0.994	0.753	−0.817
6	1/3	15.00	7.15	6.00	1.000	0.999	28.144	56.000
	2/3	6.37	4.89	6.00	1.500	0.999	27.672	24.007
	1	0.06	2.31	6.00	0.750	1.000	7.767	2.402
	2	65.28	−35.12	6.00	0.240	1.000	1.331	−3.785
	3	1547.53	−166.15	6.00	0.050	0.995	1.561	−11.397

The shaded values of b1 were manually adjusted and do not correspond to the derivative of the quadratic correlation

A limitation of the void shape evolution rule in Eq. (4.41) is that the initial void aspect ratio must be known and that it is not readily amenable to non-proportional loadings. However, it is compact, highly accurate and can be used to predict the void shape in uncoupled damage models where porosity-induced softening is minor and the void shape is only of interest for modelling coalescence. If a rate-based void shape evolution model is required, it is suggested to use the void shape evolution model from the GLD model or its variants (Pardoen and Hutchsinon 2000; Lassance et al. 2006; Scheyvaerts et al. 2010).

Care must be taken when calibrating the void aspect ratio because the trend for the aspect ratio departs from that of an isolated void prior to coalescence as shown in Fig. 4.26. Prior to coalescence, the void begins to interact with its neighbour, raising the local stress triaxiality and flattening the void. These interactions must be avoided during the calibration process as they are geometry dependent and the voids are assumed to be isolated. The limit strain in the correlation is selected to ensure interaction effects are avoided. The correlation functions for the different void geometries are compared with the unit cell data in Fig. 4.27, which shows that they accurately predict the aspect ratio to high strains while avoiding interaction effects.

Table 4.3 Parameters for the void aspect ratio for an initial porosity of 0.1 % and a hardening exponent of 0.10

Aspect ratio	T	a_2	a_l	W_0	Limit strain ε_1	R^2	Aspect ratio at strain limit W_l	Slope for extrapolation b_l
0.01	1/3	0.66	1.20	0.01	0.600	1.000	0.966	1.921
	2/3	−0.25	1.49	0.01	0.300	1.000	0.435	1.341
	1	−1.28	1.73	0.01	0.190	1.000	0.300	1.240
	2	−3.54	2.15	0.01	0.092	1.000	0.178	1.500
	3	−6.41	2.58	0.01	0.059	1.000	0.140	1.820
0.05	1/3	1.86	0.88	0.05	1.200	0.999	3.789	6.800
	2/3	0.27	1.70	0.05	0.500	1.000	0.966	1.968
	1	−1.01	2.01	0.05	0.400	1.000	0.693	1.203
	2	−5.54	2.83	0.05	0.130	1.000	0.324	1.388
	3	−26.80	5.42	0.05	0.075	0.999	0.305	1.419
1/6	1/3	2.07	1.17	0.17	1.200	0.993	4.552	6.144
	2/3	0.45	1.75	0.17	0.950	1.000	2.244	2.618
	1	−0.94	2.01	0.17	0.480	1.000	0.914	1.107
	2	−6.23	2.71	0.17	0.160	0.999	0.441	0.717
	3	−52.43	7.20	0.17	0.050	0.999	0.395	1.954
1	1/3	4.02	2.04	1.00	1.000	0.999	7.060	14.500
	2/3	1.62	2.01	1.00	1.750	1.000	9.484	7.690
	1	−0.85	1.92	1.00	0.750	1.000	1.957	0.637
	2	3.18	−1.72	1.00	0.180	1.000	0.794	−0.572
	3	48.87	−6.29	1.00	0.055	0.995	0.802	−0.914
6	1/3	14.9765	7.1021	6.00	1.100	0.999	31.934	62.000
	2/3	6.1131	5.0528	6.00	1.500	0.999	27.334	23.392
	1	−0.0903	2.4216	6.00	0.890	1.000	8.084	2.261
	2	69.2512	−33.6241	6.00	0.170	1.000	2.285	−10.079
	3	11.9209	−12.2347	6.00	0.070	1.000	5.202	−10.566

The shaded values of b1 were manually adjusted and do not correspond to the derivative of the quadratic correlation

4.6.3 *Void Coalescence*

It is of interest to evaluate the performance of the coalescence models since coalescence is predicted as a function of the void geometry. The coalescence model will only be as accurate as the sub-models that describe the void geometry. The models of Pardoen and Hutchinson (2000) and Benzerga (2002) in Eqs. (1.15) and (1.16) were evaluated using the geometry of the voids from the unit cell simulations to obtain their predictions under ideal conditions. The predictions of each model with the numerical coalescence strains are presented in Figs. 4.28 and 4.29, respectively.

Each coalescence model gives excellent agreement with the numerical coalescence strain to a surprising level of accuracy. This is not entirely unexpected since

Table 4.4 Parameters for the void aspect ratio for an initial porosity of 1 % and a hardening exponent of 0.10

Aspect ratio	T	a_2	a_1	W_0	Limit strain ε_1	R^2	Aspect ratio at strain limit W_l	Slope for extrapolation b_l
1/6	1/3	1.21	1.27	0.17	1.200	1.000	3.430	12.000
	2/3	−0.11	1.66	0.177	0.400	1.000	0.811	1.566
	1	−1.54	1.84	0.17	0.230	1.000	0.509	1.133
	2	−4.78	2.15	0.17	0.090	1.000	0.322	1.294
	3	−13.74	2.87	0.17	0.054	1.000	0.281	1.396
1	1/3	3.05	2.36	1.00	0.800	0.999	4.838	9.000
	2/3	0.64	2.42	1.00	0.900	1.000	3.693	3.565
	1	−1.20	1.95	1.00	0.375	1.000	1.564	1.053
	2	2.25	−0.77	1.00	0.080	0.999	0.953	−0.411
	3	6.40	−3.51	1.00	0.040	0.998	0.870	−3.003
6	1/3	15.13	7.32	6.00	1.000	0.999	28.453	56.000
	2/3	5.33	6.14	6.00	1.200	1.000	21.038	26.000
	1	−1.15	2.92	6.00	0.600	0.997	7.340	1.543
	2	34.27	−25.85	6.00	0.085	0.999	4.050	−20.026
	3	47.34	−26.88	6.00	0.080	1.000	4.153	−19.306

The shaded values of b_1 were manually adjusted and do not correspond to the derivative of the quadratic correlation

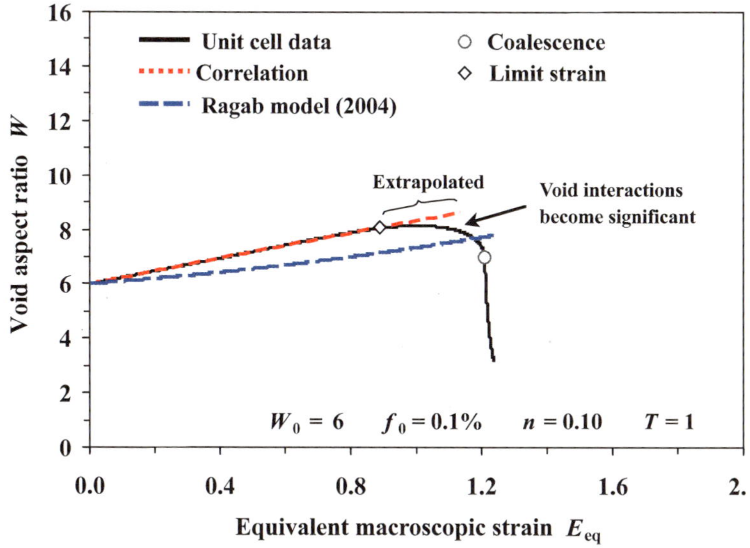

Fig. 4.26 Comparison of the void aspect correlation function with the model of Ragab (2004a) and the unit cell data. The limit strain is selected to void including the spurious interaction effects prior to coalescence

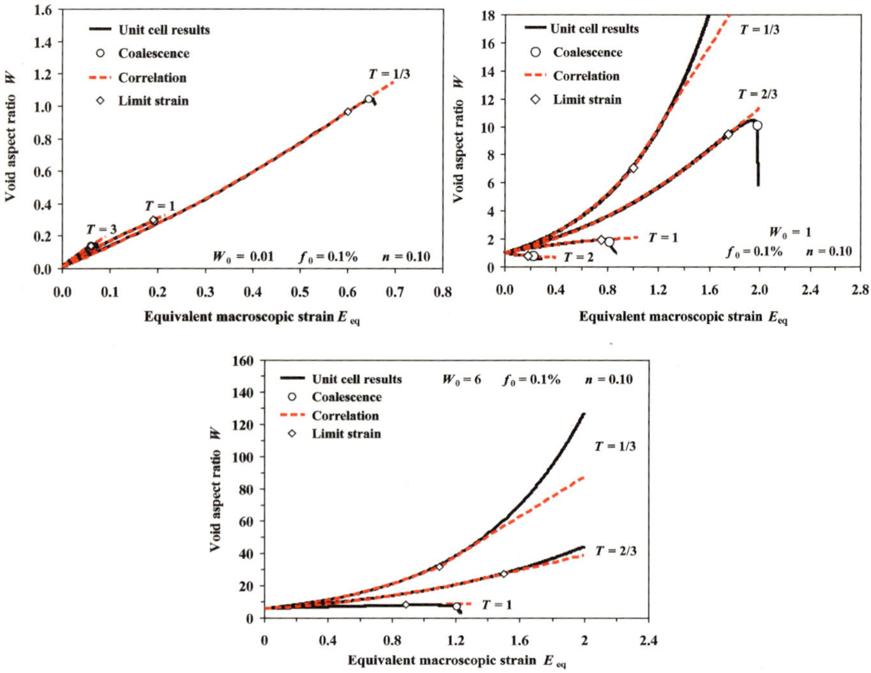

Fig. 4.27 Comparison of the void shape correlation with the unit cell data for an initially penny-shaped void (*top-left*), spherical void (*top-right*) and prolate void (*center*)

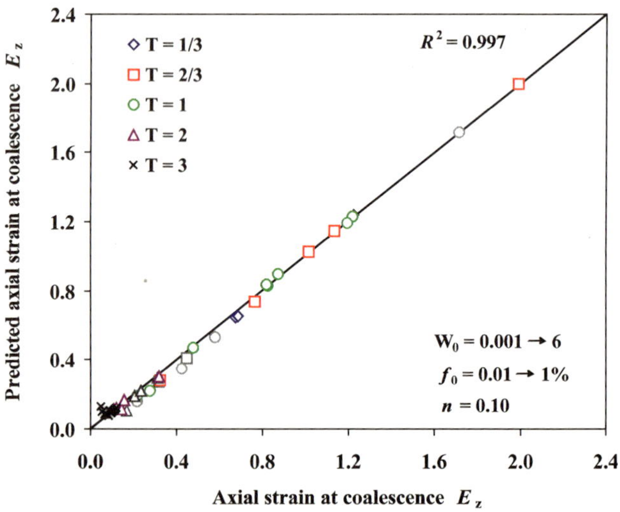

Fig. 4.28 Comparison of the numerical and predicted coalescence strains using the coalescence model of Pardoen and Hutchinson (2000) in Eq. (1.15)

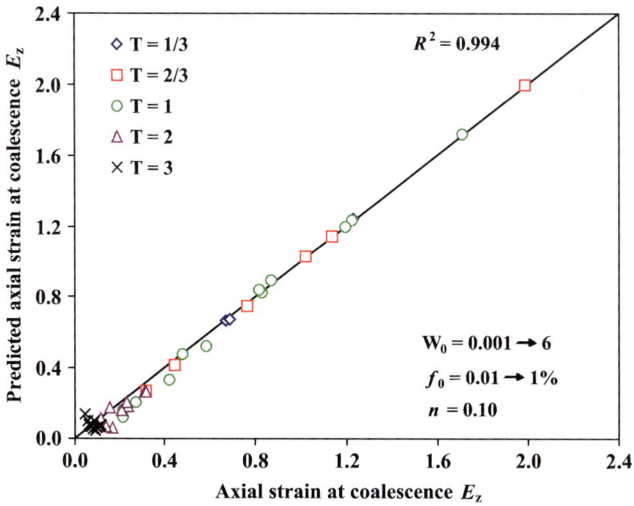

Fig. 4.29 Comparison of the numerical and predicted coalescence strains using the coalescence model of Benzerga (2002) in Eq. (1.16)

these models have been previously compared to unit cell simulations in the literature. There is no clear advantage to using either coalescence model based upon the present results. Even though the Pardoen and Hutchinson (2000) model accounts for hardening but not flat voids, and the Benzerga model (2002) assumes a perfectly plastic material, both models give very good predictions. The presence of flat voids does not translate to infinite ductilities in practice since these voids rapidly deform to shapes that are within the domain of the model. The strong agreement of both models attests to the strong underpinnings of the Thomason (1990) model that they are based upon.

The important conclusion from this validation is that the coalescence models can accurately describe void coalescence by internal necking. If the other analytical models can predict void growth and shape evolution to a similar degree of accuracy, the coalescence predictions will be sound and will agree with that of the unit cell.

4.7 Summary

A rigorous validation and calibration programme was undertaken to evaluate the analytical models used to describe void evolution and coalescence. Finite-element simulations of voided unit cells were performed to obtain benchmarks for the analytical models for a wide range of stress states. Emphasis was placed upon penny-shaped voids because they will be used extensively in the percolation model in Chap. 10 and the present evolution rules in the literature do not apply to voids of this shape. The principal contributions of this chapter are:

- The analytical models for void evolution were calibrated using the unit cell data by developing a vast library of correlation parameters.
- The semi-empirical void growth model of Ragab (2004a) was improved by calibrating the model using first-hand unit cell data and extended to the penny-shaped void regime.
- The growth and shape evolution of isolated voids within the percolation model will be representative of the actual case by virtue of the calibrated models.
- Void coalescence can be accurately predicted using the plastic limit-load model of Pardon and Hutchinson (2000) or Benzerga (2002).
- The GT yield criterion was improved by calibrating the q_2 parameter in the void growth model. Although designed for spherical voids, this calibration enables the model to give good results from penny-shaped to prolate voids

Chapter 5
Two-Dimensional (2D) Damage Percolation Modeling

As discussed in the preceding chapters, the second phase particle field in a dispersion strengthened ductile material can be described as a random distribution displaying a certain level of particle clustering (Pilkey et al. 1995). Ordinary unit cell models such as Gurson's (1977) analytical model or other numerical unit cell models (e.g. Needleman 1972; Worswick 1988) with a uniformly distributed particle/void field are unable to capture microstructures that are random and/or clustered. In order to overcome this disadvantage of unit cell models when applied to real micro-defects, Benson (1995) used a two-dimensional unit cell calculation to investigate the effect of void cluster size on ductile fracture. In his study, a unit cell with randomly distributed void clusters was modelled. Thomson et al. (1999) proposed a numerical unit cell model that contains a single particle cluster. Unfortunately, this model still possesses a periodic particle field. Real microstructures always display a random particle distribution with some superimposed degree of particle clustering. Therefore, it is doubtful that unit cell calculations are able to capture the onset of ductile fracture in real materials.

A new approach to predicting ductile damage evolution, known as the damage percolation model, has been proposed by Pilkey et al. (1998) and Worswick et al. (2001) to address this issue. In this approach, measured second phase particle fields are used as the starting point to capture the real variation in inter-particle spacing. The particle field is subjected to some uniform, remote strain field and damage nucleation and growth models are applied discretely to each particle in the field. Void coalescence is predicted as the merging of nearest-neighbouring voids and generally initiates within void clusters. Large-scale cracking is predicted once two or more clusters of voids coalesce, causing a chain reaction of profuse coalescence and gross material failure.

The damage percolation approach is the focus of this book and this chapter presents the theoretical framework for this methodology. One important issue in the damage percolation approach is the determination of the minimum particle field size or representative volume element (RVE) required to capture the bulk material response. This issue is addressed in this chapter, within the framework of a uniform

Z. Chen and C. Butcher, *Micromechanics Modelling of Ductile Fracture*,
Solid Mechanics and Its Applications 195, DOI 10.1007/978-94-007-6098-1_5,
© Springer Science+Business Media Dordrecht 2013

strain assumption. The introduction of strain gradients within the damage percolation approach is addressed in the following chapter in which the damage percolation model is coupled with a FE model.

5.1 The Damage Percolation Model

The starting point for the damage percolation model is the second phase particle field. This is usually obtained using image analysis of digital micrographs acquired directly from an optical microscope. Tessellation methods are applied to characterize second phase particle fields and the degree of clustering present in the microstructure. Matrix erosion tessellation techniques are employed to extract the degree of clustering present in the as-received alloys which contain a dispersion of Fe- and Mn-based inter-metallic particles (Pilkey 1997). The tessellated particle fields are read into the damage percolation model to simulate the development of damage under sheet forming conditions. Of particular interest are the conditions at which particle-nucleated void damage links up and the extent of linkage required to produce unstable crack growth and final fracture.

Annealed, O-temper AA5182 aluminum alloy sheets is considered in this chapter. This alloy is a candidate for use in automotive inner body panels as lightweight replacement for conventional steel. A plan view section of the as-received microstructure was prepared for metallographic examination and image analysis. The AA5182 microstructure has a dual population of Fe and Mn intermetallic particles. For simplicity, no attempt was made to distinguish between these different particle types during the acquisition and processing of the particle fields.

5.1.1 Particle Field Tessellations

A large-scale high-resolution digital image of a second-phase particle field was acquired from the planar metallographic view. This massive particle field image is roughly 5,500 × 4,250 pixels in size and has a resolution of 0.372 μm/pixel, corresponding to a physical size of about 2.0 × 1.6 mm. The constituent particles have been separated from the matrix background using standard thresholding techniques.

Relevant particle and clustering characteristics have been extracted from the large-scale high-resolution particle fields through the application of a matrix erosion tessellation algorithm. Large-scale image of the matrix erosion tessellation produced for the AA5182 microstructure is shown in Fig. 5.1.

Particle field feature data generated by the tessellation software is read directly by the damage percolation software. This model considers the nucleation and growth of voids at individual particles within the microstructure. Coalescence of individual voids to form cracks and the subsequent growth of cracks and

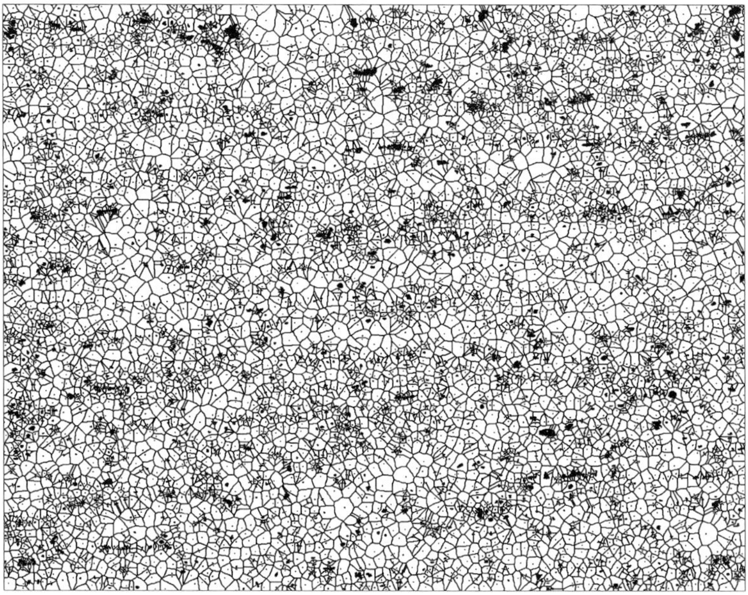

Fig. 5.1 A 5,500 × 4,250 pixel large-scale tessellated second phase particle field of Al-Mg alloy AA5182 used in the RVE study (RD, horizontal; TD, vertical) (Chen 2004)

coalescence of cracks with other cracks or isolated voids is also modelled (Worswick et al. 1998). Particle feature data utilized by the model includes:

- particle centroid coordinates
- particle principal axes (and orientation)
- nearest neighbour list
- cluster list (particles comprising each cluster)

5.1.1.1 Inter-Particle Dilational Spacing (IPDS)

The tessellation software calculates the so-called inter-particle dilational spacing (IPDS) for each tessellated particle field. In these particle clustering studies, a combination of matrix erosion tessellation and dilational counting techniques, previously employed by Shehata and Boyd (1988), were applied to the acquired particle fields (Pilkey et al. 1998). During each particle dilation step, the software records when each particle feature merges or touches another dilating particle feature. First contact or agglomeration between a particle and one of its neighbours defines then a nearest neighbour. Knowing the pixel size, the software can then determine the nearest neighbour spacing based upon the number of dilations performed. Dilational counting measures are then tabulated during the construction of a matrix erosion spatial tessellation and represented by a histogram of inter-

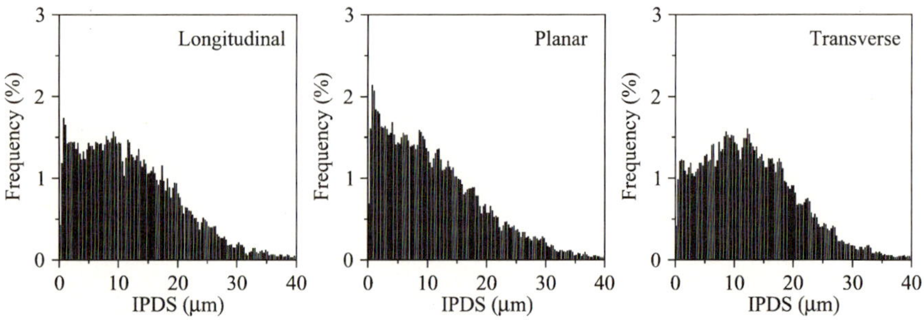

Fig. 5.2 Interparticle dilitational spacing (IPDS) of AA5182 sheets in different view planes

particle dilational spacing (IPDS) frequencies. Given that a matrix erosion tessellation algorithm involves repeated particle dilational passes, equivalent to matrix erosion passes, the number of distinct features that disappear from the particle field during each dilation pass is recorded as a frequency. The disappearance of a feature occurs when it agglomerates with another feature (i.e. dilating neighbours come in contact). At the start of the spatial tessellation process, each particle represents a feature. By recognizing that each particle dilation pass is of characteristic length in a digital image, the agglomeration frequencies can be plotted against dilational distance to produce an IPDS frequency spectrum. It follows that local peaks in the frequency of dilational merging events are indicative of characteristic spacings within the tessellated particle field. The dilating features which agglomerate at the smallest of these characteristic spacings are referred to as first order clusters, while successive IPDS peaks signify so-called second, third and higher orders of particle clusters. Figure 5.2 shows the IPDS histograms for the 1.0 mm AA5182 sheet in three view planes.

For mathematical convenience, individual particles are represented as ellipses with principal axes aligned with the sheet rolling and transverse directions. Thus any tilting of particles relative to these axes was neglected. In general, an ellipse representation of particles is considered acceptable and greatly simplifies particle interaction and void growth calculations.

5.1.1.2 Particle Aspect Ratio

Particle aspect ratio values were obtained by modelling each particle as an ellipse with equivalent second moments of area. Particles with a major axis aligned more closely to the longitudinal direction are assigned aspect ratios greater than unity. Particles with aspect ratios in excess of 2 or less than ½ are referred to here as high aspect ratio particles. Figure 5.3 shows the histograms of particle aspect ratio for the 1.0 mm AA5182 sheet in three view planes.

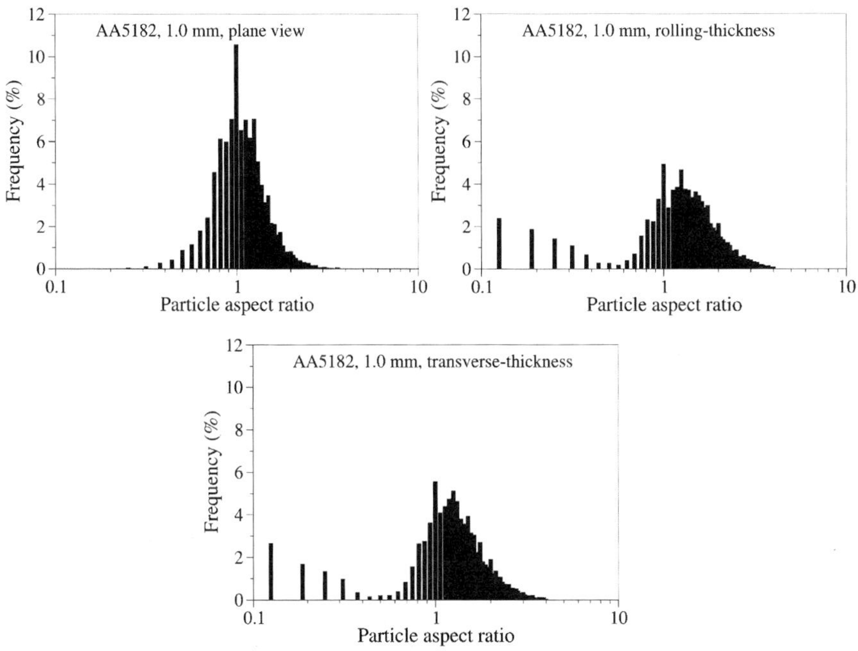

Fig. 5.3 Normalized histograms of particle aspect ratio: AA5182. Aspect ratios greater than unity indicates particles oriented longitudinally

5.1.1.3 Paricle Size

The particle size distribution is an important microstructural parameter impacting ductile fracture. Figure 5.4 shows the measured particle size distributions for a 1.0 mm AA5182 sheet. In the figure, particle size is plotted in terms of particle area in the section plane.

5.1.2 Damage Evolution Predictions

Damage evolution was treated as consisting of three distinct stages: void nucleation, growth, and coalescence. The first and last stages, nucleation and coalescence are typically the least quantified in terms of actual measurements or model predictions.

5.1.2.1 Void Nucleation

Void nucleation was treated as strain controlled and as being sensitive to particle size (Brown and Embury 1973; Teirlinck et al. 1985; Le Roy et al. 1981; Fisher and

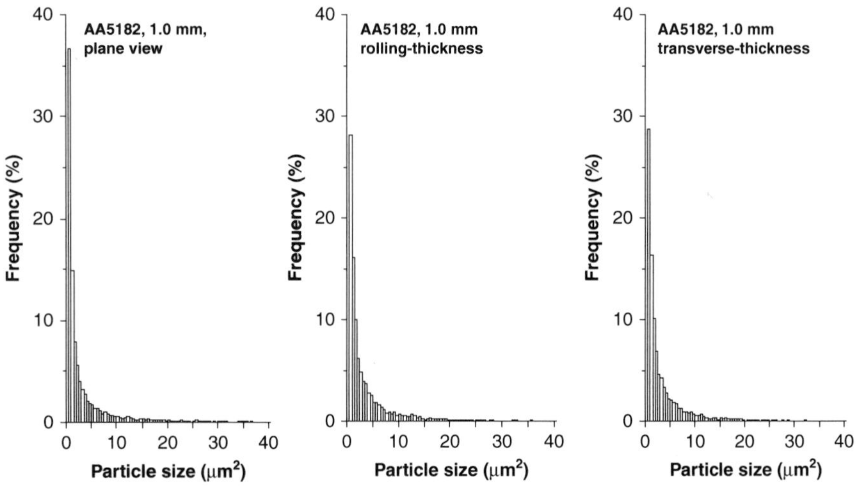

Fig. 5.4 Normalized histograms of particle size: AA5182

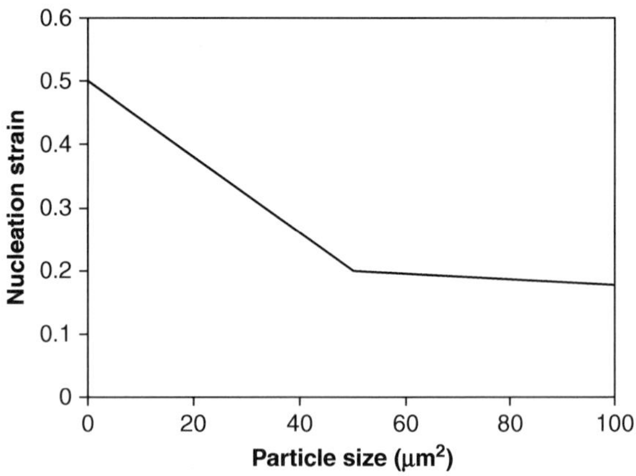

Fig. 5.5 Particle size-controlled void nucleation criterion adopted for both alloys studied (Reprinted with permission from Worswick et al. 2001)

Gurland 1981a, b). In general, nucleation is assumed to occur preferentially at larger particles and the fraction of nucleated smaller particles increases with strain. This treatment is consistent with the model suggested by Embury (1985), in which void nucleation "sweeps" through a particle population initiating preferentially at larger particles. Figure 5.5 plots the void nucleation strain as a function of particle size (area) proposed by Worswick et al. (1998, 2001) and adopted in the RVE size study for both alloys.

5.1.2.2 Void Growth

Void growth was modelled using results from unit cell calculations by Thomson et al. (1999). In that work, the critical geometric parameters governing void growth

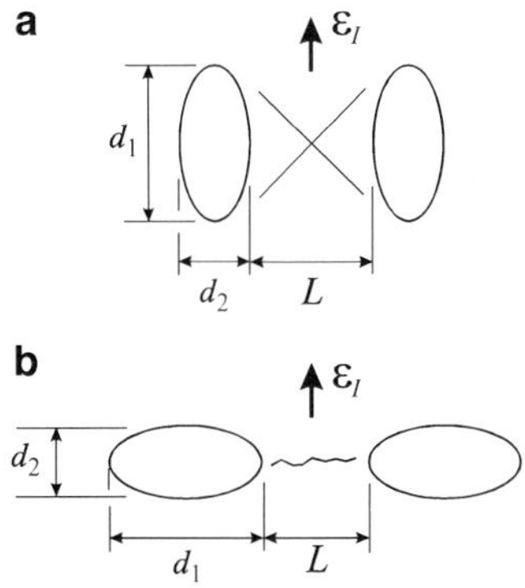

Fig. 5.6 Schematic of idealized void interaction geometry: (**a**) aligned or longitudinal case (**b**) transverse case

rate have been identified as void aspect ratio and degree of clustering. There was also a strong interaction between these geometric parameters and the strain state. Under plane strain conditions, for example, void growth was greatest when the first principal strain direction was perpendicular to the major axis of a void stringer or an elongated single void.

The current void growth treatment in the percolation damage model considers void aspect ratio and strain state, and utilizes void growth rates determined from isolated void, unit cell simulations. Clustering effects on rate of void growth are not considered at present. This simplification is somewhat justified since the behaviour of a void stringer was shown to be similar to an isolated void of similar size as the stringer. Once the flow stress in the inter-void ligament becomes saturated, the stringer "grows" much as an elongated void (Worswick et al. 1998). In the percolation damage model, the voids in a stringer-like cluster coalesce rapidly, after which they would be treated as a single larger ellipsoidal void.

Once nucleated, individual voids were assumed to grow at the prescribed rates predicted from the unit cell calculations of Thomson et al. (1999). It was assumed that after nucleation, the entire particle could be treated as void. Void growth under conditions of partial void-matrix separation should be considered in future work. Prior to nucleation, particles are assumed to deform at the bulk deformation rate of the matrix.

5.1.2.3 Void Coalescence

Void coalescence was predicted using a modified version of Brown and Embury's (1973) ligament-to-void-size-ratio criterion. Criteria based on plastic zone size, such as used by Dubensky and Koss (1987), were not employed, primarily because the strains in the matrix were very large and coalescence in the materials modelled occurs well after impingement of plastic zones surrounding neighbouring voids.

The inter-void geometry considered by Brown and Embury (1973) is depicted in Fig. 5.6a. Void coalescence is said to occur through shear band development

Fig. 5.7 Schematic of interaction geometry adopted in current work

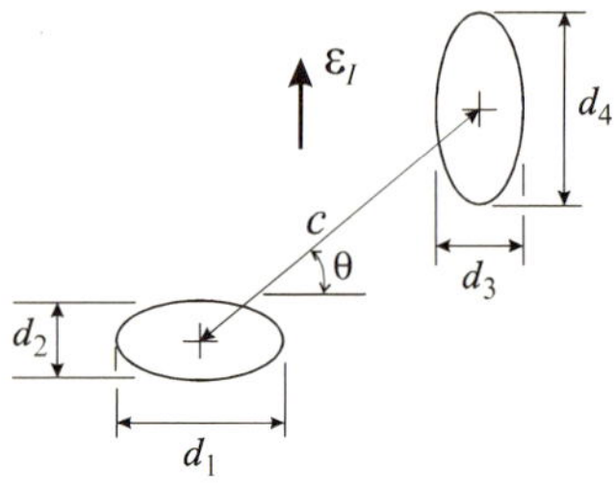

between two neighbouring elongated voids when the ratio of remaining ligament, L, to void length, d_2, drops below a critical value, γ:

$$\frac{L}{d_2} < \gamma \tag{5.1}$$

Typically, the value of γ is taken as unity or can be related to the stress state as in Weck (2006).

Brown and Embury's (1973) model was originally developed for the case of loading along the axis of elongated particles or voids as in Fig. 5.6a. This geometry applies naturally to so-called longitudinal load cases, however, the geometry for transverse loading is quite different, as shown in Fig. 5.6b. Applying Eq. (5.1) to this geometry would require that voids be positioned extremely close together before coalescence will occur since the "interaction length", L, is based on the void dimension measured along the principal straining direction. This length seems excessively small since voids elongated transverse to the ε_I-direction will experience a severe strain and stress concentration at their "poles". This effect was demonstrated by Thomson et al. (1999) using unit cell models in which inter-ligament plastic collapse appears to be the coalescence mechanism under transverse loading conditions. Based on these observations, a larger interaction distance may be more appropriate than that given by Eq. (5.1).

As a simple first step, it was decided to use

$$\frac{L}{\max(d_1, d_2)} < \gamma \tag{5.2}$$

in which the term in the denominator is the maximum of the two in-plane axes of the void. This approach extends the interaction distance for transverse loading while maintaining the Brown and Embury (1973) criterion for longitudinal loading. In all cases, γ was taken as unity, although γ may also be a function of stress or strain state as noted by Sun (1991) for triaxial stress states.

Other geometrical differences arise in comparing actual particle fields with the idealised geometries. These include differing sizes of the interacting particles as well as ligament orientations that are non-orthogonal to the ε_I -direction as depicted in Fig. 5.7.

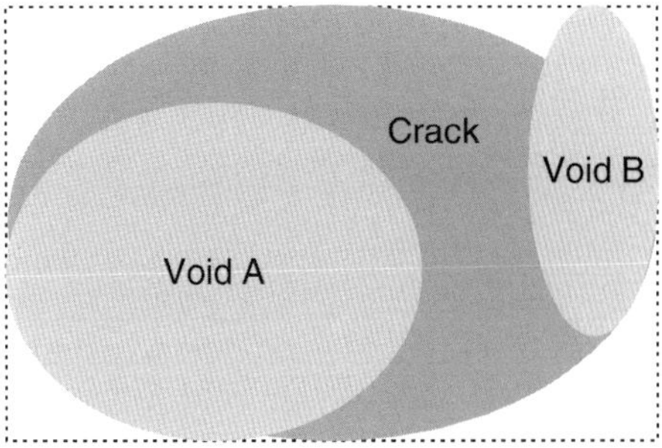

Fig. 5.8 A crack formed by coalescence of two neighbouring voids

The approach used in this work was to define an average particle size, D, as

$$D = \frac{\max(d_1, d_2) + \max(d_3, d_4)}{2} \tag{5.3}$$

and a ligament size as

$$L = c - D \tag{5.4}$$

where c is the void centre-to-centre distance and d_1, d_2, d_3 and d_4, are the void dimensions given in Fig. 5.7. Coalescence is said to occur, then, when

$$\frac{L}{D} < \gamma \tag{5.5}$$

A final restriction is placed on ligament orientation to determine whether coalescence can occur,

$$\theta < \theta_{\max} \tag{5.6}$$

where $\theta_{\max}$ is taken as 45°. In cases where θ exceeds $\theta_{\max}$, one would expect the ligament to be shielded from deformation by the voids. This requirement was enforced for the uniaxial calculations, but not for the stretch condition in which the in-plane loading is symmetric.

The search algorithm used to predict void coalescence utilizes the nearest neighbour list generated by the tessellation software. With each strain increment, Eqs. (5.3, 5.4, 5.5 and 5.6) are applied to all neighbouring pairs of nucleated voids.

5.1.2.4 Post-coalescence Treatment

Once two voids coalesce, they are treated as a single larger void or "crack", as shown in Fig. 5.8. Note that the term crack is used here for convenience and does not imply a sharp-tipped crack in a fracture mechanics sense. Propagation of damage can then proceed at three levels: (i) further coalescence of isolated voids; (ii) coalescence of isolated voids with existing cracks; and (iii) coalescence of two or more existing cracks. All of these processes can be expected to occur concurrently during continuing deformation.

Two approaches are possible to handle coalescence of voids with cracks or cracks with cracks. One approach would be to treat cracks as collections of isolated voids and to apply Eqs. (5.3, 5.4, 5.5 and 5.6) to every void-void pairing defined by the list(s) of voids in the crack(s) and their respective neighbour lists. The major drawback of this approach is that it neglects the amplification of effective void size and interaction distance occurring when one or more voids coalesce to form a larger void.

To account for this amplification effect, a different approach was used in which groups of coalesced voids were treated as a single larger elliptical void. The size of this elliptical void is such that its "bounding rectangle" encloses all voids comprising the crack (Fig. 5.8). Once established, this elliptical crack can then grow and coalesce with other voids or cracks. The significant increase in size of this crack serves to introduce the amplification of the interaction distance one would expect to occur with the onset of void coalescence.

5.2 Damage Predictions

In this section, the percolation model is demonstrated by simulating the damage evolution in AA5182 aluminum alloy sheet. Two loading conditions are considered, biaxial stretching and uniaxial stress. The uniaxial loading case is applied in two directions relative to the particle field, longitudinal and transverse. In rolling aluminum alloy sheet, second phase particles tend to elongate and align in stringers oriented along the rolling direction. Thus the longitudinal case in Fig. 5.6a corresponds to loading along the sheet rolling direction whereas the transverse case (Fig. 5.6b) represents in-plane loading perpendicular to the rolling direction. In the current study, loading was restricted to either equi-biaxial stretching ($\varepsilon_2/\varepsilon_1 = 1$) or uniaxial stress ($\sigma_2 = \sigma_3 = 0$) along the longitudinal or transverse directions. A strain step of 0.25 % is adopted in the damage percolation simulations for all three loading cases. The biaxial case corresponds to the highest triaxiality prior to necking of the sheet for which damage rates are higher. The uniaxial case is of interest since it corresponds to the stress state acting along the circumference of the stretch flange cutout (see Chap. 2). Damage variation is plotted in terms of void areal fraction, nucleated void areal fraction, areal fraction of voids in cracks and crack areal fraction to characterize the damage evolution in different stages. Also captured is the sequence of particle field evolution under various loading cases.

5.2.1 Damage Evolution

Figure 5.9 shows a typical sequence of predicted damage at various strain levels prior to fracture under biaxial stretch loading of AA5182 sheet, using a smaller

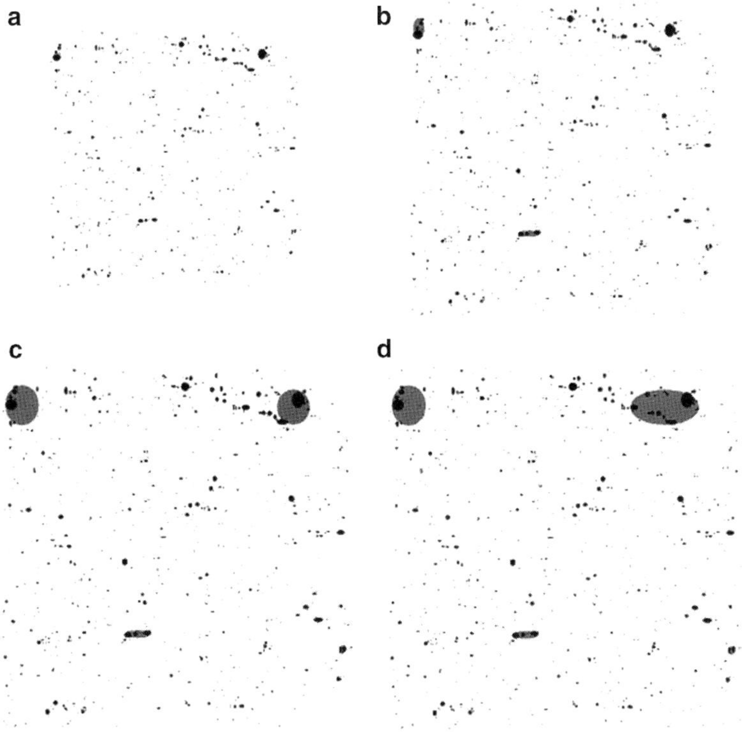

Fig. 5.9 Predicted damage for AA5182 under equi-biaxial strain conditions, 2,000 × 2,000 pixels (2,000a): (**a**) initial particle field; (**b**) 20 % strain; (**c**) 33.75 % strain; and (**d**) 34 % strain

2,000 × 2,000 pixel particle field extracted from the larger field in Fig. 5.1. The particles are shaded black and the approximate sizes of cracks are represented as grey ellipses sized to fit within the bounding box encompassing all voids within a crack. This ellipse corresponds to the "crack size" used to determine void-crack and crack-crack coalescence.

Using the adopted interaction criteria/assumptions, damage propagation becomes very dependent upon the nucleation process. Damage commences with the larger particles at a strain of roughly 19 % with local regions of void coalescence forming almost immediately at strain levels of 20 % (Fig. 5.9b). Damage progresses with nucleation of new voids which then allows formation of new cracks and growth of existing cracks, largely confined to the original particle clusters. At a much larger strain of 33.75 % (Fig. 5.9c), further void coalescence has occurred and several larger cracks are observed; however, the damage is still confined to within at most three neighbouring particle clusters. Larger cracks can only form once neighbouring cracks become large enough to bridge inter-crack ligaments (Fig. 5.9d). This process is very sensitive to void nucleation as well as void and crack interaction since the extension of cracks requires the introduction of new voids to sustain growth. At a strain of 34 % (Fig. 5.9d), the crack size becomes

Fig. 5.10 Predicted damage for AA5182 under transverse uniaxial stress conditions, 2,000 × 2,000 pixels (2,000a): (**a**) initial particle field; (**b**) 20 % strain; (**c**) 35 % strain; and (**d**) 36 % strain

critical and the linkage of four or more clusters of voids triggers a chain reaction in which crack coalescence propagates across the field without further straining (not shown).

Figure 5.10 shows the sequence of predicted damage at various strain levels prior to fracture for the same initial particle field under transverse uniaxial stress. Once again voids nucleate and coalesce within the same clusters having larger particles. However, it is observed that in Fig. 5.10, nucleation and coalescence of particles within a particle cluster on the bottom left of the field didn't happen until fracture. This behaviour is different from what was observed for the biaxial stretching case, which can be attributed to the constraint on the coalescence path given by Eq. (5.6) for uniaxial loading. This effectively reduces the number of nearest neighbours that can coalesce compared to the equi-biaxial case.

5.2.2 Predicted Damage Rates

The predicted damage-strain history for AA5182 under equi-biaxial stretch conditions using a 4,000 × 4,000 pixel particle field is plotted in Fig. 5.11. Predicted damage histories under uniaxial stress loading for AA5182 are shown in Fig. 5.12. Results are shown for both longitudinal and transverse loading in

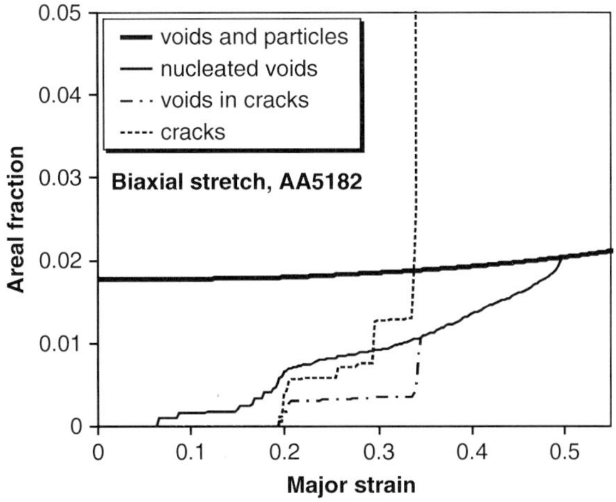

Fig. 5.11 Damage development under biaxial stretching, AA5182

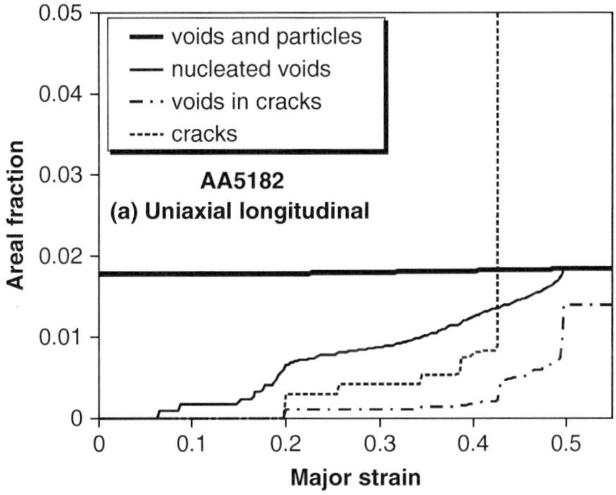

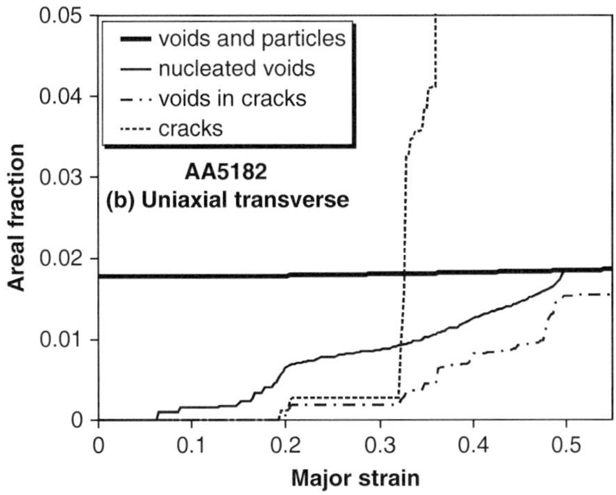

Fig. 5.12 Damage development for AA5182 under (**a**) longitudinal uniaxial stress conditions and (**b**) transverse uniaxial stress conditions

Fig. 5.12a, b, respectively. Damage development proceeds more rapidly for the transverse case. Furthermore, a much lower strain of 36.25 % is predicted for profuse void coalescence under transverse loading compared to 42.75 % for the longitudinal direction. This behaviour is attributed to the nature of the particle stringers which are aligned with the rolling direction (Fig. 5.1), providing an easier coalescence path under transverse loading conditions.

The rate of void growth is higher for the stretch cases compared to the uniaxial cases. This observation stems from the steeper slope of the solid curve in Fig. 5.11 (stretch case) compared to the uniaxial growth rates seen in Fig. 5.12. This orientation effect is attributed to the high propensity for large, oriented stringer-like clusters in AA5182 (Fig. 5.1).

5.3 Selection of Representative Volume Element (RVE)

Determination of the representative volume element (RVE) of a microstructure is essential in the micro-mechanics study of ductile fracture. As well-verified by numerous researchers (Tvergaard 1990; Meyers and Aimone 1983; Tvergaard and Needleman 1986, 1997; Needleman and Tvergaard 1991), ductile fracture is a very localized phenomenon. Therefore, it was decided to investigate the effect of RVE size, that is, the size of the region of sampled material, on the predicted onset of ductile fracture. Of particular interest was the effect of a reduction in RVE size on the predicted failure strain. Here, the RVE can be thought of as the minimum size of particle field required to obtain a repeatable prediction of damage development. In this manner, the predictions are no longer dependent upon the choice of the particle field. In addition, determination of the minimum required field size also makes the predictions more efficient, as well as the image acquisition operation which can be tedious.

The approach taken here is to study the influence of particle field size on the onset of profuse coalescence. Predictions are performed first using a large image containing many particles; these are the simulations presented in the preceding section of this chapter. Next the images are progressively subdivided into smaller images and the damage percolation simulations are repeated for each sub-image or particle field. This process of image-subdivision is repeated until the profuse void coalescence strain begins to vary significantly for the smallest image.

5.3.1 Particle Field Sizes

In order to accommodate the image sub-division process, the damage percolation software was modified to support "windowing" of the particle fields, such that smaller regions could be modelled. Given an initial point where the particle field

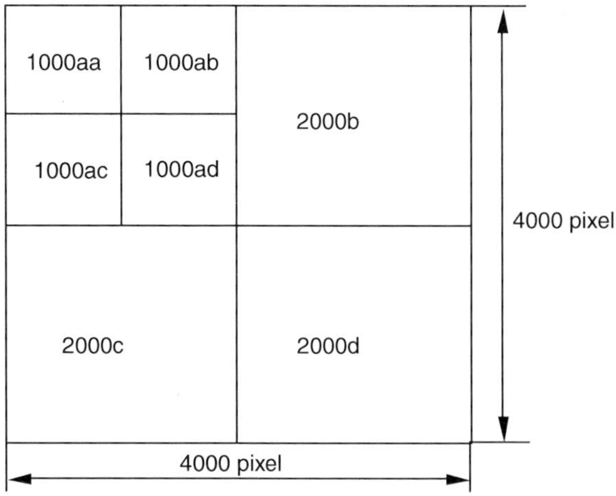

Fig. 5.13 Particle field sub-division scheme used in this study

starts and the horizontal and vertical dimensions of the field, the percolation code will read all the particle information of the field from the large field (parent field), as depicted in Fig. 5.13. Particle field subsets were taken from the large (4,250 × 5,500 pixel) image acquired for the as-received alloy. Calculations were initially performed using particle fields that were 4,000 × 4,000 pixel in size, taken from the larger acquired field. Given a pixel size of 0.372 μm, this corresponds to a total image size of 1.49 × 1.49 mm. These larger images would correspond to the largest area of interest (AOI) or representative volume element (RVE) considered in this study.

To address the effect of RVE size, the original 4,000 × 4,000 pixel particle field was divided into four equal-sized (2,000 × 2,000 pixel) sub-fields, and the damage percolation simulation was repeated for each of these sub-fields. Next, the sub-fields were further divided into 16 equal-sized (1,000 × 1,000 pixel) fields and then into sixty-four 500 × 500 pixel fields, and again used for damage percolation simulations. In principle, this sub-division process can be continued beyond this field size; however, it was determined that subdivision of the image beyond the sixty-four 500 × 500 pixel fields was not useful since catastrophic failure could not be realized for some of these smaller fields for strains in excess 100 %, which implies that the RVE should be at least larger than 500 × 500 pixels.

Table 5.1 lists the numbers of particles and initial particle areal fraction in the larger 4,000 × 4,000 pixel images and the four sub-fields (from left to right, upper to lower, denoted as a, b, c, and d). Table 5.2 gives the data for a selected 2,000 × 2,000 particle field and its four sub-fields. Both the number of particles and areal fraction vary considerably between sub-fields. The data in Table 5.2 is taken from the 2,000 × 2,000 pixel fields exhibiting failure strains closest to their 4,000 × 4,000 parent for all three loading cases.

Table 5.1 The effect of image size and particle number on the critical strain, AA5182 (4,000 pixel field) (Reprinted with permission from Worswick et al. (2001). Copyright: Elsevier)

			Critical strain		
Image size (pixel)	Number of particles	Initial particle areal fraction	Biaxial stretch	Uniaxial longitudinal	Uniaxial transverse
4,000	5,100	0.0162	0.3425	0.4275	0.3625
2,000a	1,254	0.0178	0.3425	0.4275	0.3625
2,000b	1,278	0.0167	0.395	0.4975	0.4325
2,000c	1,340	0.0152	0.4525	0.4925	0.4775
2,000d	1,228	0.0112	0.4425	0.4875	0.4925

Table 5.2 The effect of image size and particle number on the critical strain, AA5182 (2,000a) (Reprinted with permission from Worswick et al. (2001). Copyright: Elsevier)

			Critical strain		
Image size (pixel)	Number of particles	Initial particle areal fraction	Biaxial stretch	Uniaxial longitudinal	Uniaxial transverse
2,000a	1,254	0.0178	0.3425	0.4275	0.3625
1,000aa	310	0.017	0.415	0.4275	0.4375
1,000ab	348	0.0274	0.34	0.495	0.265
1,000ac	325	0.0154	0.485	0.7	0.43
1,000ad	271	0.0112	0.4925	1.3	1.05

5.3.2 Results and Discussion

The effect of the choice of representative volume element (RVE) is assessed in terms of the predicted strain to cause profuse void coalescence for different selected RVEs. For the purposes of this study, this limiting strain corresponds to the strain at which the crack areal fraction grows without further strain increment, as reflected by the steep vertical slopes seen in Figs. 5.11 and 5.12. This measure of limit strain to cause profuse coalescence is adopted to determine the influence of RVE size on the damage percolation predictions.

The limit strains at profuse coalescence for various particle fields are summarized in Tables 5.1 and 5.2. For the 2,000 × 2,000 pixel particle fields, profuse coalescence occurred in all the biaxial stretch and uniaxial cases. For the 2,000 × 2,000 fields, the critical strain level was in the range 0.34–0.44 for biaxial loading, compared to 0.43–0.50 and 0.36–0.49 for uniaxial loading in the longitudinal and transverse directions, respectively. It is worth noting that the critical strain for the 4,000 × 4,000 field was identical to those for its sub-field 2,000a which contains the particle clusters triggering the chain reaction associated with profuse void coalescence. Interestingly, this sub-field does not contain the largest number of particles when compared with its three sister-fields, but does exhibit the largest initial particle areal fraction of the four sub-fields. Figure 5.14a shows the predicted crack areal fraction using the larger AA5182 particle field (4,000 × 4,000 pixel) along with predictions from its smaller 2,000 × 2,000 pixel sub-fields (denoted as

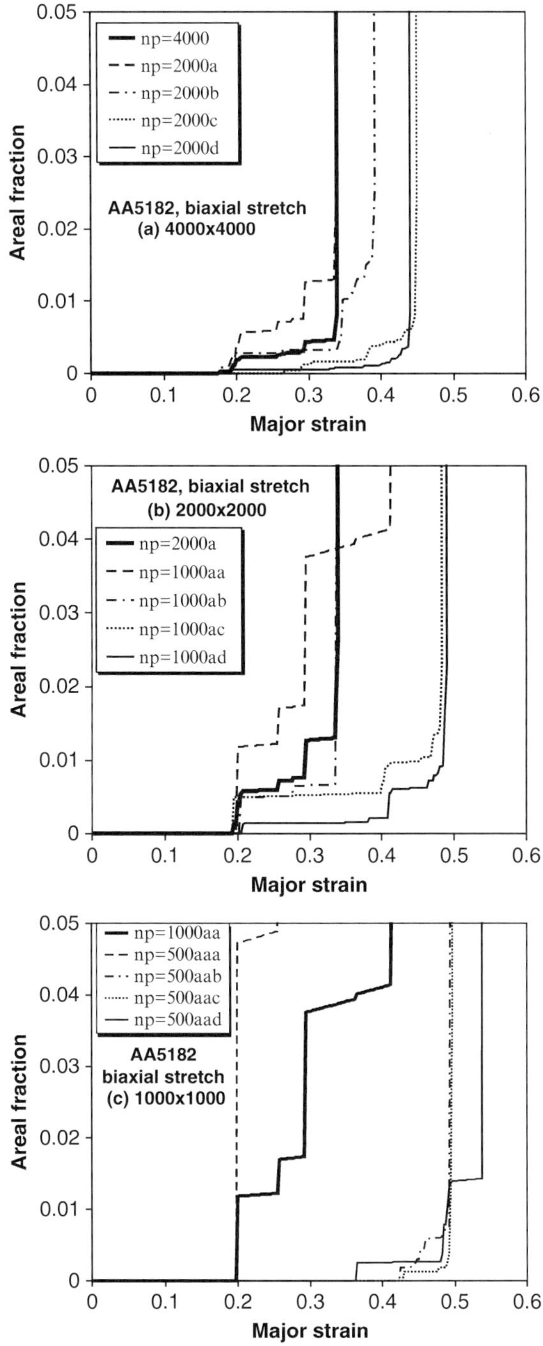

Fig. 5.14 Predicted damage history for AA5182 under equi-biaxial stretching with (**a**) 4,000 × 4,000 pixels particle field and its sub-fields (2,000a, 2,000b, 2,000c, 2,000d); (**b**) 2,000 × 2,000 pixels particle field (2,000a) and its sub-fields (1,000aa, 1,000ab, 1,000ac, 1,000ad); (**c**) 1,000 × 1,000 pixels particle field (1,000aa) and its sub-fields (500aaa, 500aab, 500aac, 500aad) (Reprinted with permission from Worswick et al. (2001). Copyright: Elsevier)

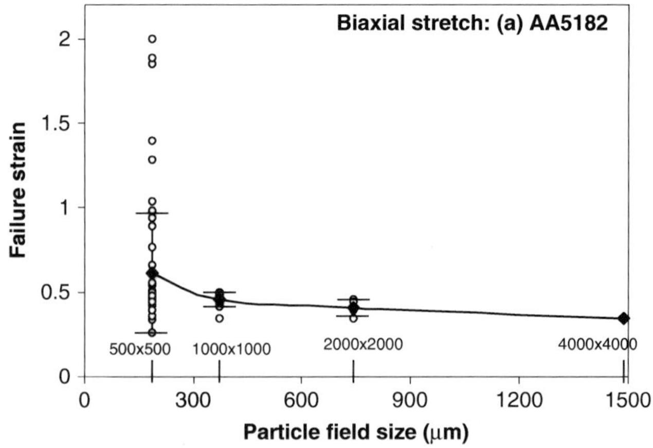

Fig. 5.15 Predicted failure strain (ductility) with RVE under biaxial stretching (Reprinted with permission from Worswick et al. (2001). Copyright: Elsevier)

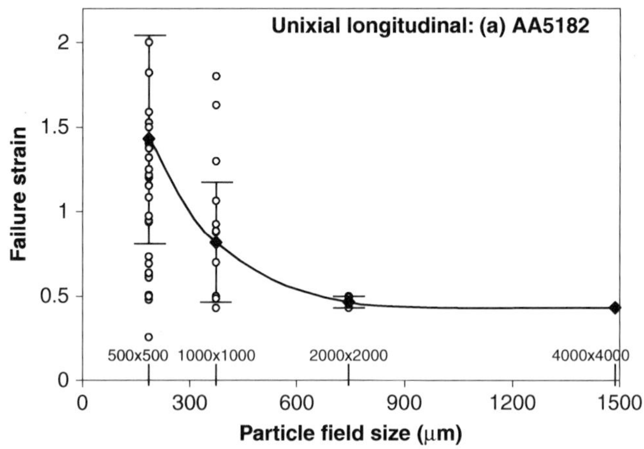

Fig. 5.16 Predicted failure strain (ductility) with RVE under longitudinal uniaxial stress condition. (**a**) AA5182 (Reprinted with permission from Worswick et al. (2001). Copyright: Elsevier)

2,000a–d in Table 5.1). The damage history of one of the smaller particle fields (2,000a) and its four sub-fields is plotted in Fig. 5.14b (Recall that this sub-field 2,000a contains the particle clusters that triggered the profuse coalescence in the $4,000 \times 4,000$ field simulation). To further illustrate the dependence of the predictions on RVE, the damage histories for a $1,000 \times 1,000$ pixel field (1,000aa) and its four sub-fields are plotted in Fig. 5.14c. It is evident that the degree of scatter in predicted failure strain increases as the size of the RVE is decreased as a natural consequence of sampling variability and the irregular particle distributions.

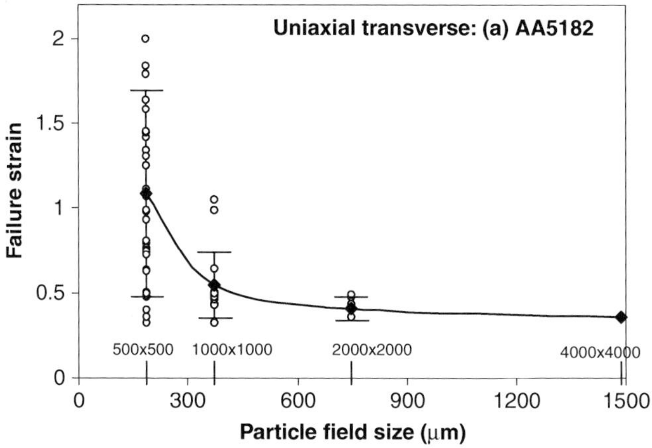

Fig. 5.17 Predicted failure strain (ductility) with RVE under transverse uniaxial stress condition (Reprinted with permission from Worswick et al. (2001). Copyright: Elsevier)

To further examine the variability in failure strain with RVE size, Fig. 5.15 plots the predicted failure strains as a function of RVE size under stretch loading (open symbols). The scatter bands in the figures correspond to the standard deviation in predicted failure strain and exhibit a very high dependence on RVE size. Note that in the calculations of variance, if the damage chain reaction did not occur when the applied strain reaches 2.0, the failure strain was taken as 2.0. Also plotted is the average value of failure strain versus RVE size (solid symbols). The degree of scatter for the predictions using the smallest RVE (500 × 500 pixels) is very high due to large sampling variability. The scatter is considerably reduced for the larger images and is similar for the 1,000 × 1,000 and 2,000 × 2,000 pixel fields. The corresponding results are plotted in Figs. 5.16 and 5.17 for longitudinal and transverse uniaxial loading, respectively. The dependence on RVE size is similar, however, the degree of scatter for the 1,000 × 1,000 pixel RVE is quite large for uniaxial loading compared to the stretch results in Fig. 5.15. This difference indicates that the rate of convergence in predicted ductility with increasing RVE size is slower for uniaxial loading. This behaviour is attributed to the constraint on coalescence given by Eq. (5.6), imposed for uniaxial loading that effectively limits the number of nearest neighbours that are candidates for coalescence compared to biaxial stretch loading for which coalescence can occur in any direction.

5.4 Summary

The validity of the damage percolation model in predicting ductile damage over the three distinct stages has been assessed with measured particle fields under various loading conditions. Damage percolation in second phase particle fields occurs as a localized process, sensitive to initial particle clustering level, rather than a uniform process, as described in unit cell micromechanical models. Although introduction of particle clusters within a unit cell approach could somehow reveal the effect of interaction between particles and clusters (Thomson 2001), percolation simulations with actual measured particle fields do supply a more complete picture as to how void damage evolves in clustered particle fields to form a macrocrack. It is found that the void nucleation process dominates ductile damage within these aluminium alloys. The representative volume element study revealed that predicted ductility, in terms of the strain to cause profuse damage, shows a satisfactory convergence, for particle fields at least 2,000 $\times$ 2,000 pixels (0.75 $\times$ 0.75 mm) in size.

Chapter 6
Two-Dimensional (2D) Damage Percolation/ Finite Element Modeling of Sheet Metal Forming

Experimental evidence and numerical simulation have established that ductile damage critically limits the formability of sheet metals (Gelin 1998; Hu et al. 2000; Tang et al. 1999). To accurately predict formability and to optimize material processing to achieve enhanced formability, it is important to understand how heterogeneously distributed micro-defects affect the macromechanical behaviour of sheet metal. Therefore, it is of both theoretical and practical interest to investigate damage evolution during sheet metal forming.

This research seeks to address the role of ductile damage in determining sheet metal formability. Of particular interest is the variable nature of material microstructures, for example the statistical variation in second phase particle size and spacing. It is well known that damage will first nucleate within closely spaced regions of second phase particles, known as particle clusters. It is considerably less well established how ductile damage then propagates outwards from particle clusters to establish macro-cracking and at what level damage becomes critical. Pilkey (1997) has conducted experimental research to examine the effect of particle clustering on ductile fracture of aluminum alloy sheets under various loading conditions. Detailed two-dimensional statistical data identifying various particle distribution properties has been acquired for Al-Si alloy sheets and then correlated with various formability trends (Pilkey 1997). The challenge exists, however, to properly include detailed statistical measures of microstructure in models of ductile fracture. Although there is an enormous collection of literature on the mechanical modelling of ductile fracture, none of the models available so far is able to properly include measured statistical particle fields.

Following the previous chapter, this chapter describes the extension of the percolation model to deal with ductile fracture in sheet metal forming. Strong strain gradients and heterogeneous second phase particle fields are the major concerns in this study. To achieve this goal, the damage percolation model is coupled with a Gurson (1977) based finite element (FE) model so that heterogeneity of microstructure and high strain gradients in sheet metal forming can be implemented using a coupled model.

Z. Chen and C. Butcher, *Micromechanics Modelling of Ductile Fracture*,
Solid Mechanics and Its Applications 195, DOI 10.1007/978-94-007-6098-1_6,

© Springer Science+Business Media Dordrecht 2013

6.1 Stretch Flange Experiment

Automotive aluminum-magnesium alloys, AA5182 of 1.6 and 1.0 mm thicknesses, were employed in the present research. The nominal chemical compositions and the mechanical properties are listed in Tables 6.1 and 6.2. Stretch flange forming operations are commonly used in the automotive industry to fabricate inner door panels (Fig. 6.1). A blank with an inner cutout is clamped against a die while a punch moves downwards to form the flange through expansion of the cutout. A backup punch holds the sheet against the punch to make the bend at the punch profile radius. Drawbeads are often introduced to limit the rate of cutout expansion.

The repeated bending and unbending of the sheet, as it flows through the drawbeads can lead to damage initiation followed by necking at the cutout edge or tearing at the punch nose and around the profile radius. The amount of damage induced by the drawbeads in the sheet metal is of particular interest.

Table 6.1 Nominal chemical composition of AA5182, wt % (Finn MJ, private communication, 1999)

Si	Fe	Mg	Mn	Cu	Ti
0.08	0.3	4.6	0.33	0.04	

Table 6.2 Mechanical properties of AA5182 from uniaxial tensile tests (Finn MJ, private communication, 1999)

Gauge (mm)	σ_0 (MPa)	E (GPa)	ν
1.6	117.34	71.71	0.33
1.0	120.17	71.71	0.33

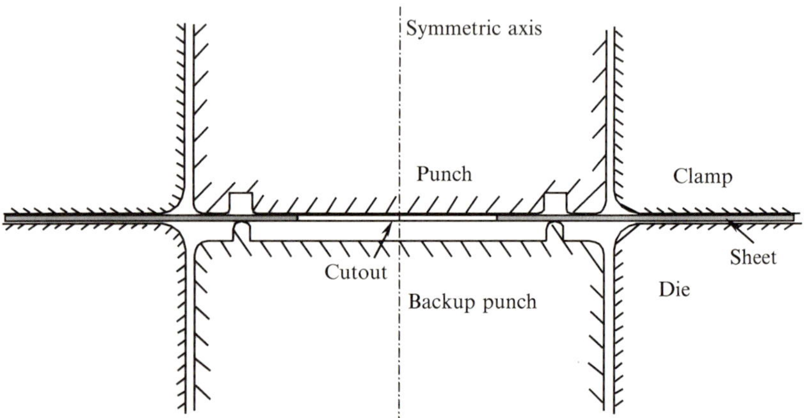

Fig. 6.1 Stretch flange forming tooling schematic

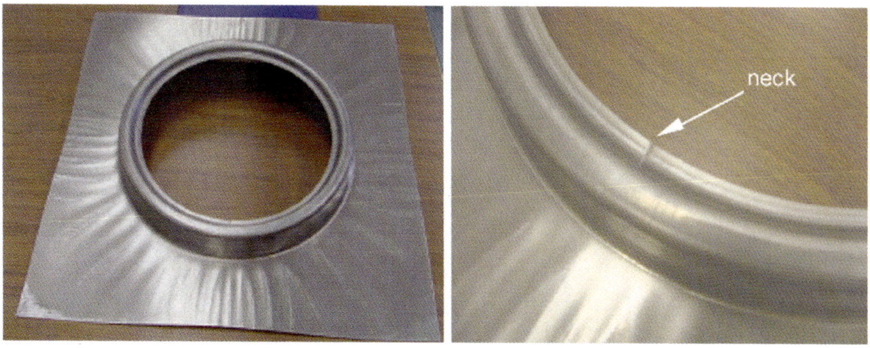

Fig. 6.2 A stretch flange sample fractured by inner edge necking (Reprinted with permission from Chen et al. (2005). Copyright: Elsevier)

A triple-action servo-hydraulic press, developed at the University of Waterloo (Cinotti 2003), was used in the stretch flange experiments. Figure 6.1 is a schematic of the z-flange tooling, which incorporates mating drawbeads on the main and backup punches. Drawbeads are used in commercial stretch flange operations to control or limit the rate of cutout expansion. Of interest in the current research is the level of damage induced by the bending-unbending of the sheet as it passes through the drawbeads. Detailed dimensions of the forming apparatus are given in Chen (2004). To capture the initiation of fracture during the stretch flange forming operation, a Piezotron® acoustic emission (AE) Sensor from Kistler Instrument Corp., was adapted to this tooling set by Orlov (private communication, 2003). The sensor is attached to the main punch to detect the acoustic pulse associated with the onset of cracking. The AE sensor is especially well suited for measuring AEs above 50 kHz.

Two different sets of experiments were conducted. The first is a crack detection test, where the punch motion is stopped immediately once the AE exceeds the threshold for crack initiation, as determined by relevant calibrating tests (Orlov O, private communication, 2003). Sheet specimens with cutout radii in the range of 88–98 mm, in increments of 2 mm, were tested to failure. The main purpose of this test is to investigate the effect of cutout size on the formability of stretch flanges. The second set of experiments comprised of a series of interrupted tests. For this part of the experiment, only the smallest (88 mm) and the largest (98 mm) cutout radii were considered. For each cutout size, the forming operation was interrupted at four different levels of punch depth, namely 25, 50, 75, and 90 % of the punch depth to fracture. The recovered samples were sectioned and their damage levels measured using metallographic, optical microscopy and image analysis methods. Damage development and strain path during the forming process is thereby captured within stretch flange samples.

A series of blanks were pressed in the first set of experiments in order to determine the limit punch depth that could be successfully formed without splitting as a function of the cutout radius. Two fracture modes were observed: one designated inner edge necking (Fig. 6.2) and the other as circumferential cracking

Fig. 6.3 A stretch flange sample fractured by circumferential cracking (Reprinted with permission from Chen et al. (2005). Copyright: Elsevier)

(Fig. 6.3). Fracture by inner edge necking can be arrested immediately after crack initiation (Fig. 6.2) using the AE sensor to stop the test. Crack arrest is possible due to the high strain gradient in the radial direction. Regions of lower strain in the outer area away from the cutout edge resist propagation of the neck and/or crack. However, fracture due to circumferential cracking at the punch nose propagates rapidly (Fig. 6.3) because of the axisymmetric nature of the crack orientation and high stresses in the sidewall. From the measured strain path of the interrupted stretch flanges, it is found that the radial necking-induced inner edge cracking is from the elevated circumferential uniaxial strains at the cutout, which is not thought to be damage-controlled. On the other hand, circumferential cracking at the punch nose is due to the biaxial stretching strain state which greatly promotes damage development and eventually causes ductile fracture. Details of strain path measurement can be found in Chen (2004). Here we focus on the damage-induced circumferential cracking at the punch nose.

The main punch displacement, backup punch force and clamping force were recorded by the load cells and displacement sensors mounted on the tooling and transferred to a PC for further analysis. Multi-channel servo-controllers were employed for signal conditioning and servo-control of the actuators. Further details of the apparatus and tooling are given by Cinotti (2003).

The percolation damage calculations presented in the previous chapter have served to demonstrate the nature of damage progression within a real second phase particle field; however, one limitation of these calculations is the assumption of a uniform strain field. In this chapter, modifications to the damage percolation approach are presented to allow consideration of non-uniform strain fields. The approach taken is to utilize a "loose coupling" of the percolation code with an elastic–plastic FE code that utilizes a Gurson-based damage model to describe the material constitutive response.

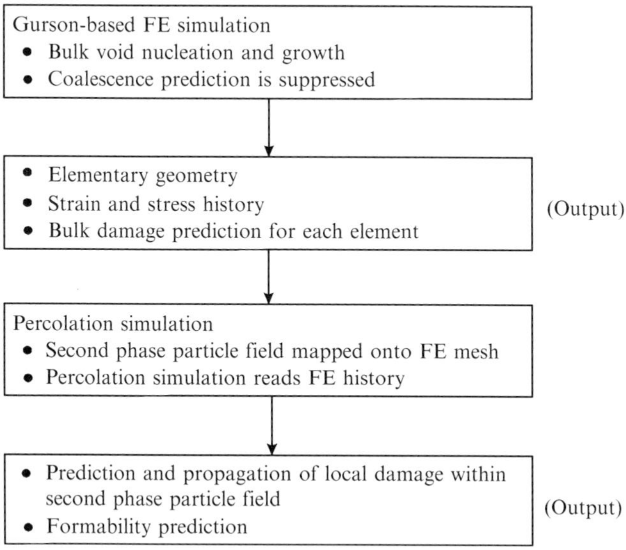

Fig. 6.4 A schematic of the one-way coupled FE-damage percolation model

The advantages to adopting such an approach are as follows:

- The FE code can readily handle the complex boundary conditions associated with metal forming operations, thus real strain gradients can be introduced into the percolation simulations;
- Bulk damage and constitutive softening predictions can be obtained using a Gurson-based constitutive model;
- The percolation code can utilize the FE stress, strain and bulk damage predictions and then predict the percolation of damage through the measured second phase particle field.

The approach in developing this "loose coupling" is to employ the FE model to handle the prediction of strain gradients and damage-induced softening, while the percolation calculations predict damage evolution within the tessellated second phase particle field, based upon deformation and bulk damage history from the FE calculation. Figure 6.4 outlines the nature of the coupling, which is essentially one-way with the FE history read in by the percolation code. Generally, the physical size of the simulation domain for the percolation code is much smaller; thus only a small region of the FE mesh, referred to as the mesh area of interest, or AOI, is considered in the percolation model.

The major changes required in the damage percolation model are to read in: (i) the initial mesh geometry for the AOI; and (ii) the nodal displacement history as well as element stress, strain and porosity histories. To accommodate the real deformation field, the measured second phase particle fields are mapped onto the mesh area of interest. Each particle is assigned a "parent element" and will experience the deformation and bulk damage history of the parent. The percolation code has been further modified to support these coupled calculations in terms of the void nucleation, growth, coalescence and post-coalescence treatments, as outlined in Sects. 5.2, 5.3 and 5.4. Prior to discussing these modifications, the Gurson-Tvergaard-Needleman (GTN) constitutive model is presented, since it is adopted in the FE part of the coupled simulations.

6.2 GTN-Based Damage Model

The well-known Gurson-Tvergaard-Needleman (GTN) constitutive model (Gurson 1977; Tvergaard 1981; Tvergaard and Needleman 1984) is employed in the FE part of the coupled model. Since void coalescence and final failure are to be treated in the damage percolation part of the model, the usual void coalescence treatment in the GTN model has been suppressed. Thus, the GTN-based calculations are limited to predict the nucleation and growth of the bulk porosity (damage) and the damage-induced constitutive softening.

The GTN-damage model (Gurson 1977; Tvergaard 1981; Tvergaard and Needleman 1984) is based upon the pressure-sensitive yield function proposed by Gurson (1977):

$$\Phi = \left(\frac{\Sigma_{eq}}{\bar{\sigma}}\right)^2 + 2f^* q_1 \cosh\left(q_2 \frac{3\Sigma_m}{2\bar{\sigma}}\right) - 1 - q_1^2 f^{*2} = 0 \tag{6.1}$$

where f^* is the effective void volume fraction, Σ_{eq} and Σ_m are the von Mises effective stres s and hydrostatic stress, respectively, and $\bar{\sigma}$ is the matrix flow stress. The von Mises effective stress, Σ_{eq}, is defined by:

$$\Sigma_{eq} = \sqrt{\frac{3}{2}\Sigma'_{ij}\Sigma'_{ij}} \tag{6.2}$$

where Σ'_{ij} is the deviatoric stress component. The "q" coefficients are "calibration coefficients" introduced by Tvergaard (1981) to better represent void interaction effects in plastically deforming materials. The function f^* was introduced by Tvergaard and Needleman (1984) to account for the onset of void coalescence,

$$f^* = \begin{cases} f & \text{if } f \le f_c \\ f_c + \frac{f_u^* - f_c}{f_f - f_c}(f - f_c) & \text{if } f > f_c \end{cases} \tag{6.3}$$

in which, f_c is a critical porosity value at which void coalescence commences, f_f is the porosity value at failure and $f_u^* = 1/q_1$. It is seen from Eq. (5.3) that fracture through void coalescence is assumed to initiate once the porosity reaches a critical value, f_c. In the current study, void coalescence is suppressed by specifying large values of f_c and f_f. Thus $f^* = f$ for all of the coupled simulations.

The rate of increase of void volume fraction is due to the growth of existing voids and the nucleation of new voids:

$$\dot{f} = \dot{f}_{growth} + \dot{f}_{nucleation} \tag{6.4}$$

Assuming the material is plastically incompressible, the growth of existing voids is related to the hydrostatic component of macroscopic plastic strain by:

$$\dot{f}_{growth} = 3(1-f)\dot{E}_{\mathrm{hyd}} \tag{6.5}$$

The contribution of void nucleation will be material dependent. For the aluminum-magnesium alloys in this study, a plastic strain-controlled nucleation rule is adopted, that assumes that voids nucleate at second phase particles and there is a normal distribution of nucleation strain for the total population of particles (Chu and Needleman 1980),

$$\dot{f}_{nucleation} = A\dot{\bar{\varepsilon}}^p \tag{6.6}$$

in which

$$A = \frac{f_N}{S_N\sqrt{2\pi}} \exp\left[-\frac{1}{2}\left(\frac{\bar{\varepsilon}^p - \varepsilon_N}{S_N}\right)^2\right] \tag{6.7}$$

where $\bar{\varepsilon}^p$ is the effective plastic strain representing the microscopic strain-state in the matrix material. The term f_N represents the volume fraction of void-nucleating particles, while ε_N and S_N are the average and standard deviation of the strains at which particles nucleate voids. In the coupled calculations, values of ε_N and s_N were determined such that the void nucleation rate using Eqs. (6.6) and (6.7) closely matched those for the percolation model. Further elaboration of the nucleation model is given in Sect. 6.3.1.

The effective plastic strain in the matrix material is calculated according to the principle of work equivalence,

$$\dot{\bar{\varepsilon}}^p = \frac{\Sigma_{ij}\dot{E}^p_{ij}}{(1-f)\bar{\sigma}} \tag{6.8}$$

where $\dot{E}^p_{ij}$ is the rate of macroscopic plastic strain. The plastic flow in the material is taken from the uniaxial stress versus effective plastic strain curve obtained by uniaxial tensile tests.

6.3 Coupled Percolation Model – Damage Predictions

The percolation damage algorithm, as adapted to perform coupled predictions, largely follows that described in Chap. 4 of this thesis. Modifications are required, however, to handle the deformation and state information generated by the GTN-based damage predictions outlined in the previous section. The following sections describe the changes required in percolation model of Chap. 5 to accomplish this "one-way loose-coupling".

6.3.1 Void Nucleation

Void nucleation due to particle decohesion or cracking is directly related to the local plastic strain in the vicinity of a particle. Thus, in coupled model, it is assumed that once the equivalent plastic strain of its parent element attains the nucleation strain, i.e.

$$\bar{\varepsilon}^p \geq \varepsilon_n \tag{6.9}$$

the particle is nucleated and treated as a void, where ε_n is the nucleation strain of the specific particle.

At the current stage of implementation of the coupled calculations, the void nucleation treatment differs somewhat between the two parts of the model. The nucleation strains in the percolation code were treated in the same manner as was presented in Chap. 4; however, the predicted nucleation strain as a function of particle size was adjusted to better match recent results by Winkler (2003) for the aluminum-magnesium alloy AA5182. In that work, the nucleation parameters in Eqs. (5.6) and (5.7) were determined for 1.6 mm AA5754 and AA5182, to match measured bulk damage levels in the stretch flange samples (Chen 2004). The AA5182 samples exhibited high levels of damage and a nucleation strain of 0.2 was recommended by Winkler (2003). Note that a standard deviation (s_N) of 20 % was assumed for both alloys (Winkler 2003).

In the current research, detailed measured data concerning the particle nucleation strain as a function of particle size was not available, although such measurements are underway in related work (Winkler 2003). Instead, the form of the assumed relationship between nucleation strain was adjusted such that: (i) the average bulk nucleation strain matched the values determined by Winkler (2003); and (ii) the slope of the porosity evolution versus strain matched that predicted by the GTN model for the assumed values of s_N. Figure 6.5 shows the assumed dependence of nucleation strain as a function of particle size for both alloys. The bold curves were those matching the average nucleation strains determined by Winkler (2003). Recognizing that this description of nucleation strain is somewhat subjective and requires future validation, it was elected to consider two additional

Fig. 6.5 Void nucleation strain versus particle size adopted in the percolation simulations corresponding to the nucleation strain (ε_N) used in the GTN-based FE calculation

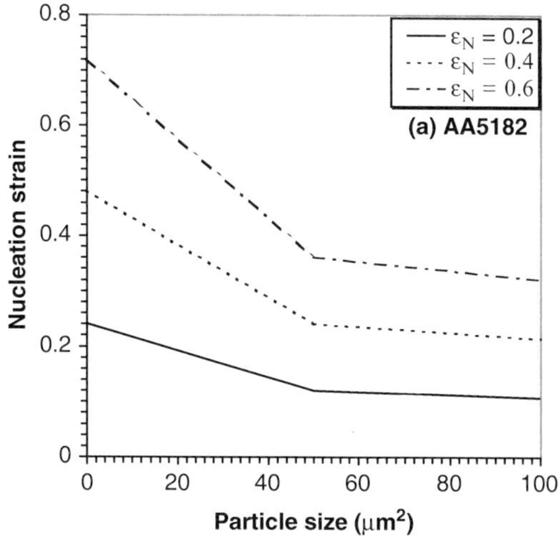

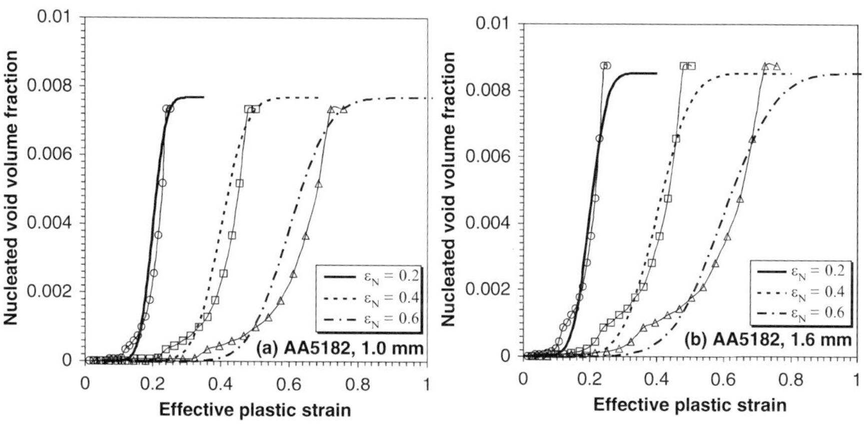

Fig. 6.6 Predictions of nucleated void volume fraction using the various void nucleation strains plotted in Fig. 5.2 (open symbols). Curves are predicted using corresponding values for ε_N and values for S_N in Table 5.1 (Reprinted with permission from Chen and Worswick (2008). Copyright: Elsevier)

levels of average nucleation strain to evaluate the effect of nucleation strain in a parametric fashion. Levels of $\varepsilon_N = 0.4$ and 0.6 were selected as being intermediate to the levels determined by Winkler. The nucleation curves using these levels are also plotted in Fig. 6.5 and were considered in the coupled percolation simulations.

By combining the functional dependence of particle nucleation strain on particle size in Fig. 6.5 with the measured particle size frequencies, it then becomes possible to predict the volume fraction of nucleated particles as a function of strain. This data is plotted for each alloy and thickness in Fig. 6.6 (open symbols).

Table 6.3 Controlling parameters in the GTN-based material model

f_N	**0.00768**	**0.00853**
f_0	0	0
ε_{N1}	0.2	0.2
ε_{N2}	0.4	0.4
ε_{N3}	0.6	0.6
s_N	15 %	15 %

In an effort to "synchronize" the Gurson nucleation prediction with the percolation model, values of s_N were determined for each sheet material to match the percolation predictions in Fig. 6.6 for each level of ε_N (curves). It is seen from the figures that the correspondence between the two nucleation treatments is reasonably good; thus the bulk nucleation behaviours will be similar in the two models. Differences will occur, however, on an element-by-element basis due to the random placement of particles mapped onto the FE mesh. Better matching of the nucleation treatment is possible, but requires tighter two-way coupling of the models and was considered beyond the scope of the current work.

Table 6.3 summarizes the controlling parameters used in the GTN-based damage simulations, corresponding to the curves in Fig. 6.6. Note that the standard deviation of void-nucleation strain, s_N, was reduced to 15 %, for the 1.0 mm sheets to reflect the effect of the rolling process. The thinner 1.0 mm sheet has a larger population of small particles compared to the 1.6 mm sheet. Note also that the value of f_N selected for each sheet was the higher of the measured particle areal fractions in the rolling- and transverse-through-thickness planes to facilitate a conservative prediction of flange formability. An initial porosity of zero ($f_0 = 0$) was chosen for both alloys based on metallographic analyses by Pilkey (private communication, 2001).

6.3.2 *Void Growth*

Void growth in the coupled model is treated using the predicted evolution of porosity within the GTN-based FE calculations. Within the percolation code, voids are assumed to enlarge during the deformation process according to

$$a^{(i)} = a^{(i-1)}\left(\frac{1+f^{(i)}}{1+f^{(i-1)}}\right)^{1/3} \tag{6.10}$$

where $a^{(i)}$ is the semi-axis of a void at the ith step and $f^{(i)}$ the porosity predicted for the corresponding parent finite element. This approach enforces equal rates of void growth between the GTN-based FE and percolation predictions.

Equation (6.10) implies that void expansion is isotropic and that details of local void geometry are neglected, such as the void aspect ratio and cluster effects addressed by Thomson et al. (1999). In addition, void growth under conditions of partial void-matrix separation hasn't been considered in the current study and the entire particle was treated as a void once nucleated.

Despite these simplifications, this void growth treatment is robust and can handle large changes in strain path using the isotropic Gurson void growth assumptions. Note that Eq. (5.10) naturally handles void closure effects that may occur during bend-unbend operations, for example.

6.3.3 Void Coalescence

Prediction of void coalescence is performed exclusively within the percolation damage code. Void coalescence using the so-called critical void volume fraction criterion (Eq. 6.3) within the GTN-based damage model is suppressed. Void coalescence was predicted using the modified version of Brown and Embury's (1973) ligament to void size ratio criterion, as described previously in Sect. 5.1.2.3.

In real forming operations, the strain field and strain path in the part are complex; hence, restrictions on ligament orientation for coalescence expressed by Eq. (5.6) were eliminated in the coupled model. The search algorithm used to predict void coalescence utilizes the nearest neighbour list generated by the tessellation software. At each strain increment step, Eqs. (5.2), (5.3), (5.4), (5.5) and (6.9), (6.10) are applied to all neighbouring pairs of nucleated voids. Note that the neighbouring particle search list spans across neighbouring elements in the finite element mesh.

6.3.4 Post-coalescence Treatment

The void-crack and crack-crack coalescence treatments are the same as described in Sect. 5.1 for the uniform strain model. In the coupled model, however, cracks are assigned to a parent element based on the location of their centroids. At each step, crack growth is predicted based upon the hydrostatic plastic strain component of its parent element with

$$a_m^{(i)} = a_m^{(i-1)} \left(\frac{1 + \varepsilon_{kk}^{(i)}}{1 + \varepsilon_{kk}^{(i-1)}} \right)^{1/3} \tag{6.11}$$

where $a_m^{(i)}$ is the semi-axis of a crack and $\varepsilon_{kk}^{(i)}$ is the hydrostatic strain at the ith step.

6.4 Particle Field Mapping

In coupled model, one of the most significant changes in damage percolation code is the mapping of the particle field onto the FE mesh AOI. Figure 6.7 shows a typical mesh AOI and mapped second phase particle field for 1.6 mm AA5182. This mapping requires the assignment of a parent element to each second phase particle.

A straightforward search algorithm is used to assign each particle a parent element. Damage evolution within the particle sub-field of each element is governed by the stress, strain and porosity history of the parent. For each time step, particle, void and crack positions are updated through linear interpolation of the nodal displacements of their parent elements, which can be expressed as

$$x_k = \sum_{j=1}^{4} N_j(s,t) X_k^j \tag{6.12}$$

where: N_j are the standard shape functions for a linear iso-parametric finite element; $j = 1, 2, 3, 4$ refers to nodal label; x_k and X_k^j are the respective coordinates of the centroid of a crack or particle and the nodes of its parent element; and s and t are the local coordinates of the crack or particle centroid within its parent element.

Cracks formed by coalescence of voids and/or cracks are also assigned to parent elements once coalescence occurs. A search algorithm is used to assign a parent element to each crack. Once a new crack is formed, the coordinates of its centroid can change and shift from one element to another due to coalescence of additional voids or cracks situated amongst different elements.

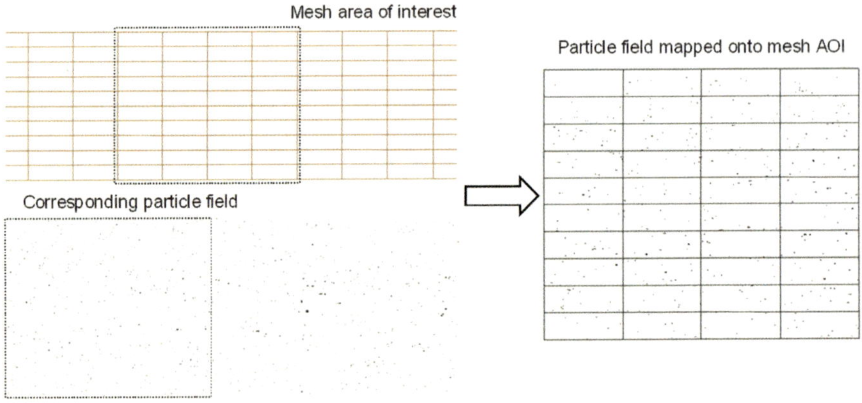

Fig. 6.7 A particle field is mapped onto the mesh area of interest from a larger tessellated second phase particle field of AA5182 sheet (Reprinted with permission from Chen et al. (2005). Copyright: Elsevier)

6.5 Coupled Model – Mesh and Particle Fields

6.5.1 FE Mesh

To save computation cost, axisymmetric solid continuum elements were employed in the FE part of the coupled model to model the blank and the rigid tooling because of the axisymmetric nature of the load and geometry. Calculations were performed using a commercial explicit dynamic FE code, LS-DYNA (Hallquist 2006). This code was linked with a user-supplied constitutive model embodying the Gurson constitutive model, developed by Worswick and Pelletier (1998).

The mesh adopted for the tooling and blank is shown in Fig. 6.8. Four-node, linear quadratic elements were used, with ten and six elements through the thickness of 1.6 mm and 1.0 mm blanks, respectively. From the predicted contours of porosity in the blank, it was found that maximum porosity always occurred in the area initially located right above the top of the male drawbead. Therefore, a finer mesh is employed in that area and this region was adopted as the mesh AOI for the subsequent damage percolation simulation. To further examine the bending/unbending effect on damage development during stretch flange forming, the top and bottom elements, designated elements A and B, were used to track damage histories.

The tooling elements were modelled as rigid and penalty function-based contact interfaces were defined between the tooling surfaces and the blank (Hallquist 2006).

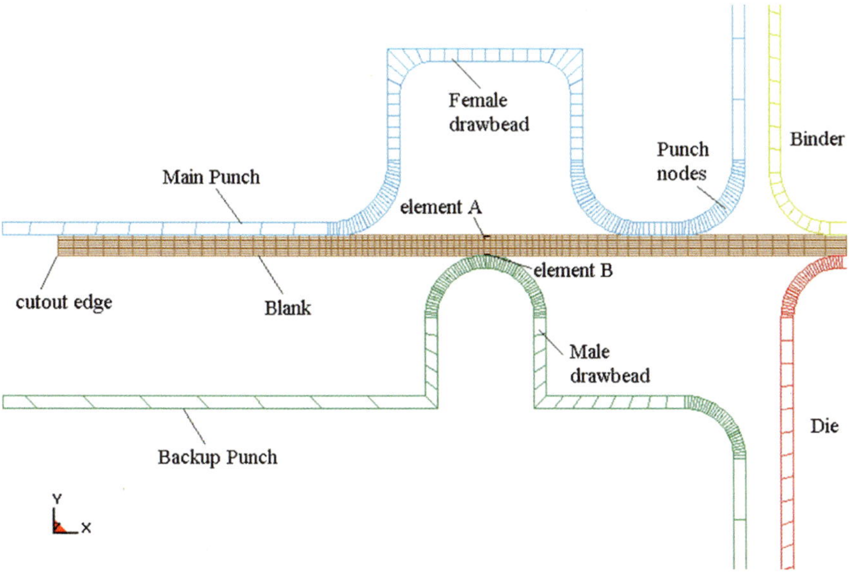

Fig. 6.8 Axisymmetric FE mesh used to model the stretch flange forming, 1.6 mm gauge, 92 mm cutout is shown (Reprinted with permission from Chen et al. (2005). Copyright: Elsevier)

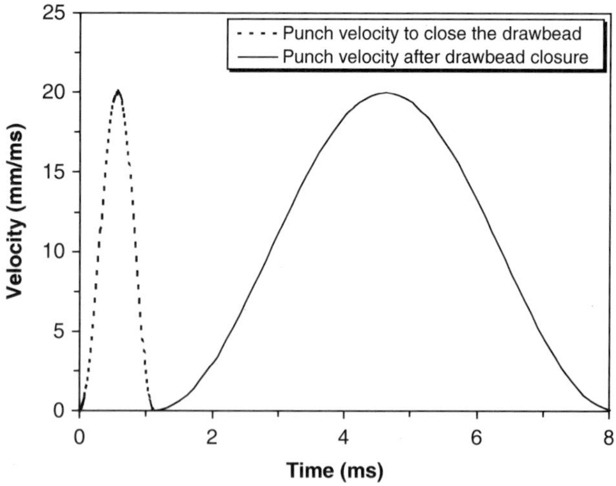

Fig. 6.9 Velocity histories of the main punch and backup punch in stretch flange forming (Reprinted with permission from Chen et al. (2005). Copyright: Elsevier)

A coefficient of friction of 0.074 (Draw-sol) was selected at all the interfaces between the blank and the punches for AA5182 based upon the drawbead simulator test data (Chen 2004). The blank interfaces with the die and binder were assigned a friction coefficient of 0.15 (dry friction) in order to minimize draw-in.

6.5.2 Boundary Conditions – Tooling Motion

A clamping force of 378 kN is applied on the binder to hold the blank during the forming operation. Before the bead is closed, a prescribed sinusoidal velocity (Fig. 6.9) is used to move the backup punch upwards to close the bead. After the drawbead is closed, the kinematic constraint on the backup punch is removed and replaced with a clamping force of 222.5 kN that holds the bead closed during the main punch motion. Once the backup punch force is applied, the main punch moves downward, expanding the stretch flange cutout. The velocity history prescribed for the main and backup punches are displayed in Fig. 6.9. A sinusoidal velocity profile is adopted to control or limit inertial effects. The forming operation is terminated once the main punch reaches its limit position.

6.5.3 Second Phase Particle Fields

1.0 and 1.6 mm AA5182 sheets were considered in stretch-flange forming simulations. The second-phase particle field information described in Chap. 5 was adopted in the simulation. Table 6.4 lists relevant data from the measured particle fields used in the simulations.

Table 6.4 Particle field used in the coupled FE/damage percolation modelling

Material	AA5182_1.0	AA5182_1.6
Particle areal fraction (%)	0.768	0.8531
Number of particles	8,486	13,038
Particle field size (mm)	4.4 × 1.0	4.4 × 1.6

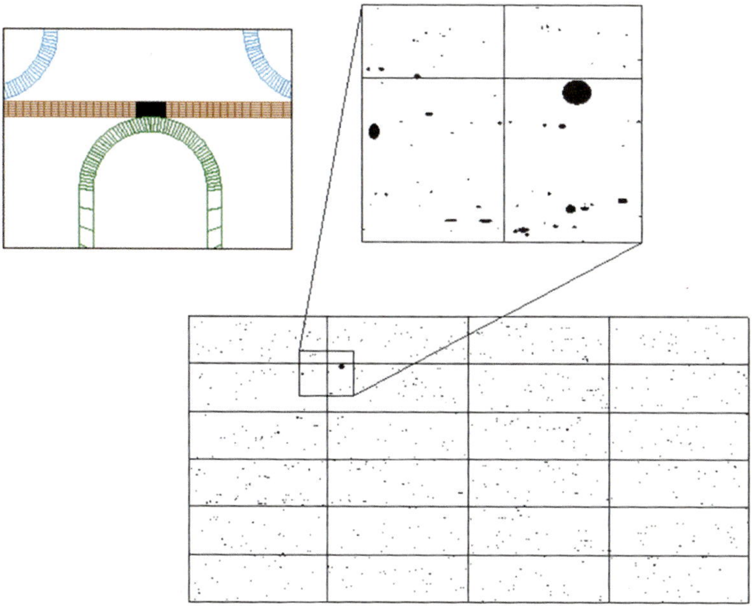

Fig. 6.10 Initial position of the mesh AOI and the corresponding mapped particle field, 1.0 mm AA5182

In the damage percolation simulations of stretch flange forming, particle fields of roughly 14,300 × 7,000 pixels (2.0 × 1.0 mm) and 14,300 × 11,400 pixels (2.0 × 1.6 mm) were separated from the original large particle fields and mapped onto a mesh area of either 24 elements (4 × 6) or 40 elements (4 × 10), respectively, located immediately above the male drawbead. This region was the most severely damaged area in the coalescence-suppressed GTN-based FE simulation. Damage percolation simulations were performed for this area using the coupled approach. Figs. 6.10 and 6.11 show the mesh areas of interest and the corresponding mapped particle fields used for 1.0 mm and 1.6 mm AA5182.

6.6 GTN-Based FE Results

In developing a coupled simulation of a forming operation it is first necessary to identify the worst-damaged area in the part; this becomes the damage AOI for the percolation simulation. In order to locate this region, several considerations are

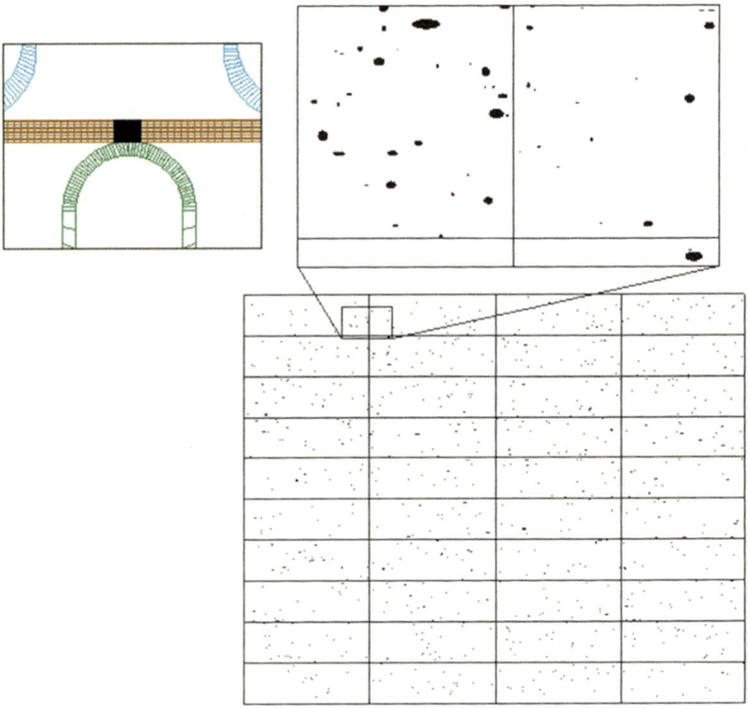

Fig. 6.11 Initial position of the mesh AOI and the corresponding mapped particle field, 1.6 mm AA5182 (Reprinted with permission from Chen et al. (2005). Copyright: Elsevier)

important. First, the GTN-based FE simulation is conducted to ascertain the potential most-damaged areas. Then damage percolation simulations are performed using the element and nodal data for each region. It is important to consider several candidate regions since the region of highest damage may differ between the GTN-based and percolation damage-based calculations. Finally, the region to fail at the lowest punch depth is taken as the most damaged area for the coupled model.

Such a procedure was used to determine the most damaged area for the coupled model. For all the blanks and materials considered in this study, it was found that the area initially located immediately above the top of the drawbead was the most damaged region. Therefore, this region was adopted as the mesh AOI for all of the models presented here.

6.6.1 Porosity Predictions

The predicted porosity levels are strongly related to sheet thickness. Figures 6.12 and 6.13 show a sequence of deformed contour plots of porosity for the 1.0 and 1.6 mm AA5182 samples with an 88 mm radius cutout, using a nucleation strain of

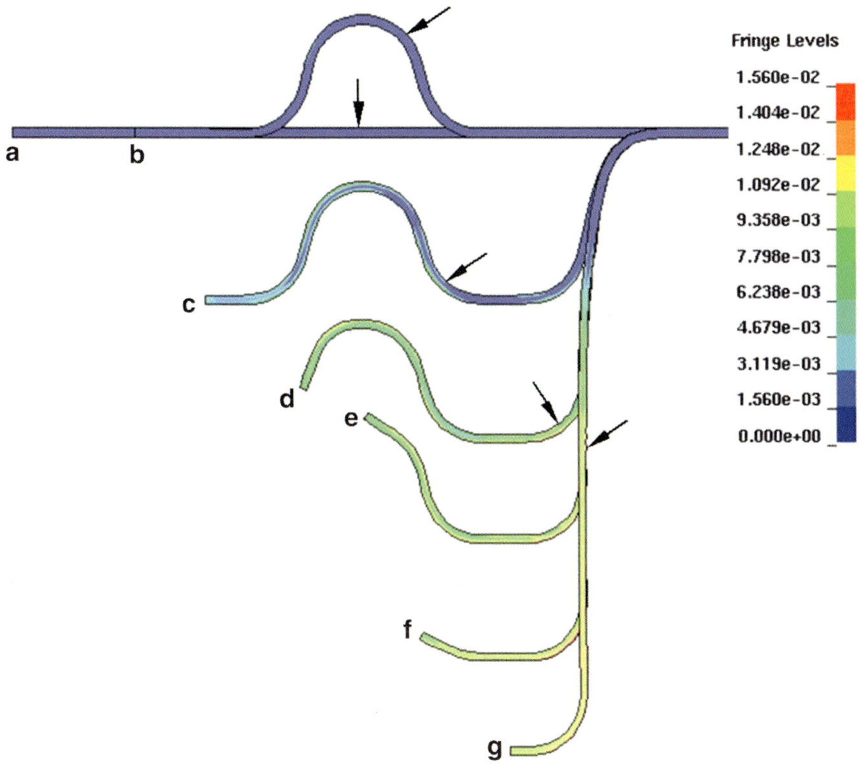

Fig. 6.12 Porosity contours within a deformed AA5182, 1.0 mm stretch flange with an 88 mm cutout using void nucleation strain, $\varepsilon_N = 0.2$ (coalescence-suppressed GTN-based FE calculation): (**a**) initial position; (**b**) drawbead closure at $t = 1.125$ ms; (c) $t = 3.6$ ms (punch depth = 16.4 mm); (**d**) $t = 4.4$ ms (punch depth = 30.1 mm); (**e**) $t = 4.88$ ms (punch depth = 40.1 mm); (**f**) $t = 5.5$ ms (punch depth = 51.7 mm); (**g**) $t = 6.1$ ms (punch depth = 60.9 mm). The arrows serve to locate the AOI

0.2. Much lower levels of void nucleation occur during drawbead closure for the 1.0 mm sheet due to the lower bending strain within the drawbead. With punch movement, the maximum porosity increases to 1.6 % for the 1.0 mm sheet compared to 2.1 % for the 1.6 mm sheet.

The progression of damage development for a higher nucleation strain is shown in Fig. 6.14. These are the sequences of porosity contours for the 1.6 mm AA5182 sample with an 88 mm radius cutout, considering nucleation strain of 0.6. Comparison of this figure to Fig. 6.13 demonstrates that the level of the void nucleation strain greatly affects the predicted level of damage in the samples. Higher void nucleation strain levels retard void initiation and result in lower porosity levels. As the void nucleation strain increases from 0.2 to 0.6, the maximum porosity in the deformed part decreases from 2.1 to 1.4 %, as indicated in the figures. In addition, the void nucleation strain has strong impact on damage development during

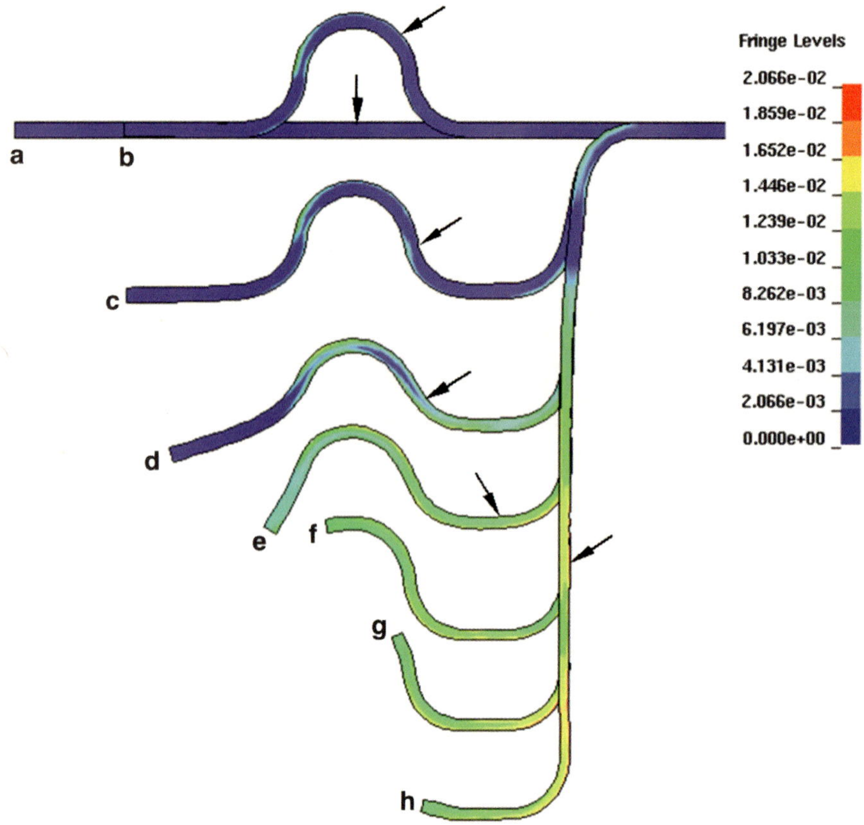

Fig. 6.13 Porosity contours within a deformed AA5182, 1.6 mm stretch flange with an 88 mm cutout using void nucleation strain, $\varepsilon_N = 0.2$ (coalescence-suppressed GTN-based FE calculation): (**a**) initial position; (**b**) drawbead closure at $t = 1.125$ ms; (**c**) $t = 3.6$ ms (punch depth = 16.4 mm); (**d**) $t = 4.4$ ms (punch depth = 30.1 mm); (**e**) $t = 4.88$ ms (punch depth = 40.1 mm); (**f**) $t = 5.5$ ms (punch depth = 51.7 mm); (**g**) $t = 6.1$ ms (punch depth = 60.9 mm); (**h**) $t = 8.1$ ms (punch depth = 70 mm). The arrows serve to locate the AOI (Reprinted with permission from Chen et al. (2005). Copyright: Elsevier)

drawbead closure (Figs. 6.13b and 6.14b). For a nucleation strain of 0.2, significant void nucleation is predicted during drawbead closure. However, for a nucleation strain of 0.6, the predicted void nucleation during drawbead closure was very low. For 1.0 mm sheet, damage development within the stretch flanges exhibits similar dependence on nucleation strain (not shown).

Figure 6.15 summarizes graphically the predicted peak porosity values for all of the stretch flange cases modelled. For both thicknesses, the peak porosity is strongly dependent on the nucleation strain level. Higher nucleation strain leads to lower peak porosity. The peak porosity in the 1.6 mm sheet does not show a dependence on cutout size, within the range considered. On the other hand, the peak porosity of the 1.0 mm sheet does indicate a mild, cutout-size dependence at a nucleation strain level of 0.2; that is, larger cutouts induce lower peak porosities. The 1.0 mm sheet exhibits much lower peak porosity in comparison to the 1.6 mm sheet.

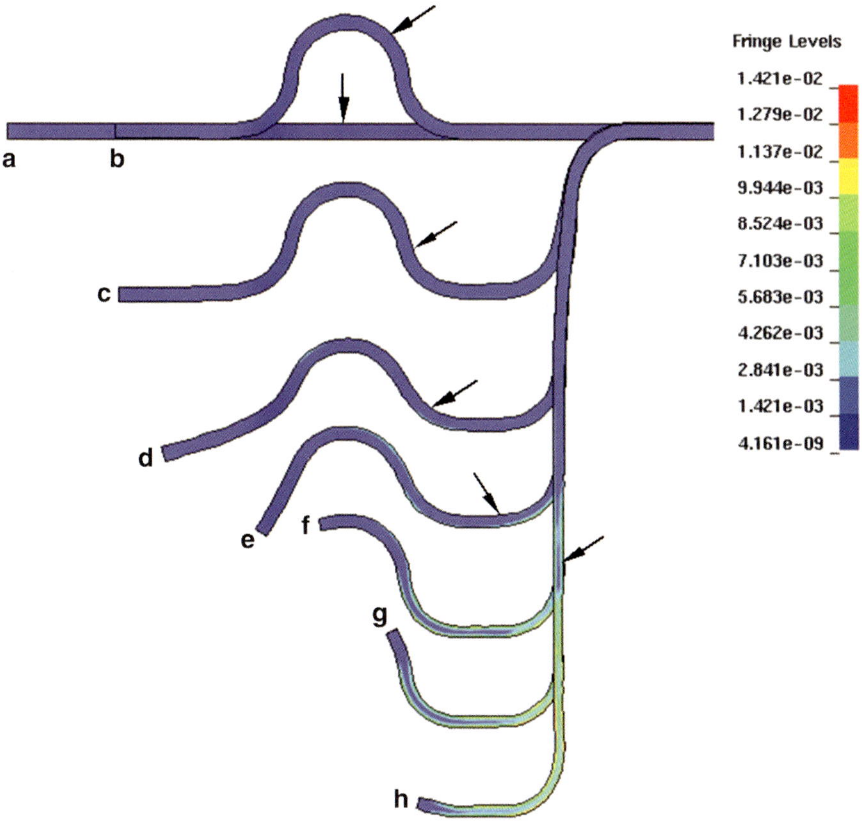

Fig. 6.14 Porosity contours within a deformed AA5182, 1.6 mm stretch flange with an 88 mm cutout using void nucleation strain, $\varepsilon_N = 0.6$ (coalescence-suppressed GTN-based FE calculation): (**a**) initial position; (**b**) drawbead closure at $t = 1.125$ ms; (**c**) $t = 3.6$ ms (punch depth = 16.4 mm); (**d**) $t = 4.4$ ms (punch depth = 30.1 mm); (**e**) $t = 4.88$ ms (punch depth = 40.1 mm); (**f**) $t = 5.5$ ms (punch depth = 51.7 mm); (**g**) $t = 6.1$ ms (punch depth = 60.9 mm); (**h**) $t = 8.1$ ms (punch depth = 70 mm). The arrows serve to locate the AOI

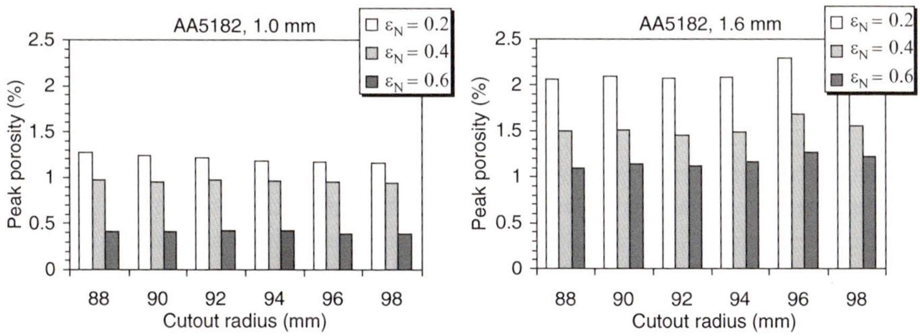

Fig. 6.15 Peak porosity as a function of flange cutout radius

indicates the radial failures at the cutout edge and the lighter shading corresponds to circumferential failures at the punch profile radius. Note that the maximum punch depth considered was 70 mm; thus, the predictions plotted at this level have not failed. Several observations can be made:

- The dependency of predicted punch depth on nucleation strain is very strong and overwhelms any effect due to cutout radius for the range considered;
- The scatter in the experiments is very large and of similar magnitude to the range of the limit punch depth in the models.

In spite of the experimental scatter, there exists a qualitative agreement between the predicted and measured limit punch depths. The results indicate that a nucleation strain in the range 0.25–0.5 would lead to predicted limit punch depths in general accord with the observed failures for the model AA5182. These values are consistent with current results by Winkler (2003).

Further refinement of the material damage parameters is not thought possible at this time, pending improved repeatability of the stretch flange experiments and/or more in-depth metallographic studies of nucleation behaviour for these alloys (Winkler 2003).

6.8 Summary

Stretch flange formability, in terms of limit punch depth to fracture, is predicted through the coupled model as a function of cutout radius and void nucleation strain. The experiments and models have shown that the limit punch depth to fracture is not particularly sensitive to the variation of cutout size within the range considered. The predicted damage rate, however, did decrease with an increase in cutout size. Increased sheet thickness leads to an increase in damage rate in the models, favouring circumferential cracking at the punch nose. This trend was reflected in higher predicted damage rates and a transition to punch nose failures for the thicker AA5182 samples.

The void nucleation strain has a strong impact on the predicted damage rate and formability. Higher nucleation strain leads to higher predicted formability and lower damage rate. Comparison of the predicted formability and damage rate with experimental results suggests a nucleation strain of 0.2–0.5 for these alloys.

Chapter 7
Two Dimensional (2D) Damage Percolation with Stress State

Butcher and Chen (2009b, c) extended the 2-D damage percolation model used by Chen (2004) to directly incorporate the stress state, material softening and a coalescence model linking the void geometry with the stress state via the plastic limit-load criterion. Unlike the prior percolation model, the stress state is directly determined from the GT yield surface by performing a dynamic homogenization at each time step to calculate the equivalent void in the material to account for softening. In the previous damage percolation models (Worswick et al. 2001; Chen 2004; Chen et al. 2005), void nucleation and coalescence were modeled using only geometric considerations and the effect of stress state was not considered. Void growth and shape evolution are strain-controlled and were reasonably well represented in the percolation model for well-defined stress states but a simplified coalescence rule was employed that did not account for the stress state. In addition, the fracture predictions of the percolation model are extremely sensitive to the void nucleation rule. In continuum modeling, void nucleation is often represented using a bulk averaged criterion. Obviously, this averaged criterion is unsuitable for percolation modeling since nucleation occurs at the individual particle scale.

A phenomenological nucleation criterion was developed where void nucleation is related to the particle morphology, hydrostatic stress and shear loading. The nucleation criterion was calibrated by subjecting three particle fields from an aluminum-magnesium alloy to different loading conditions to achieve agreement with experimental forming limit data. More importantly, the model was able to provide predictions for the average area fraction and size of the void nucleating particles that are in good agreement with the available experimental data.

Z. Chen and C. Butcher, *Micromechanics Modelling of Ductile Fracture*,
Solid Mechanics and Its Applications 195, DOI 10.1007/978-94-007-6098-1_7,
© Springer Science+Business Media Dordrecht 2013

7.1 Development of a Phenomenological Void Nucleation Criterion for Percolation Modeling

Void nucleation primarily occurs at second phase particles via particle cracking or separation of the particle-matrix interface (debonding). A realistic nucleation criterion should account for many factors such as the nucleation mechanism (cracking or debonding), particle morphology (size, shape, clustering and volume fraction), and stress state. Additional factors which can influence void nucleation are the strain rate, temperature and level of pre-strain in the material (Horstemeyer et al. 2003). Here, a phenomenological nucleation model is developed where the nucleation strain is assumed to be a function of the particle morphology and stress state:

$$\varepsilon_{\mathrm{N}} = \varepsilon_{\mathrm{N}_0} g(d) \cdot h(f_{\mathrm{p}}) \cdot s(T, \mu_L) \tag{7.1}$$

where $\varepsilon_{\mathrm{N}_0}$ is the nucleation strain in pure shear which is scaled by weighting functions $g(d)$, $h(f_p)$,and $s(T, \mu_L)$, related to the particle diameter, d, area fraction of second phase particles, f_{p}, and stress state defined by the stress triaxiality, T, and Lode parameter, μ_L. The influence of particle clustering and aspect ratio on void nucleation is not accounted for in this model and no assumptions are made regarding the nucleation mechanism. This nucleation rule is therefore a lumped-parameter model that accounts for the various types of nucleation mechanisms.

7.1.1 Particle Size and Area Fraction Functions

In general, the strain required to nucleate a void is inversely related to particle size with larger particles nucleating voids at lower strains (Embury 1985; Dighe et al. 2002; Shabrov and Needleman 2002). The particle size-related nucleation parameter suggested by Horstemeyer et al. (2003) was adopted for the size-related weighting function:

$$g(d) = \frac{1}{\sqrt{d}} = \left(\frac{4}{\pi} A_p\right)^{\frac{-1}{4}} \tag{7.2}$$

where d is the particle diameter and A_{p} is the particle area in the plane of deformation.

Although it is tempting to assume that the materials with higher volume fractions of particles will experience more nucleation events, the opposite trend has been observed experimentally (Gangalee and Gurland 1967; Mazinani and Poole 2007). The probability of nucleation tends to be higher for materials with a smaller particle content because the particles are more widely distributed throughout the material and will experience a higher load. Materials with a high volume fraction of particles tend to have a measure of connectivity and are therefore able to share the load with their neighbours, decreasing the average stress in the particles. Here, we assume that the nucleation strain increases with the area fraction of second phase particles, such that

$$h(f_{\mathrm{p}}) = {f_{\mathrm{p}}}^{1/3} \tag{7.3}$$

where the exponent of one-third is selected based upon the work of Gangalee and Gurland (1967) who used the ratio of $\sqrt{d}/{f_{\mathrm{n}}}^{1/3}$ to characterize nucleation for a range of particle sizes and area fractions in aluminum-silicon alloys. The product of the size and area fraction functions, $g(d) \cdot h(f_{\mathrm{n}})$, results in the inverse of this ratio, implying that the nucleation strain decreases with particle size and increases with the second phase particle content. Smaller particles nucleate at higher strains because they contain a lower probability of containing an internal defect that would promote particle cracking and they also tend to be spherical and tightly bounded to the matrix material. Large particles tend to be irregularly shaped and contain defects such as surface cracks that form during solidification, promoting nucleation early in the deformation process. The nucleation model of Horstemeyer and Gokhale (1999) also includes a dependence upon $\sqrt{d}/{f_{\mathrm{n}}}^{1/3}$ which acts as a scaling factor for the nucleation rate.

7.1.2 *Stress State Dependence Function*

It is well known that the nucleation and fracture strains decrease with increasing stress triaxialitybut that stress triaxiality cannot uniquely define the stress state since it is related to only two of the three invariants of the stress deviator tensor: the equivalent and mean stresses. Consequently, for a given triaxiality, the severity of shear loading is unknown. The shear or deviatoric stress state can be characterized by incorporating a dependence upon the third stress invariant using the Lode parameter. Recent works have shown that void growth, shape change and coalescence are related to both the stress triaxiality and severity of shear loading (Zhang et al. 2001; Xue 2007; Barsoum and Faleskog 2007a, b; Scheyvaerts et al. 2011). The stress state can be described using the stress triaxiality, T, in addition to the Lode parameter, μ_{L}, defined as follows

$$\mu_{\mathrm{L}} = \frac{2\sigma_2 - \sigma_1 - \sigma_3}{\sigma_1 - \sigma_3} = \frac{3(\sigma_2 - \sigma_{\mathrm{hyd}})}{\sigma_1 - \sigma_3} \tag{7.4}$$

where $\sigma_1, \ \sigma_2, \ \sigma_3$ are the principal stresses in descending order. The stress state for proportional, plane-stress loading can be easily estimated using for a von Mises material or damage-based material with a small volume void volume fraction, the stress triaxialities corresponding to uniaxial tension, plane strain and equal-biaxial tension are: $1/3$, $1/\sqrt{3}$ and $2/3$, respectively. Lode parameters of -1, 0 and 1 correspond to uniaxial tension, generalized shear and equal-biaxial tension, respectively, with the values in between representing a state of combined tension and shear. The stress state is determined in the percolation model using the Gurson-Tvergaard (GT) material model to account for material softening. However, the difference between the stress state obtained using the GT model and the von Mises model is negligible since the failure porosity in all of the particle fields is less than 1 % in AA5182 sheet and the triaxiality is also low (less than one). Overall, material softening is included in the percolation model for completeness but it is not significant in the present application.

While the influence of shear on void nucleation has often been neglected, the nucleation model of Horstemeyer and Gokhale (1999) included a dependence upon the third stress invariant which predicted a higher nucleation rate in combined tension and torsion compared to pure tension. Furthermore, Dighe et al. (2002) experimentally observed a higher degree of particle cracking in torsion than in tension for an Al-Si alloy. Recent x-ray microtomography studies of dual phase steels in uniaxial tension by Maire et al. (2008) observed that the nucleation strain could be well described by the stress triaxiality. For a general material, the nucleation rule used by Maire et al. (2008) can be written as

$$\varepsilon_{\mathrm{N}} = \varepsilon_{\mathrm{No}} \exp(-kT) \tag{7.5}$$

where $\varepsilon_{\mathrm{No}}$ is the nucleation strain corresponding to pure shear ($T = 0$) and k is the triaxiality scale factor identified through calibration with experiment data. However, since many different shear stress states can be obtained for the same triaxiality, the scale factor would almost certainly change if it was identified in a state of combined tension and shear rather than uniaxial tension. Therefore, we propose a phenomenological dependence of the stress triaxiality parameter on the Lode parameter

$$k(\mu_L) = 2 - \beta|\mu_L| \tag{7.6}$$

where β is an adjustable parameter. The value of 2 in Eq. (7.6) was arbitrarily selected so that $k = 1$ in axisymmetric tension if $\beta = 1$. For simplicity, no distinction was made between the Lode parameters in uniaxial and equal-biaxial tension (Xue 2008; Butcher and Chen 2009a, b, c). The stress state weighting function, $s(T, \mu_L)$ in Eq. (7.1) is

$$s(T, \mu_L) = \exp[-T(2 - \beta|\mu_L|)] \tag{7.7}$$

From inspection of Eq. (7.7), β is a function of the stress triaxiality and the absolute value of the Lode parameter and should be identified in at least two different stress states.

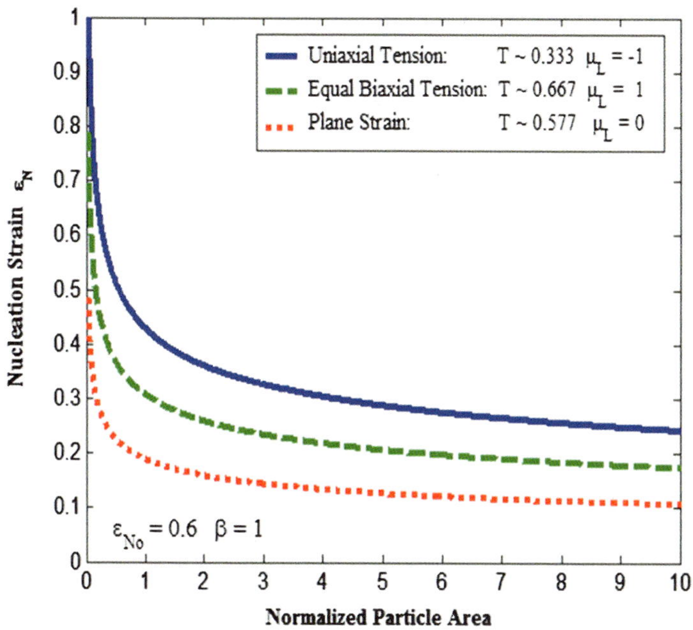

Fig. 7.1 Variation of the nucleation strain with particle size and stress state

7.1.3 Application of the Nucleation Criterion to Various Particle Fields

Since the nucleation model will be applied to different particle fields, the particle morphology should be measured relative to the average morphology. Therefore, the nucleation model is expressed as

$$\varepsilon_{\mathrm{N}} = \varepsilon_{\mathrm{No}} \left(\frac{f_p}{f_{\text{p-avg}}} \right)^{\frac{1}{3}} \cdot \left(\frac{A_p}{A_{\text{p-avg}}} \right)^{\frac{-1}{4}} \cdot \exp[-T(2 - \beta|\mu_L|)] \tag{7.8}$$

where $A_{\text{p-avg}}$ and $f_{\text{p-avg}}$ are the average particle area and area fraction for all particle fields. The nucleation parameters $\varepsilon_{\mathrm{No}}$ and β are identified through calibration of the percolation model with experimental forming limit data but could have been determined using torsion and tensile test data. The general trend for the variation of the nucleation strain with particle size and stress state is presented in Fig. 7.1.

The nucleation model reflects the contribution of shear loading because the nucleation strain is lower in plane strain than in equal-biaxial tension even though the triaxiality is smaller for plane strain. A similar trend is observed in forming limit diagrams where the limit strain decreases from uniaxial tension to plane strain (triaxiality and shear loading increasing). The forming limit strain then increases as an equal-biaxial condition is approached (triaxiality increasing, shear loading decreasing). A combined tensile and shear stress state will also promote void coalescence (Barsoum and Faleskog 2007a, b).

7.2 Percolation Modeling of Ductile Fracture

The principal techniques and methodologies used in the 2D percolation model are similar to the previous model described in Chap. 6. The main modeling assumptions used in this incarnation of the model are as follows:

- The particle field is two-dimensional and the voids and particles represent cylinders in the through-thickness direction
- The two dimensional particle field is in a plane stress state and deforms homogenously
- Local stress and strain heterogeneity within void and particle clusters are neglected
- The stress state within the particles is not considered
- Particles, voids and cracks are elliptical; they do not rotate during deformation
- No distinction is made between nucleation via particle cracking, debonding or partial debonding; the initial size of the nucleated void is equal to the particle size
- Voids are assumed to grow as isolated voids
- Void coalescence occurs by necking failure of the inter-void ligament. Ligament shearing or combined necking and shearing are not considered
- The void nucleation criterion is evaluated for each particle. When the plastic strain reaches the nucleation strain the particle is replaced with a void equal to the particle dimensions.
- The GT yield function used in the FE-percolation model of Chen (2004) is used to directly compute the stress in the particle field without the loose-coupling method between the FE simulation and the percolation model

7.2.1 Void Coalescence

The coalescence rule used in the previous percolation models was a modified Brown and Embury (1973) criterion independent of the stress state. The incorporation of the stress state into the percolation model enables a physically sound treatment of coalescence since it is strongly related to both the microstructure and applied stress. The 2D variant of the plastic limit-load coalescence criterion of Thomason (1990) is adopted in the present work since coalescence is based upon necking failure of the inter-void ligament and is related to the stress state and void geometry. Coalescence occurs when the following condition is satisfied

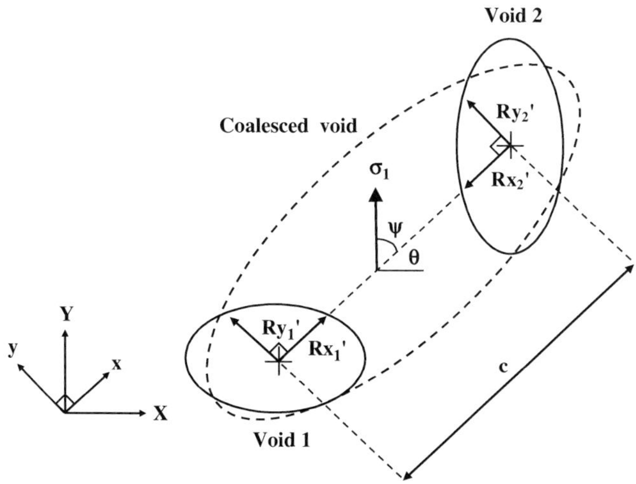

Fig. 7.2 Schematic of void interaction geometry

$$\frac{\sigma_1}{\bar{\sigma}} \geq \frac{2}{\sqrt{3}}\left(\sqrt{1+\frac{1}{4\tan^2\psi}}+\frac{\chi^{-1}-1}{4W}\right)(1-\chi) \qquad (7.9)$$

where ψ is the angle of the maximum principal stress relative to the material ligament; W is the void aspect ratio (R_y/R_x), and χ is the void spacing ratio defined as the ratio of the lateral void radius to half of the lateral void spacing. However, the void distribution in a real material is not periodic as idealized in the plastic limit-load criterion. Therefore, we suggest that the geometrical parameters in Eq. (7.9) be determined within a local Cartesian coordinate system x'–y', where x' is in alignment with the ligament orientation, as shown in Fig. 7.2. Accordingly, the void aspect and spacing ratios are defined as

$$W = \frac{R'_{y1}+R'_{y2}}{R'_{x1}+R'_{x2}} \quad \chi = \frac{R'_{x1}+R'_{x2}}{c} \qquad (7.10, 7.11)$$

where c is the center-to-center distance between two voids.

The x'-axis of the local coordinate system, x'–y', is in alignment with the ligament orientation defined by the angle, θ, with respect to the horizontal direction. Equation (7.9) is evaluated at each time-step for all neighbouring pairs of nucleated voids.

7.2.2 *Profuse Coalescence and Failure of the Particle Field*

The onset of profuse void coalescence signaling failure of the particle field is easily identified as the voids rapidly link-up throughout the particle field to form a single void which encompasses the entire field. Failure of the particle field was identified by homogenizing all of the individual voids into an equivalent void and evaluating the coalescence model in Eq. (7.9) at each time-step. Since the porosity increases so rapidly at the onset of profuse coalescence, the choice of global failure criterion was not an important factor. Assuming a global critical porosity of 1 % as suggested by Chen (2004) for AA5182 sheet led to similar results identified by using Eq. (7.9) and by visual means. Ultimately, the coalescence criterion in Eq. (7.9) was adopted since it does not introduce any additional parameters into the model.

7.3 Particle Field Tessellations

Three tessellated AA5182 particle fields of 2,000 × 2,000 pixels or 0.75 mm × 0.75 mm were extracted from a large-scale, plane-view image and designated as P1, P2, and P3 as shown in Fig. 7.3. The particle aspect ratio and spacing ratio are calculated using Eqs. (7.10) and (7.12), respectively. The second phase particles exhibit significant clustering as evidenced by the large average particle spacing ratio and coalescence can be expected to occur shortly after a void is nucleated next to an existing void. The particle field data is summarized in Table 7.1 and the size distributions are shown in Fig. 7.4.

7.4 Calibration of the Nucleation Model

The nucleation parameters in Eq. (7.8) are identified parametrically by comparing the fracture predictions of the percolation model for each particle field with the experimental forming limit curve data of Chen (2004) and Wu et al. (2003) for

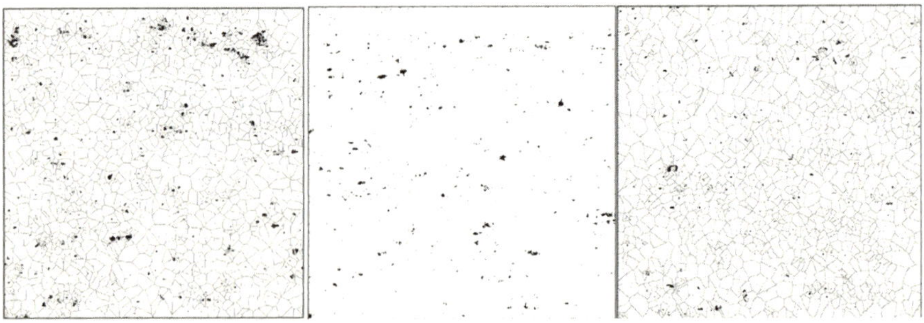

Fig. 7.3 Tessellated AA5182 particle fields: P1 (*left*), P2(*center*) and P3 (*left*). Each particle field is composed of 2,000 × 2,000 pixels obtained from a plane-view large scale image. Rolling direction is horizontal and the transverse direction is vertical

Table 7.1 Particle field information for AA5182 sheet

Particle field	Initial particle area fraction f_n	Particle area A (μm^2)	Particle aspect ratio λ_p	Particle spacing ratio χ_p
P1	0.0177	7.876	1.20	0.618
P2	0.0142	5.416	1.25	0.562
P3	0.0079	4.208	1.24	0.531
Average	0.0133	5.833	1.23	0.570

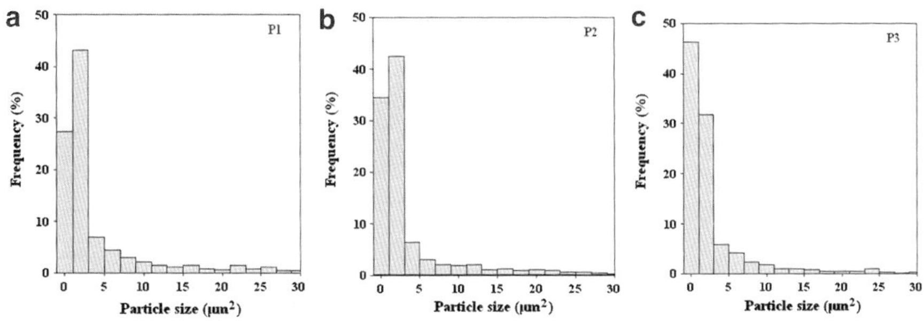

Fig. 7.4 Particle size distribution for particle fields: (**a**) P1, (**b**) P2, and (**c**) P3 obtained from AA5182, 1.5 mm sheet, plane-view

AA5182 sheet. Each particle field was subjected to a range of proportional loading where the loading condition is characterized using the ratio of the minor to major strain ($\varepsilon_2/\varepsilon_1$). The strain ratio is varied from −0.5 (uniaxial tension) to 1.0 (equal-biaxial stretching) in 0.125 increments. The limit strain is detected at the onset of profuse coalescence by evaluating Eq. (7.9) for the entire particle field. For each strain ratio, the particle field was loaded in both the rolling and transverse directions for a total of six simulations per strain ratio. The particle field is enforced to deform homogeneously and remain rectangular throughout the deformation process so that it representative of a unit cell. An alternative approach to predicting the forming-limit diagrams could be performed within the M-K framework (Marciniak and Kuczynski 1967) where the presence of a material imperfection results in the development of a localized deformation band and subsequent failure. In this approach, the existence of void clusters would hasten band localization.

7.4.1 Comparison of Predicted and Experimental Forming Limits

The forming limit predictions for the particle fields in both the rolling and transverse directions are compared with the experimental data in Fig. 7.5. The forming limit data obtained using the percolation model is in general agreement with the experiment

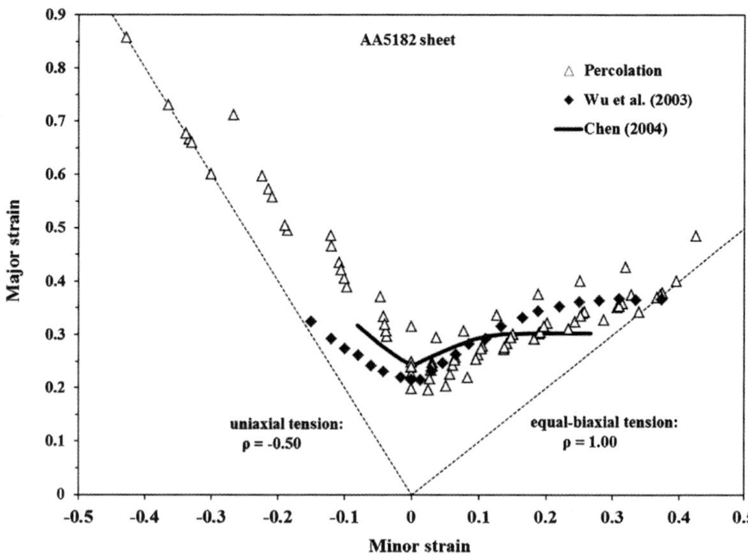

Fig. 7.5 Comparison of the experimental forming limit data with the values obtained using the calibrated percolation model. The experimental data of Chen (2004) is for exactly the same material as the percolation model, while the data of Wu et al. (2003) is for the sheet from the same material supplier (Reprinted with permission from Butcher and Chen (2009b). Copyright: Elsevier)

with nucleation parameters of $\varepsilon_{No} = 0.85$ and $\beta = 1.80$. The best performance of the model occurs in the biaxial stretching condition (right-hand side of FLC) while overestimating formability in the uniaxial stretching condition (left-hand side of FLC). The variation between particle fields and loading direction is greatest in uniaxial tension because the coalescence path is more constrained. Consequently, some particle fields will possess a more favorable particle distribution which hinders coalescence in one-direction and are thus able to continue stable void growth and nucleation to higher plastic strains. The large strains obtained in uniaxial tension can be attributed to the enforced homogeneous deformation of the particle field, thereby suppressing necking. Despite this restriction, these values are not entirely unreasonable since true strains of 0.55 in uniaxial tension have been observed in stretch flanging of this alloy (Worswick et al. 2001).

The limit strains are lowest for particle field P1 and highest for field P2. It is not surprising that particle field P1 experiences the lowest fracture strains as it contains many clusters of relatively large particles compared to fields P2 and P3 as seen in Fig. 7.3. Overall, the scatter in the predictions of the percolation model is reasonable considering that the distribution of the particles in each field is very different despite their similarities in their average statistics such as the aspect ratio and spacing. Generally, a particle field with an excessive amount of particle clustering will result in fracture at lower strains since it possesses a higher number of fracture paths.

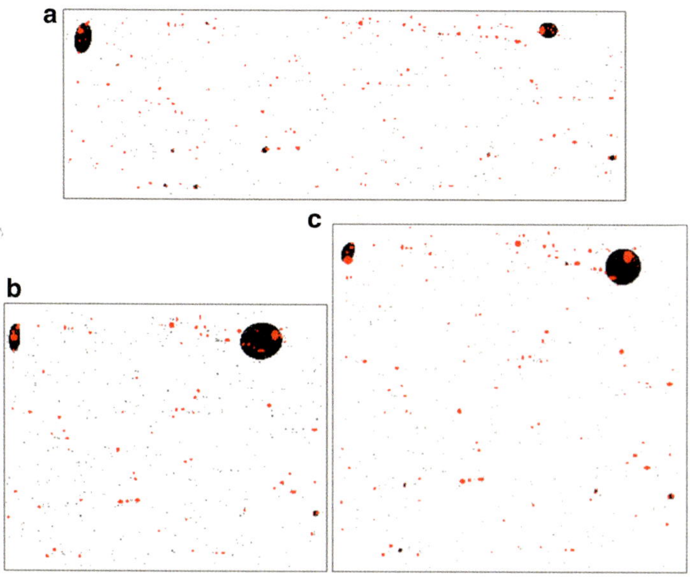

Fig. 7.6 Predicted damage prior to profuse coalescence in particle field P1 under: (**a**) uniaxial tension at 66 % major strain (**b**) plane strain at 23.9 % major strain, and (**c**) equal biaxial tension at 37.8 % major strain. The red ellipses indicate nucleated voids. The rolling direction is horizontal and load is applied in the same direction for uniaxial tension and plane strain conditions

7.4.2 Void Nucleation

Damage initiation in AA5182 particle field P1 prior to profuse coalescence is presented in Fig. 7.6 for uniaxial tension, plane strain and equal-biaxial tension. The grey ellipses are second-phase particles; the red ellipses represent particles which have nucleated voids and the black ellipses are coalesced voids. For each loading condition, void nucleation and localized coalescence occurs at large particles which are located in clusters. Void coalescence occurs rapidly in the clusters to form large local voids which quickly grow and link-up among different clusters throughout the material causing failure. Similar behaviour is observed in particle fields P2 and P3 where coalescence is ultimately controlled by the nucleation of large particles within two or three clusters.

The trends for void nucleation for each particle field subjected to uniaxial, plane strain and equal-biaxial tension are presented in Fig. 7.7. The nucleation trend for each particle field is the average nucleation response of the rolling and transverse directions. For each particle field, the area fraction of particles which nucleate voids is normalized by the field's particle area fraction.

The nucleation trends are similar for each particle field and loading condition with equal-biaxial tension having a slightly higher nucleation rate than uniaxial tension. The overall behavior is similar for both the uniaxial and equal-biaxial cases

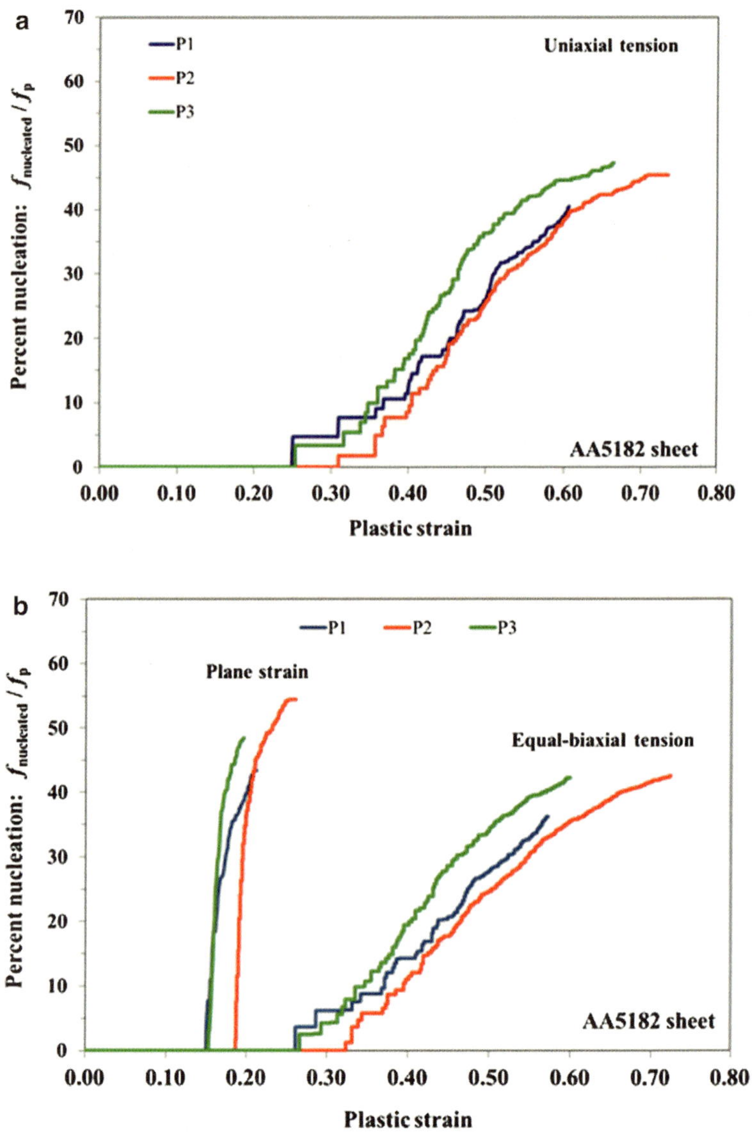

Fig. 7.7 Average percentage of particles which nucleated voids in: (**a**) uniaxial tension, (**b**) plane strain and equal-biaxial tension

where nucleation increases proportionally with plastic strain once a threshold strain of 0.25–0.32 is reached. Overall, the influence of the stress triaxiality is small within the range of triaxiality considered (1/3 to 2/3). Conversely, the influence of the Lode parameter on void nucleation is dramatic as demonstrated in the plane

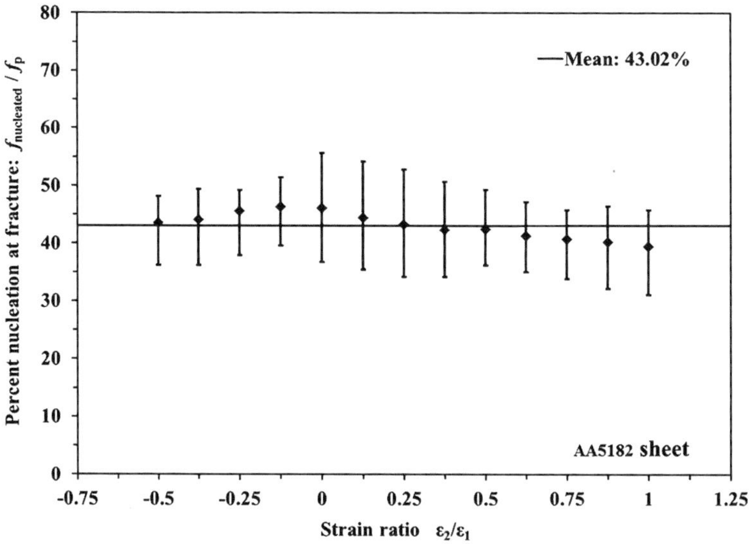

Fig. 7.8 Area percentage of particles which have nucleated voids at fracture. A total of sixmeasurements were obtained for each loading condition and are expressed as a band for clarity

strain loading condition where shear effects can be considerable. For a thin sheet (plane stress) subjected to a plane strain loading, the Lode parameter reaches a value of 0 with a stress triaxiality of about 0.577. This corresponds to a stress state of pure shear with a hydrostatic stress component. This stress state is more severe than simple shear where the Lode parameter is also 0 but there is no superimposed hydrostatic stress ($T = 0$) to promote void growth and coalescence. The nucleation model reflects the severity of the stress state in plane strain loading and thus nucleation occurs at lower plastic strains. For plane strain loading, nucleation is negligible in the particle fields until a critical plastic strain has been reached, where a rapid burst of nucleation occurs, triggering widespread coalescence and fracture. This trend mimics the behaviour of AA5182 where nucleation is minimal until the development of shear bands in the material that promote localized nucleation and coalescence (Hadianfard et al. 2008).

On average, 43 % of the second-phase particles nucleate voids regardless of the loading condition as shown in Fig. 7.8. It is important to note that this result is based upon area fractions and thus reflects that nucleation predominantly occurs at large particles. Physically, it is not likely that 43 % of the particles nucleate a void; rather it is the nucleation from the large particles that inflates the overall particle fraction. The average area fraction of void nucleating particles in AA5182 sheet was calculated to be 0.57 % which is in reasonable agreement with the value of 0.77 % determined in the stretch flanging operations of Chen (2004).

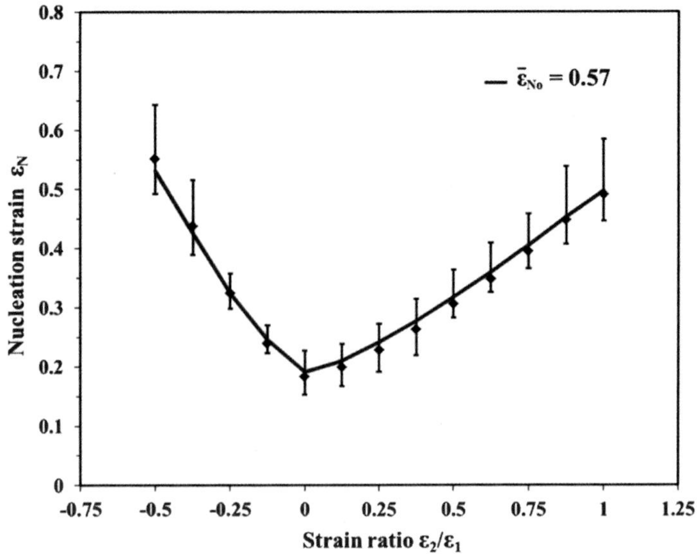

Fig. 7.9 Variation of the nucleation strain with proportional loading. A total of six measurementswere obtained for each loading condition and are expressed as a band for clarity

7.4.3 Average Nucleation Strain

The mean nucleation strain under various proportional loading histories is presented in Fig. 7.9. The heterogeneous nucleation model for the nucleation strain in Eq. (7.8) can be transformed to a bulk criterion since the average particle field will have a volume fraction of $f_p = f_{\text{p-avg}}$ with an average particle size of $A_p = A_{\text{p-avg}}$. The nucleation strain parameter, ε_{No}, will be changed to a new effective value denoted $\bar{\varepsilon}_{\text{No}}$, but the stress dependence parameter, β, remains the same value of 1.80 identified from the particle field analysis. The average nucleation strain is expressed as

$$\varepsilon_{\text{N}} = \varepsilon_{\text{No}} \left(\frac{f_{\text{p}}}{f_{\text{p-avg}}} \right)^{\frac{1}{3}} \left(\frac{A_p}{A_{\text{p-avg}}} \right)^{\frac{-1}{4}} \exp[-T(2 - \beta|\mu_L|)]$$
$$\rightarrow \bar{\varepsilon}_{\text{No}} \exp[-T(2 - \beta|\mu_L|)] \quad (7.12)$$

The effective nucleation scale parameter is easily identified through calibration with the mean nucleation strain obtained from the particle fields. A value of $\bar{\varepsilon}_{\text{No}} = 0.57$ is found to give good agreement with the percolation data (Fig. 7.9). It is difficult to compare the predicted nucleation strains with the values obtained in the literature since the nucleation strain is typically considered independent of the stress state and is therefore specific to the test conditions in which it was determined. Overall, the reported nucleation strains for AA5182 in the literature cover a very large range from 0.20 to 0.70 (Lievers et al. 2004).

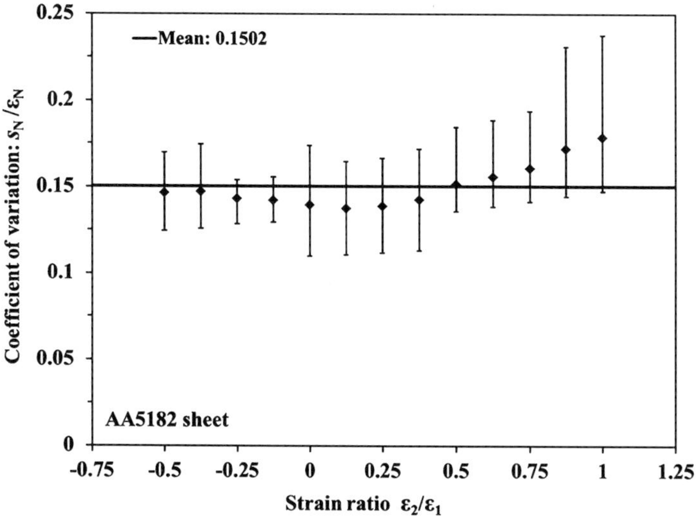

Fig. 7.10 Influence of proportional loading on the coefficient of variation for nucleation in the model AA5182-O sheet. A total of six measurements were obtained for each loading condition and are expressed as a band for clarity

It is of interest to evaluate the standard deviation of the nucleation strain for each particle field for comparison with the values reported in the literature. Since different nucleation strains and standard deviations have been reported for this alloy, the standard deviation is normalized by the mean nucleation strain to compute the coefficient of variation (CV). Strictly speaking, the nucleation model used in the percolation model is independent of the loading direction since the particle aspect ratio is neglected. However, nucleation and thus the CV are related to the loading direction through the coalescence model. The coalescence path is dependent upon the loading direction and the same particle field can exhibit a higher degree of nucleation if coalescence is delayed and a higher strain is reached. The CV does not appear to exhibit a strong dependence upon the stress state as shown in Fig. 7.10. Overall, the mean CV is 15.02 % and this value is in general agreement with the work of Orlov (2006), Lievers et al. (2004) and Chen (2004) who suggested respective values of 9.38, 14.1 and 20 % for the same material.

7.4.4 *Average Size of Damaged Particle*

The average size of the particle that nucleates a void (damaged particle) can be extracted from the percolation model for each particle field and loading condition. As shown in Fig. 7.11, the average diameter of the damaged particles exhibits no clear dependence upon the loading condition. The mean diameter of the damaged

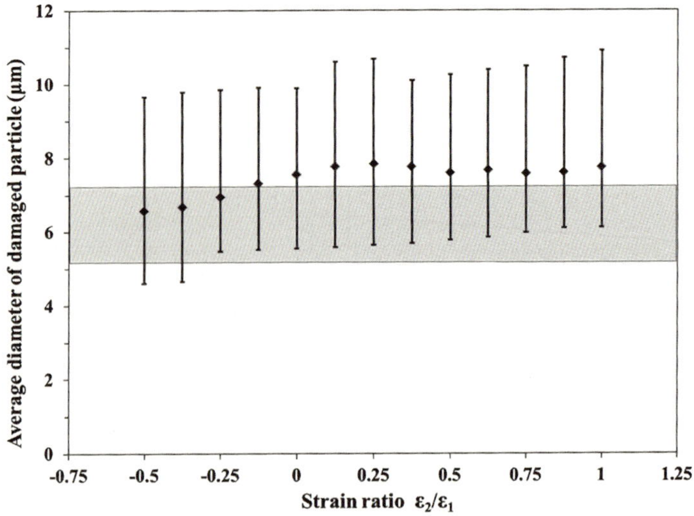

Fig. 7.11 Comparison of the predicted average diameter of damaged particles with the experiment data of Hadianfard et al. (2008). The experimental range for the diameter of damaged particles is presented as a shaded band. A total of six measurements were obtained for each loading condition and are expressed as a band for clarity

particles is 7.43 μm and is in general agreement with the recent experimental work of Hadianfard et al. (2008) for the same model material. The experimentally observed range for the diameter of the damaged particles in the quasi-static loading regime is presented as a shaded band in Fig. 7.11. This result is certainly encouraging as the proposed nucleation criterion in Eq. (7.8) gives good predictions for the size of the particle which nucleates a void in this alloy. It is important to note that only three distinct particle fields have been evaluated in the present work with two measurements obtained for each loading condition (rolling and transverse directions). Future work should include a larger number of particle fields to assess variability due to different particle distributions.

7.5 Calibration of a Continuum-Based Nucleation Rule Using the Percolation Model

The damage percolation model can provide great insight into the role of the microstructure on ductile fracture but its practical application is limited since finite-element simulations of a sheet metal forming process requires a continuum nucleation model. However, the percolation model can be used to achieve physically meaningful values for the nucleation strain by synchronizing void nucleation between the heterogeneous and continuum nucleation models for the model material, AA5182 sheet. As such, void nucleation within the local microstructure can be related to the overall material behavior using traditional micromechanical constitutive models.

7.5.1 Continuum Nucleation Model

The strain-controlled void nucleation model of Chu and Needleman (1980) in Eq. (1.5) is once again employed to predict nucleation. It is presented in this section for the sake of continuity and has the form:

$$\dot{f}_{\text{nucleation}} = \frac{f_{\text{n}}}{s_{\text{N}}\sqrt{2\pi}} \exp\left[-\frac{1}{2}\left(\frac{\bar{\varepsilon}^{\text{p}} - \varepsilon_{\text{N}}}{s_{\text{N}}}\right)^2\right]\dot{\bar{\varepsilon}}^{\text{p}} \tag{7.13}$$

where ε_{N} and s_{N} are the average and standard deviation of the nucleation strain; $\dot{\bar{\varepsilon}}^{\text{p}}$ is the equivalent plastic strain increment; f_{n} is the volume or area fraction of the particles that nucleate voids.

7.5.2 Synchronization of the Void Nucleation Criteria

The parameters required to be identified in the continuum nucleation model in Eq. (7.13) are the fraction of nucleating particles and the average nucleation strain and its standard deviation. The previous percolation simulations suggest that on average, 43 % of the area fraction of second phase particles nucleate voids regardless of the stress state so that $f_{\text{n}} = 0.43 f_{\text{p-avg}}$ or 0.572 % in the continuum model. Similarly, a good estimate for the standard deviation of the nucleation strain was found to be 15 % of the average nucleation strain.

The reduced form of the heterogeneous nucleation model in Eq. (7.12) is substituted into the continuum nucleation model in Eq. (7.13). The resulting nucleation criterion is

$$\begin{aligned} \dot{f}_{\text{nucleation}} &= \frac{0.015213}{\varepsilon_{\text{N}}} \exp\left[-\frac{1}{2}\left(\frac{\bar{\varepsilon}^{\text{p}} - \varepsilon_{\text{N}}}{0.15\varepsilon_{\text{N}}}\right)^2\right]\dot{\bar{\varepsilon}}^{\text{p}} \\ \varepsilon_{\text{N}} &= \bar{\varepsilon}_{No} \exp[-T(2 - 1.8|\mu_L|)] \end{aligned} \tag{7.14}$$

The only unknown parameter is the effective nucleation strain, $\bar{\varepsilon}_{\text{No}}$, is identified through calibration with the nucleation trends for the three particle fields in uniaxial tension, plane strain, and equal-biaxial tension. In this approach, all of the nucleation parameters in the continuum nucleation model have been identified based on

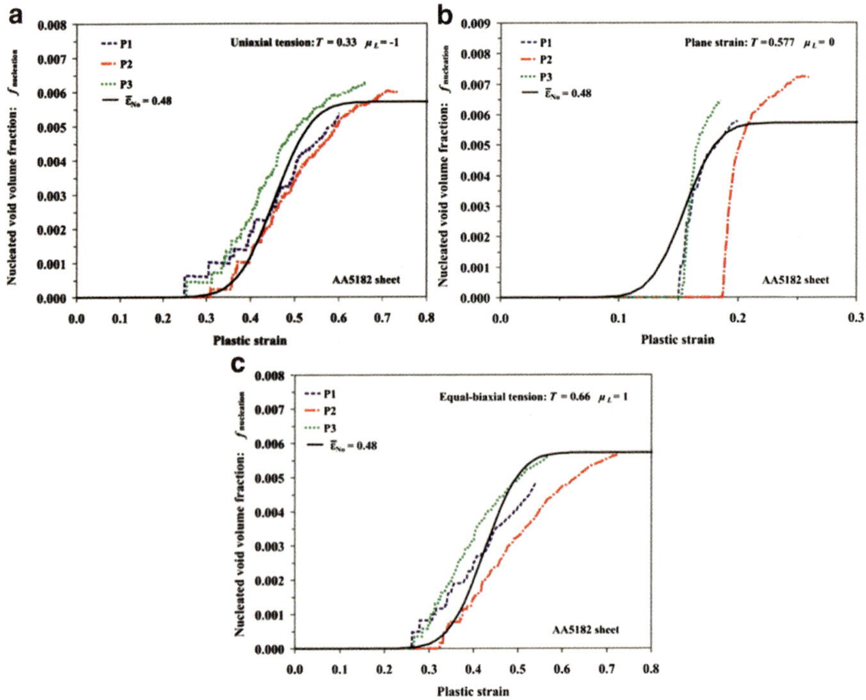

Fig. 7.12 Comparison of the predicted nucleated void volume fraction of the synchronized continuum model with nucleation in the three particle fields subjected to (**a**) uniaxial tension, (**b**) plane strain, and (**c**) equal-biaxial tension (Reprinted with permission from Butcher and Chen (2009b). Copyright: Elsevier)

the particle distribution of the material to ensure physically sound estimates of void nucleation for the stress states commonly encountered in sheet metal forming.

The synchronized continuum nucleation model with $\bar{\varepsilon}_{No} = 0.48$ provided the best agreement with the nucleation trends obtained in the particle fields as shown in Fig. 7.12. The synchronized continuum nucleation model is able to characterize nucleation over a range of stress states unlike the traditional Chu and Needleman (1980) model in Eq. (7.13) where the nucleation strain is independent of the stress state. The synchronized continuum model in Eq. (7.14) yields nucleation strains of 0.45, 0.15 and 0.42 for uniaxial tension, plane strain and equal-biaxial tension, respectively. The Chu and Needleman nucleation model could have been calibrated using the particle field data but nucleation strain would not be transferable to other stress states. For example, a nucleation strain of 0.435 in the Chu and Needleman model would give good agreement with the particle field data in both uniaxial and equal-biaxial tension but would poorly represent nucleation in plane strain where the calibrated nucleation strain should be about 0.15.

Overall, the synchronized continuum nucleation model can give very good estimates for void nucleation within the measured particle distributions, and it is transferable among the range of stress states encountered in sheet metal forming.

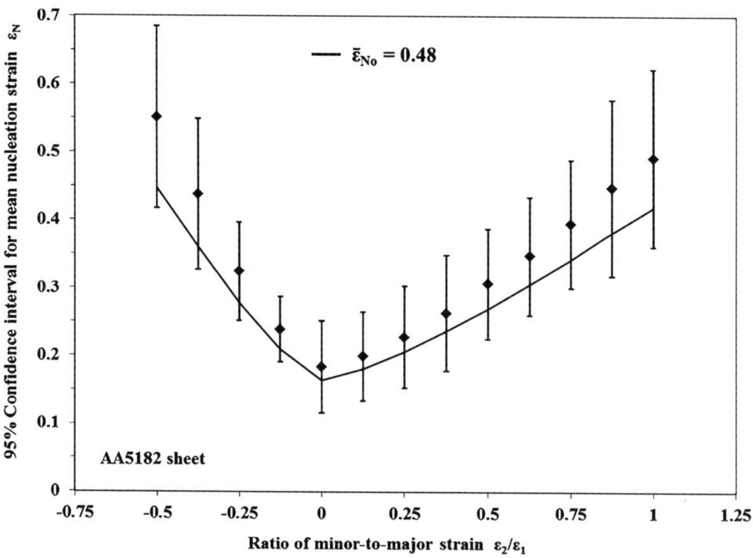

Fig. 7.13 Comparison of the mean nucleation strain obtained using the synchronized continuum model with the results of the heterogeneous model for the three particle fields for different loading conditions

The continuum nucleation model gives good agreement with the average nucleation strain obtained from the particle fields from uniaxial tension to equal-biaxial tension as shown in Fig. 7.13. The synchronized continuum model could be incorporated into finite-element simulations to provide realistic estimates of void nucleation in practical sheet metal forming operations of the model alloy.

7.6 Summary

A phenomenological void nucleation criterion was developed and implemented into a damage percolation model where nucleation occurs at the individual particle scale as a function of the particle morphology and stress state. The stress state is characterized by incorporating a dependence upon the stress triaxiality and severity of shear loading using the Lode parameter. The nucleation rule has been applied to the percolation modeling of AA5182 sheet to recreate an experimental forming limit data with nucleation values of $\varepsilon_{No} = 0.85$ and $\beta = 1.80$. The calibrated nucleation rule gives good agreement with the experimental size for void nucleating particles. Void nucleation within measured particle distributions was also used to quantify nucleation at the continuum scale by synchronizing a well-known nucleation model with the results from the particle field simulations for a variety of stress states.

Chapter 8
Three-Dimensional Particle Fields

There are two distinct challenges in the development of a multi-scale damage percolation model: (i) the development and application of the micromechanical models used to predict damage initiation and evolution and (ii) obtaining the experimental particle distributions for the model to use. While the development of the micromechanical models and implementation of the percolation model is a significant endeavour, obtaining the experimental particle distributions is equally as challenging, and arguably more tedious.

Obtaining the required experimental particle distributions remains a difficult process, especially for 3-D fields, and will often not be feasible due to cost considerations or access to the required equipment. For most materials, the basic statistics that describe the particle distribution can be reliably estimated from optical microscopy such as the average particle size, shape, spacing and volume fraction. In light of this assumption, it is of great interest to develop a particle field generator that can create representative particle fields using the experiment data (if available) or from assumed statistical distributions. The incorporation of a particle field generator into the percolation model enables fracture to be treated in a stochastic manner where the variation in the microstructure will lead to a range of predicted fracture strains.

The performance of a particle field generator rests upon knowledge of the probability density distributions for the governing variables. There are two commonly used methods to measure a particle distribution: x-ray microtomography (3-D) or from high resolution optical images (2-D). An image tessellation algorithm is then applied to extract the particle and void information such as the centroid, semi-axes, orientation and nearest neighbour information.

X-ray microtomography is the obvious choice for obtaining a three-dimensional particle distribution if the particles in the material of interest are on the order of 1 μm or larger. X-ray tomography studies of ductile fracture have long suffered from issues of resolution but recent advances have improved the resolution to about 0.7 μm at the state-of-the-art synchrotron in Grenoble, France (Maire et al. 2006; Orlov 2006). However, access to synchrotrons capable of achieving this degree of resolution is limited. As a result, obtaining 2-D images from metallographic

Z. Chen and C. Butcher, *Micromechanics Modelling of Ductile Fracture*, Solid Mechanics and Its Applications 195, DOI 10.1007/978-94-007-6098-1_8,
© Springer Science+Business Media Dordrecht 2013

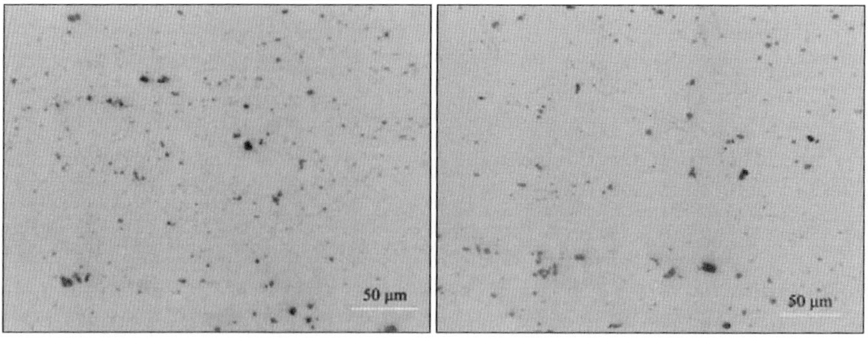

Fig. 8.1 In-plane images of the particle distribution for AA5182 sheet for a thickness of 1.0 mm (*left*) and 1.6 mm (*right*). The rolling direction is *horizontal* and the transverse direction is *vertical*

samples is the standard method of measuring a particle distribution as it is inexpensive, accessible, and has very good resolution of 0.139 µm per pixel (Chen 2004). A three-dimensional distribution can be obtained by successive sectioning of the sample and assembling the 2-D images but this is an extremely time-consuming process and may not be appropriate for materials with fine particles (Mangan et al. 1999). A typical 2-D image obtained using the sectioning and polishing method is presented in Fig. 8.1. Similarly, 3-D microtomography images are shown in Fig. 8.2 to describe the break-up of particles in AA5182 during the rolling process and in a subsequent tensile test.

This section will discuss the basic procedure for generating the particle field within the percolation model followed by sections discussing the various techniques and modeling treatments of the algorithm. Finally, the particle field generator will be used to generate representative particle fields of an aluminum-magnesium alloy and validated using the experimental results in the literature.

8.1 Particle Field Generator

The procedure for generating a particle field begins with the finite-element model. First, the algorithm identifies the elements that have been set as percolation-type elements from the input file generated using the finite-element software, LS-DYNA. The node information of these elements is used to identify the corners of the volume to determine a single element block that is termed the 'global percolation volume'. The volume is then populated with objects based upon their packing densities and arranged into clusters. The element volume is then decomposed into its constituent elements and the individual objects are assigned to their parent element as shown in Fig. 8.3. These steps are performed automatically at the start of a finite-element simulation as part of the pre-processing routine of the percolation model. The subsequent sections will discuss the decisions and modeling treatments required in each stage of the particle generation process, starting with the sampling techniques used to define the objects.

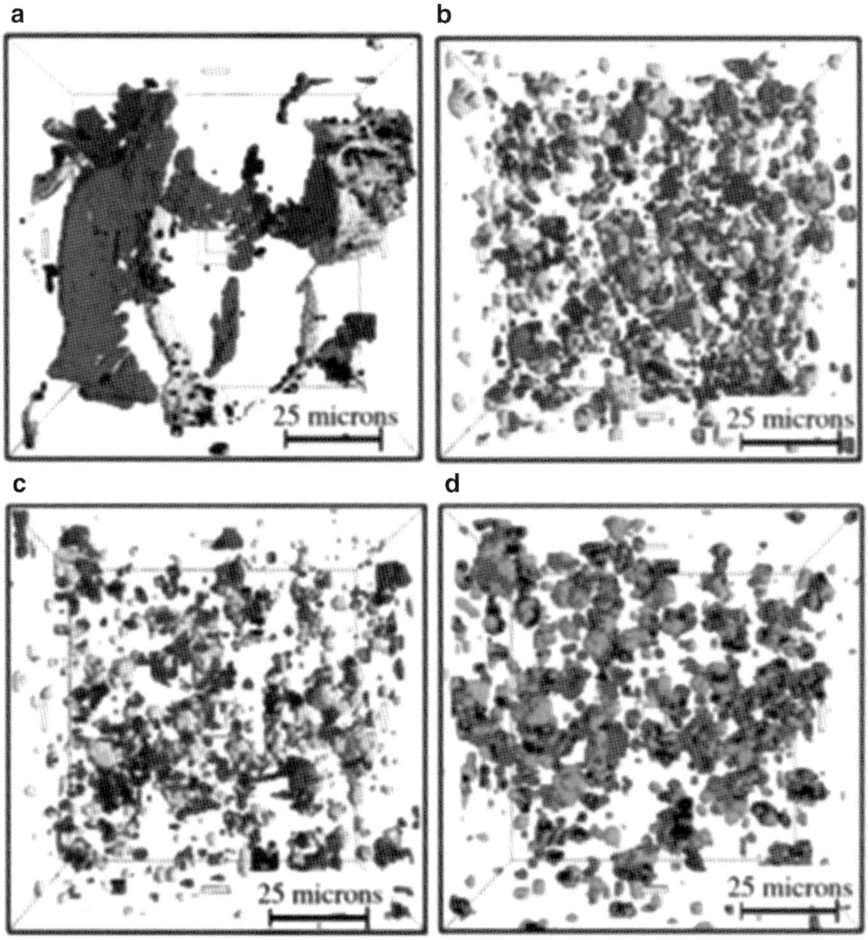

Fig. 8.2 3D view of intermetallic phases and voids in 1 mm thick AA5182 sheet obtained by Maire et al. (2006): (**a**) as cast state; (**b**) the hot rolled state; (**c**) cold rolled state; and (**d**) microstructure at the end of a tensile test. *Light grey spots* show the iron-rich particles, *dark grey spots* are Mg2Si-particles and *black spots* are pores (Reprinted with permission from Maire et al. 2006. Copyright 2006 Elsevier)

8.1.1 Particle Field Basics

Only the basics of the particle field generator developed by Butcher (2011) will be described for brevity and the interested reader is also referred to the thesis of Orlov (2006). The particle field may contain any number of particle types in addition to the voids. No restrictions are made on the properties of the particles and voids except that they must be ellipsoidal. Since the voids and particles are modeled as ellipsoids and generated using the same procedure, they will simply be referred to

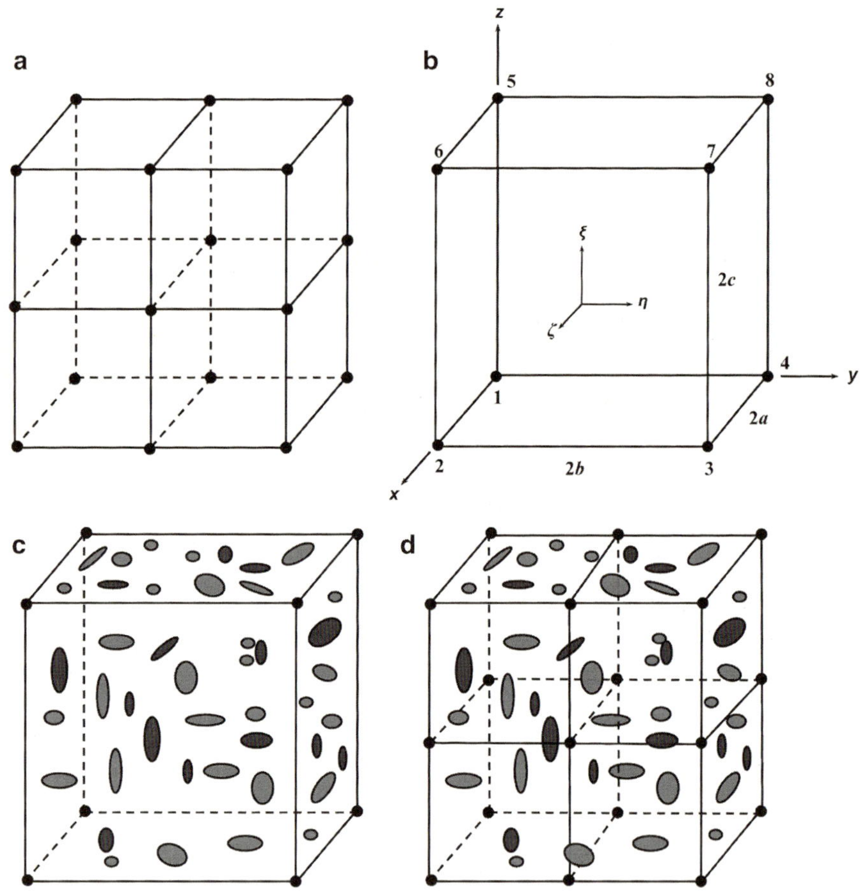

Fig. 8.3 Procedure for generating a particle field within a block of percolation elements (**a**) percolation elements identified from the global finite element mesh, (**b**) percolation elements combined to create a global element to define volume for particle field generation, (**c**) particles and voids are generated within the volume and (**d**) particles and void are assigned to parent elements

as *objects* in the proceeding sections. The procedures in this chapter are presented only for a 3-D brick element to provide a general 3-D solution but readily reduce to 2-D elements by constraining particle generation to one plane.

8.1.2 Generation of Random Variables

The particle field generator algorithm recreates the particle and void distributions through knowledge of the probability density functions (PDF) for the relevant variables such as the object semi-axes (radii), orientation angles and inter-object

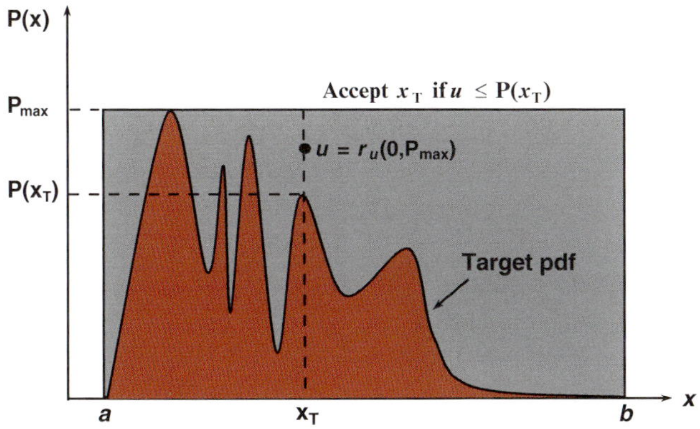

Fig. 8.4 Schematic of the rejection-sampling algorithm used to determine a random variable, x, from a target probability density function (PDF)

spacings. If the experimental distribution is known, the random variable can be generated from this distribution using a rejection-sampling technique. This method provides greater confidence in the resulting particle field since the variables of interest will have the same statistical distributions as experimentally observed. The generation of random variables from an assumed distribution such as the normal distribution is a trivial exercise and only the rejection-sampling method will be described here.

8.1.2.1 Rejection-Sampling Algorithm

A simple rejection-sampling algorithm has been adopted to obtain a random variable, x, from its experimental PDF. First, the curve data that defines the density function is pre-processed to determine maximum probability in the curve, P_{max}, as well as the left and right bounds for the variable, defined as a and b. The operation of the rejection-sampling algorithm is as follows:

- Randomly generate a trial value for the variable, x_T, that lies within $[a, b]$.
- Evaluate the probability at the trial value, $P(x_T)$, by interpolating the density curve data.
- Generate a random number, u, that lies within the range $[0, P_{max}]$.
- Accept x_T if $u \geq P(x_T)$. Otherwise, reject x_T and repeat the process with a new trial x_T.

A graphical representation of the rejection-sampling algorithm is presented in Fig. 8.4. Naturally, the resolution of the generated distribution improves with the number of samples taken. It is difficult to specify the minimum number of particles and voids to be generated in a representative volume element (RVE) because the rejection-sampling algorithm is a function of the experimental PDF. If the particle distribution exhibits multiple modes, the size of the RVE must be made larger to increase the number of particles/voids in the element. An adaptive sampling technique may also be required for modelling complex distributions since the standard algorithm becomes computationally expensive. Conversely, well-behaved distributions can be modelled with a smaller number of samples. It is recommended that a parametric study be performed for each material of interest to establish a minimum RVE size.

8.1.3 Object Generation

8.1.3.1 Number of Objects

The number of each type of object to be generated within the volume is determined from the measured object densities, ρ_i, that are a measure of the number of objects per unit volume. This metric is chosen because it is easily measured experimentally compared to using volume fractions such as the porosity for the voids. This metric will lead to slight variations in the global volume fractions of the objects since the sizes of the objects will vary according to their distributions. Once the volume of the element block, V_G, has been computed, the number of objects of the i-th object type, N_i, to be created is

$$N_i = \rho_i(\#objects/vol)V_G \tag{8.1}$$

The use of object densities also simplifies the implementation of the algorithm because the generation process can be terminated upon reaching the desired number of objects. Alternatively, if a target volume fraction of an object is used to control the generation process, an iterative procedure would be required to continually remove and create objects to meet this target.

Generation of Object Semi-axes

All objects are general three-dimensional ellipsoids with semi-axes R_1, R_2, R_3. Since the density functions for these radii must be obtained experimentally, they correspond to a specific direction. For sheet metals, these are the rolling, transverse and short transverse or through-thickness directions. These axes could also correspond to the directions of anisotropy in a general material. The directions for the rolling, transverse and thickness directions are denoted as $\boldsymbol{r}$, $\boldsymbol{t}_r$, and $\boldsymbol{t}_t$, respectively. After the radii have been generated, they are sorted and reclassified into the traditional form for ellipsoids with a, b, c, as the semi-axes, where $a > b > c$.

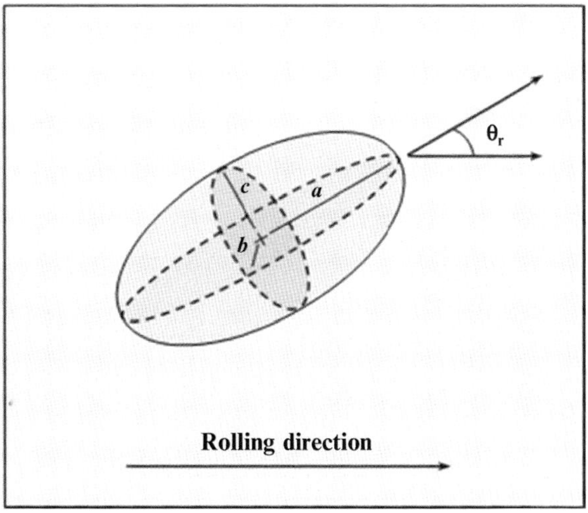

Fig. 8.5 Orientation of an object relative to the rolling direction by a solid angle, θr

Generation of Object Orientation

Unlike previous percolation models (Worswick et al. 2001; Chen 2004; Orlov 2006), no restrictions are placed upon the orientation of the objects within the field. In sheet materials, the particles and voids tend to naturally align themselves with the rolling direction. To account for the orientation of the objects, a preferential direction can be specified along with the density function for the solid angle, θ_r, from which an objects semi-major axis deviates. An example of this effect is shown in Fig. 8.5.

8.1.4 Generation of Objects Within Clusters

8.1.4.1 Object Generation Sequence

The sequence for generating the objects of a specific type can be random or follow a specific order, such as first generating all of the particles of type 1 followed by all of the particles of type 2 and then the voids. The object generation sequence plays a role in determining the degree of clustering in the material by controlling the population of available neighbours. If the object type is randomly selected, the degree of clustering of the objects will be higher as the pool of potential neighbours is kept to the current number of objects. Early in the generation process, the number of available neighbours is small and the same neighbours could be selected repeatedly, leading to excessive clustering. Conversely, if an entire population of particles is generated before a population that requires neighbours such as voids, the voids will not be as near each other as they have selected their neighbouring particle from a larger pool of potential neighbours.

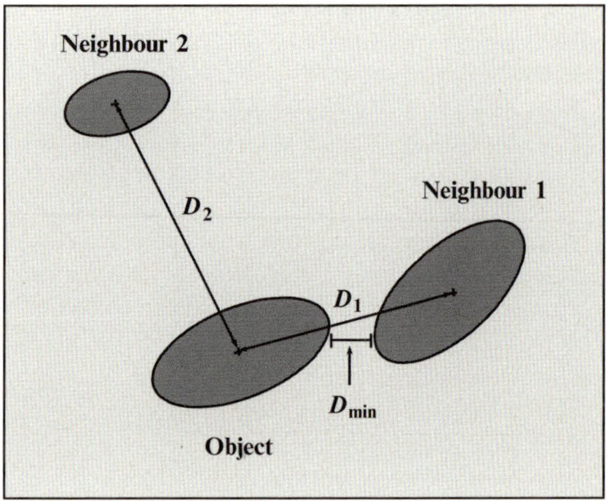

Fig. 8.6 Schematic showing the possible measures used to calculate the inter-object spacing

8.1.4.2 Distance Between Objects

Accurate knowledge of the density distributions for the distance between object types is critical to recreate the spatial distribution of the objects within the material. Experimentally, the distance between objects is commonly measured using the inter-particle dilatational spacing (IPDS) (Worswick et al. 2001; Chen 2004; Orlov 2006). This value is determined by calculating the distances between a particle or void and all of its surrounding neighbours. The minimum distance is called the IPDS and the process is repeated for all of the particles within the microstructure to develop the density function. Naturally, an IPDS distribution exists for each object and neighbour combination in the material such as IPDS distribution for voids and particles of type 1, or the distribution between particles of type 1 and type 2 and so forth. In the present work, the IPDS will be simply referred to as the inter-object spacing, denoted D_s, to provide a general term that can refer to voids or particles.

There is some uncertainty surrounding the calculation of the inter-object spacing. Typically, the spacing is taken as the minimum of all of the center-to-center distances between an object and all of its neighbours. Using this definition, $D_s = \min(D_1,D_2)$ in Fig. 8.6. However, by inspection of Fig. 8.6, it can be seen that the minimum spacing, D_{min}, between two objects can be very different compared to the center-to-center distance. The center-to-center spacing becomes a better estimate of the minimum spacing when the objects are far apart and have similar orientations.

8.1.4.3 Clusters

The voids and second-phase particles/inclusions in sheet materials are preferentially aligned in the rolling direction. Consequently, the rolling process creates aligned clusters or 'stringers' where neighbouring voids and particles are linearly aligned in

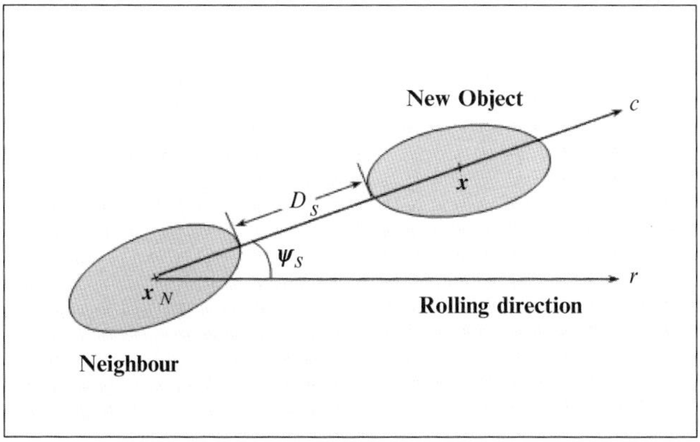

Fig. 8.7 Schematic showing the creation of a stringer-type cluster

the rolling direction. In this work, we will consider two types of clusters: general and stringer. A general cluster is formed by simply placing one object (void or particle) next to an existing object at a distance specified from the density function for the inter-object distances between the two object types. Stringer-type clusters are similar to the general type with a constraint on the location that an object can be placed next to a neighbour as shown in Fig. 8.7. An object will be placed within a general cluster when a random number from 0 to 1 is greater than the probability of clustering, P_c, such that

$$r_c(0,1) \geq \mathrm{P}_c \tag{8.2}$$

If the object has been selected to be within a cluster, then the cluster will be of the stringer-type if

$$r_s(0,1) \geq \mathrm{P}_s \tag{8.3}$$

where P_s, is the probability the object will be in a stringer-type cluster.

Object is not in a cluster: $r_c(0,1) < \mathrm{P}_c$

If the trial object does not lie within a cluster, the centroid of the object, $\boldsymbol{x}$, is randomly selected within the percolation volume.

Stringer-type cluster: $r_c(0,1) \geq \mathrm{P}_c$ and $r_s(0,1) \geq \mathrm{P}_s$

The following process is used to create a stringer-type cluster:

- An existing object is randomly selected to be the neighbour of the trial object being generated and is located at $\boldsymbol{x}_N$.
- The center-to-center vector between the two objects is denoted as $\boldsymbol{c} = (l_c, m_c, n_c)$ and the rolling direction vector is defined as $\boldsymbol{r} = (l_r, m_r, n_r)$.
- The probability density function for the stringer solid angle, ψ_s, between $\boldsymbol{c}$ and $\boldsymbol{r}$ is known for the two object types.

- The probability density function is known for the inter-object distance, D_s, between the two object types in the stringer.
- The center-to-center vector is determined that satisfies the stringer angle requirement.
- The minimum distance is computed between the ellipsoid and its neighbour. An iterative process is performed until the minimum distance along this vector is D_s. This process ensures that the two objects do not overlap and identifies the centroid, $\boldsymbol{x}$, of the object.
- If no suitable neighbour is found after a set number of attempts, a new trial object is created with a different orientation and dimensions. The neighbour search is then resumed.

General cluster: $r_c(0,1) \geq P_c$ and $r_s(0,1) < P_s$

If the new object is located within a general cluster, the process is the same as for the stringer cluster. The only difference between the models is in computing the center-to-center vector between the two objects. In this case, the position of the trial object, $\boldsymbol{x}$, relative to its neighbour is

$$\boldsymbol{x} = \boldsymbol{x}_N + \frac{\boldsymbol{c}(r_1(-1,1), r_2(-1,1), r_3(-1,1))}{|\boldsymbol{c}|} D_s \tag{8.4}$$

Where $\boldsymbol{c}$ is composed of three random numbers generated from a uniform distribution.

8.1.5 Particle Properties

The properties of the particles can also be assigned if their probability density functions are known. This is an important feature of the generator as the composition of the particles certainly varies in a real material and will have a significant effect on void nucleation and ultimately fracture. The distributions for the elastic modulus, E, Poisson's ratio, v, and the yield strength of the particles, σ_y, can be varied in the current version of the generator.

8.1.6 Object Constraints

An object will be rejected from the particle field if any of the following constraints are violated:

8.1.6.1 The Centroid of the Object Lies Inside the Percolation Volume

The centroid of an object, $\boldsymbol{x}$, lies within the global element block when the following constraint is true for each face:

$$\boldsymbol{n}_i \cdot \boldsymbol{x} - d_i < 0 \tag{8.5}$$

where $\boldsymbol{n}_{\mathrm{i}}$, are the outward facing normal vectors for the planes that define the element faces and d_{i} is a constant of the plane.

8.1.6.2 The Object Does Not Touch or Overlap with Any Other Object

No object is allowed to touch or overlap with another object. This constraint is enforced by evaluating the minimum distance between an object and every other object in the particle field. The objects are touching if the minimum distance is 0 or the minimum distance algorithm does not converge.

8.1.6.3 Minimum Distance Between Objects

The minimum distance between two arbitrary ellipsoids is a non-linear optimization problem that requires an iterative solution. In this work, we adopt the geometric-based algorithm of Lin and Han (2002) as it is simple, efficient and has very good convergence properties.

8.1.6.4 The Object Does Not Touch or Intersect with Any of the Facing Planes That Define the Percolation Volume

The minimum distance between an object and a plane must be determined to test if an object has intersected the boundary of the element block. This problem could be simplified to the case of testing for an intersection between a plane and ellipsoid but the minimum distance is preferred as it provides a more general solution. If the generated particle field were to be meshed in a finite-element model for homogenization studies, a minimum distance between the ellipsoids and the cell walls would have to be enforced (Pierard et al. 2007). The minimum distance between a plane and an ellipsoid can be solved by adapting the method of Lin and Han (2002). In this case, the equation of the second ellipsoid is replaced with the equation of the plane and the algorithm converges rapidly.

8.1.6.5 Neighbour Constraints

The type of permissible neighbour can be specified for each object. This information is used in modeling the clusters/nearest neighbours. The type of neighbour that an object may have is important for determining the spacing between the objects for accurate representation of object clusters. The algorithm allows for a probability distribution for the spacing between each type of neighbour in the material. This is important as the distribution for the spacing between voids and different particle types can be used in the model if they are known.

8.1.7 Identification of the Parent Element

Each object can belong to only a single parent element. An object may be intersected by the boundary between two percolation elements, but it belongs to the element in which its centroid resides. An object with its centroid at $\boldsymbol{x} = (x, y, z)$ within the global volume is located within its parent element when the following three constraints are satisfied:

$$-1 \leq \frac{x - x_c}{2a} \leq 1 \qquad -1 \leq \frac{y - y_c}{2b} \leq 1 \qquad -1 \leq \frac{z - z_c}{2c} \leq 1 \tag{8.6}$$

where x_c, y_c, z_c denote the center of the parent element that has side lengths of $2a$, $2b$, $2c$. These constraints are evaluated for each element within the element block until Eqs. (8.6a, b, c) are satisfied for the current object.

8.1.8 Numerical Implementation

The particle field generator algorithm was written in C++ and directly ported into the percolation code. A Matlab code was developed for visualization of the particle fields. The particle generator was ported into the finite-element percolation code so that a new particle field could be automatically generated at run-time. This feature provides the foundation for developing a heuristic version of the model where a large number of simulations can be performed to capture the statistical variation in the material behaviour. The algorithm for the particle field generation process is shown in Fig. 8.8.

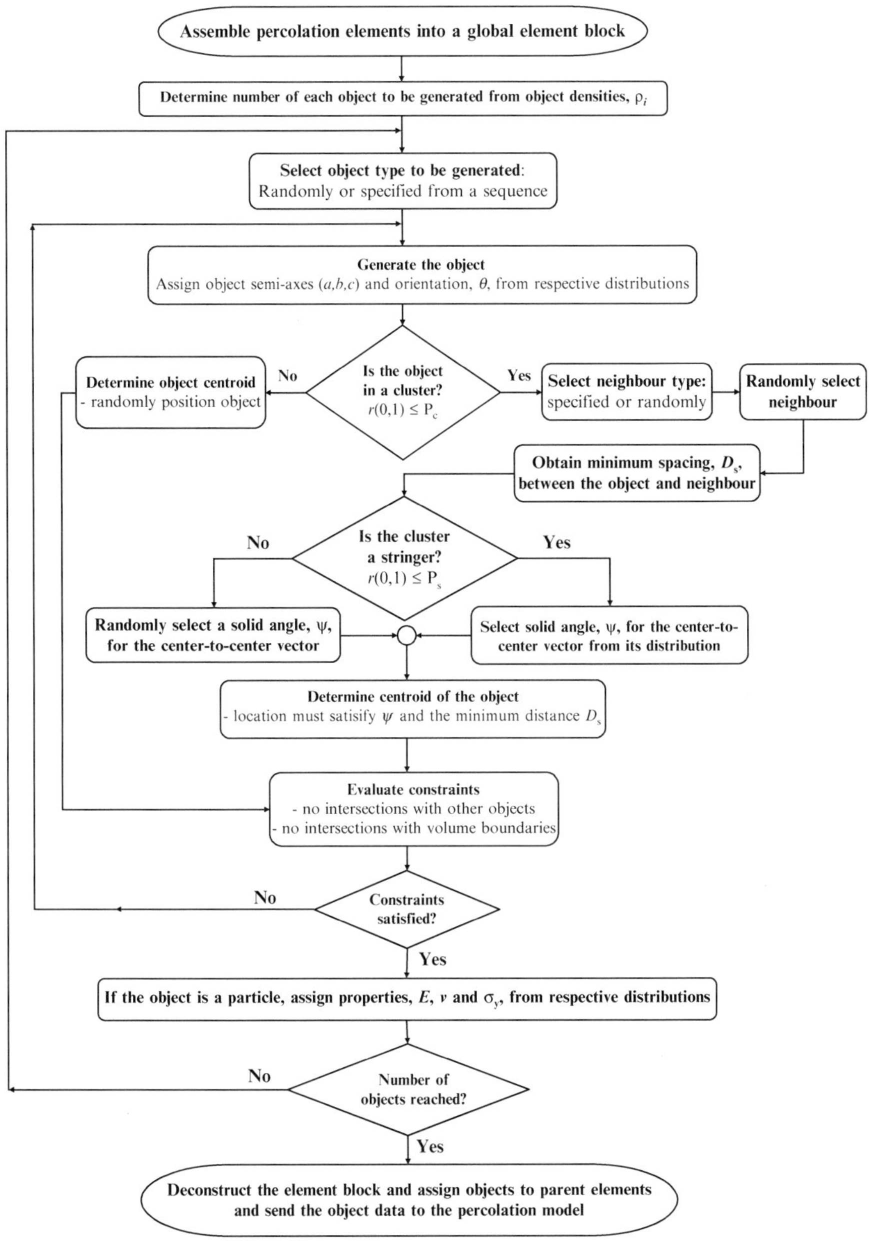

Fig. 8.8 Algorithm for the particle field generator

8.2 Application of the Particle Field Generator to an Al-Mg Alloy

Fortunately, a high-resolution x-ray microtomography study of the material of interest, AA5182 alloy, was performed by Orlov (2006). These results will be employed in the present work to generate the 3-D particle distributions for the percolation model. The resolution of the pixels within the voxel images was 0.7 μm which can capture most particles in the AA5182 sheet. Particles that are too small to be detected at this resolution are not expected to have a significant influence on the fracture strain predictions of the model because it is the large particles that will primarily nucleate voids. An experimental study of this alloy by Hadianfard et al. (2008) found that the average diameter of a broken particle (void nucleation) ranged from about 4.5 to 7.25 μm. From this study, we can conclude that the resolution used by Orlov (2006) of 0.7 μm was sufficient to capture the particles that contribute to the fracture mechanism.

8.2.1 Material Characterization

As-cast 5xxx series aluminum-magnesium alloys contain coarse intermetallic particles that may be larger than 100 μm in size. These particles then break up during the rolling process into particle fragments of 1–10 μm in size (Maire et al. 2006) as illustrated in Fig. 8.2. There are two main types of intermetallic particles in AA5182 sheet: iron-rich and Mg_2Si particles. The chemical composition of the alloy is presented in Table 8.1. The particles are oriented and elongated along the rolling direction and exhibit a strong degree of clustering. The initial voids in the material are located within particle clusters and transverse to the rolling direction, suggesting they were formed by particle cracking during the rolling process.

The tessellated particle fields were extracted from the voxel images by Orlov (2006) using a matrix erosion technique that approximated the voids and particles as ellipsoids that are aligned along the rolling, transverse or through-thickness directions of the material. The majority of the second-phase objects in AA5182 are Fe-rich (83 %), with about 10 % Mg_2Si particles located near Fe-rich particles with the remaining 7 % being voids. The object densities for the 1 mm cold rolled AA5182 sheet of Orlov (2006) are presented in Table 8.2 with the volume fractions in Table 8.3. The proximity of the voids to the particles was measured by the number of particle-void interfaces (PVI).

Overall, the percentages of the voids that shared an interface with the Fe-rich and Mg_2Si particles were reported as 46 and 54 %, respectively. This result indicates that the probability of a void nucleating during the cold rolling process is approximately equal for both particle types. This is a fortuitous result as no special measures must be taken in the particle generation process to position voids adjacent to a preferential particle type.

Table 8.1 Chemical composition of the AA5182-O alloy

Element	Si	Fe	Mg	Mn	Cu
Wt %	0.08	0.21	4.6	0.33	0.04

Table 8.2 Object densities (number per mm^3) in AA5182 sheet (Orlov 2006)

Fe-rich	Mg_2Si	Voids	Fe-rich PVI	Mg_2Si-PVI
659,127	77,045	55,889	28,103	33,439

Table 8.3 Volume fractions of the constituents in AA5182 sheet (Orlov 2006)

Fe-rich	Mg_2Si	Voids	Matrix
0.00483	0.000485	0.000529	0.994156

The sample volume was 70 × 70 × 70 μm

8.2.2 *Parameters Used in the Particle Field Generation Process for AA5182*

To generate particle fields that are in accordance with the experimental results of Orlov (2006), certain considerations must be made so that the predicted distributions are consistent with the measurement techniques used. For example, the objects must be aligned in the material directions since orientation was not considered in the image tessellation process. The following criteria and parameters were used in generating the particle fields.

- Volume of the particle field is 200 × 200 × 200 μm.
- All objects are aligned with the rolling, transverse and thickness directions. No orientation angle.
- Rejection-sampling will be used to generate the object semi-axes.
- All clusters are of the stringer type with a randomly selected solid angle of −20° to 20° relative to the rolling direction.
- Fe-rich particles are generated first followed by the Mg_2Si particles and then the voids.
- The Fe-rich particles are randomly distributed within the volume.
- The probability of an Mg_2Si particle being in a cluster is 80 %. The spacing between an Mg_2Si particle and its neighbouring particle obeys a normal distribution with a mean of 0.7 μm and a standard deviation of 3 μm. These values were suggested by Orlov (2006).
- All voids are assumed to be within a cluster due to the assumption that the voids were created by the particle cracking during the rolling process.
- The center-to-center spacing between a void and its neighbouring particle is selected from a uniform distribution from 1 μm to 4.5 μm. These values were determined parameterically to give agreement with the experimental spacing distribution.

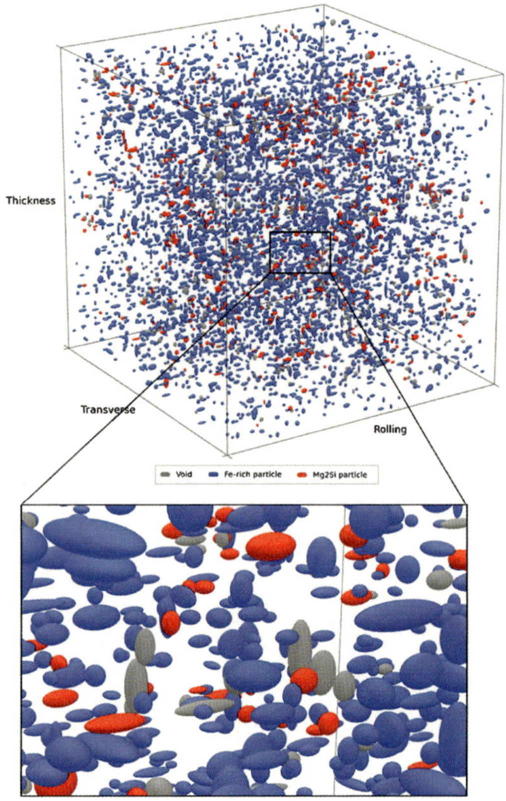

Fig. 8.9 Generated particle field of AA5182 with a volume of 200 μm × 200 μm × 200 μm

8.2.3 *Particle Field Generation Results*

A typical particle field with a volume of 200 × 200 × 200 μm is shown in Fig. 8.9 where the clustering of the voids and particles is evident along with their preferential orientation along the rolling direction. The number of voids, Mg_2Si and Fe-rich particles in this volume are 447, 616, 5,273, respectively. The number of objects to be created is important for the rejection-sampling algorithm to recreate the distributions for the object dimensions/properties. The resulting distribution will converge to the target distribution provided a sufficient number of random samples are taken.

8.2.4 *Object Dimensions*

The generated distributions for the semi-axes of the Fe-rich particles, Mg_2Si particles and voids are presented in Figs. 8.10, 8.11 and 8.12. The frequency distributions of the Fe-rich partcle dimensions are in excellent agreement with its

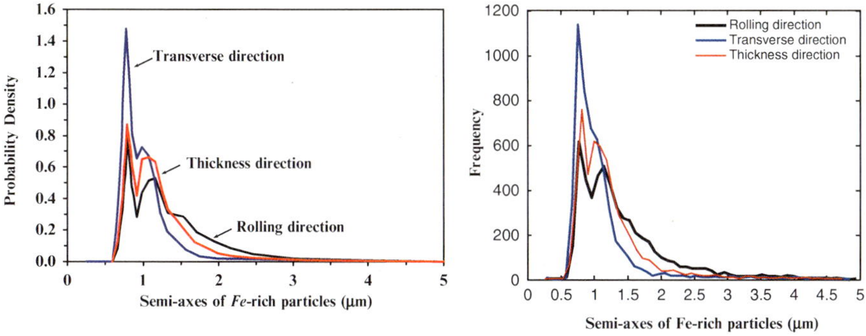

Fig. 8.10 Experimental PDF adapted from Orlov (2006) (*left*) and the generated frequency distributions for the semi-axes of the Fe-rich particles (*right*)

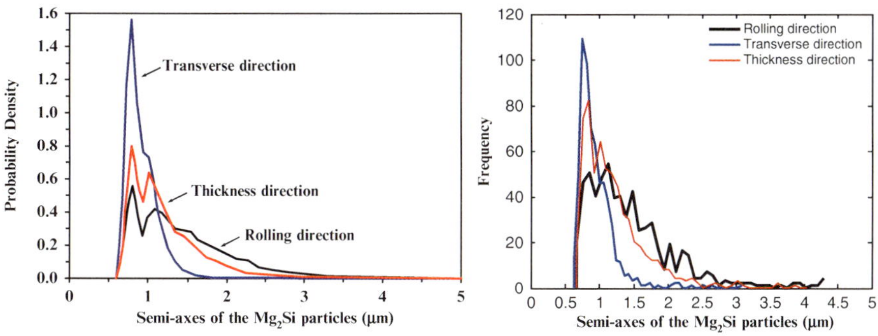

Fig. 8.11 Experimental PDF adapted from Orlov (2006) (*left*) and the generated frequency distributions for the semi-axes of the Mg2Si particles

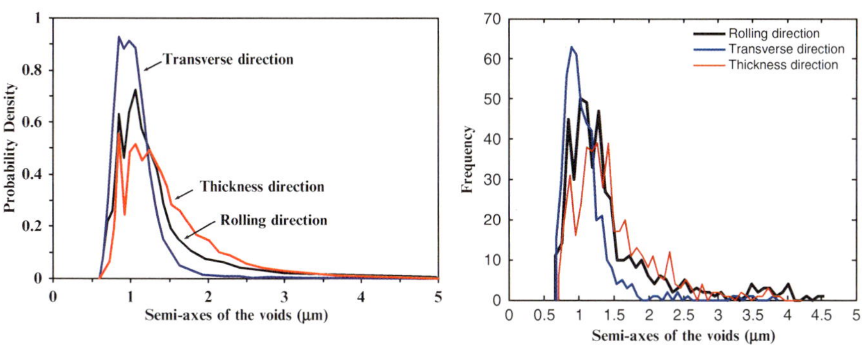

Fig. 8.12 Experimental PDF adapted from Orlov (2006) (*left*) and the generated frequency distributions for the semi-axes of the voids

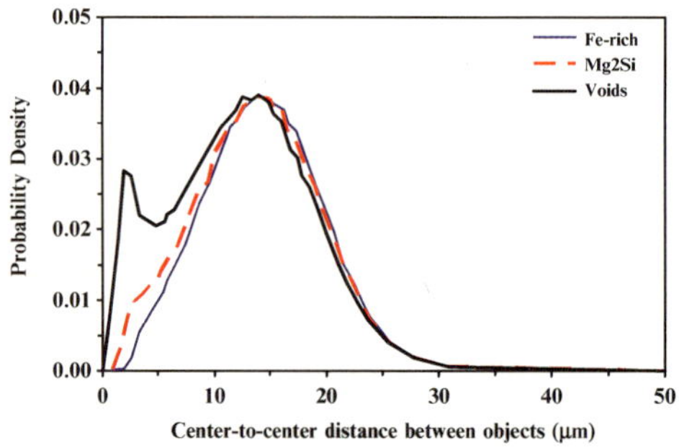

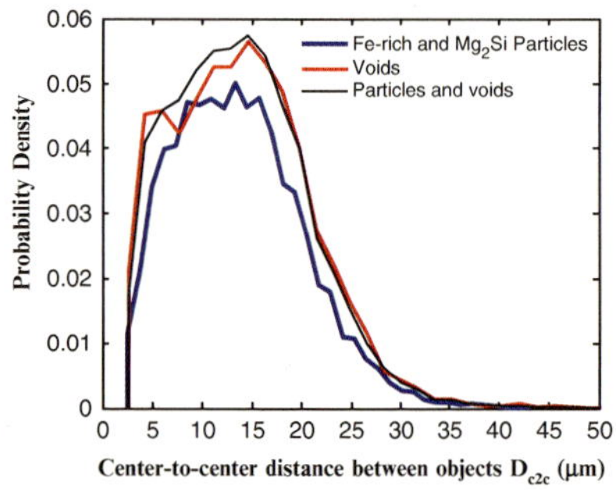

Fig. 8.13 Experimental PDF adapted from Orlov (2006) (*left*) and the generated distribution for the minimum center-to-center distance between objects

PDF distributions shown in Fig. 8.10 since 5,273 samples were taken from the target distribution in the sampling algorithm. The distributions for the Mg_2Si particles in Fig. 8.11 exhibit more scatter since only 616 particles were generated. However, the general trends are in good agreement with the experimental distribution in Fig. 8.11a and is an acceptable representation. Finally, for the voids in Fig. 6.12, the scatter is again larger compared to the Mg_2Si particles since only 447 voids are created. Nevertheless, the general trends for the void distribution can be seen when compared to the target distribution in Fig. 8.12a, especially in the transverse direction. Overall, the rejection-sampling technique employed in the present work can successfully recreate the target distribution provided a sufficient number of samples are taken. For the model alloy, decent results can be obtained with 400–600 objects and excellent agreement at 5,000 samples.

8.2.5 Spatial Distribution of the Objects

The ability of the particle field generator to recreate the spatial distribution of the objects is a true test of the performance of the algorithm since it is not a straightforward process as generating the object dimensions. The spatial distribution of the object is not explicitly controlled and is a function of the clustering parameters (probability of clustering, cluster type, allowable neighbours in the cluster), the inter-object spacing distribution functions and even the sequence that the different object populations are generated. To obtain object spacing values in accordance with the experimental results of Orlov (2006), the dimensions of the particle field must match the experimental size of $200 \times 200 \times 200$ μm.

The probability density distributions for the spatial distributions are shown in Fig. 8.13 and are in good agreement with the experimental distributions in Fig. 5.2. A parametric study was performed to determine the inter-object spacings for the voids and their neighbouring particles since the voids exhibit a bi-modal distribution with one strong peak at 2–4 μm and another at about 15 μm, which is the average value for all object types. A uniform distribution with a range of 1–4.5 μm enabled the distribution of the voids to exhibit the characteristic bi-modal distribution. The Fe-rich particles were generated randomly within the material and requires no assumed distribution for the particle spacing. A preliminary study for the distribution of the spacings of the Fe-particles was performed but was insensitive to the assumed values due to the large number of particles, 5,273, in the volume so that the resulting distribution for the spacings was random. For the Mg_2Si particles, a probability of clustering of 80 % with the spacing distribution following a normal distribution with a mean and standard deviation of 0.7 and 3 μm, respectively. These values were suggested by Orlov (2006). Finally, the stringer angle was varied as well but exhibited no significant influence on the resulting object-spacings. So it was assumed to vary from -20° to 20° relative to the rolling direction as this is a typical range observed in sheet metals.

Overall, the spatial distributions of the particles and voids in an AA5182 alloy can be recreated using the proposed algorithm, albiet to the accuracy of experiment data considered. It is important to mention that the clustering criterion used by Orlov (2006) is only a first-order approximation to the actual clustering in the material and could be improved in future work. The use of the minimum center-to-center object spacing provides only a rough estimate of the true clustering behaviour as it neglects the dimensions of the objects and therefore, their true spacings. The absolute minimum distance between the objects should be used in conjunction with the center-to-center distance to quantify the object spacing using a dimensionless spacing ratio. Additionally, the average spacings surrounding the objects should also be reported to obtain a radial spacing distribution. Therefore, although the particle field generator in the present work provides good agreement with the experimental distributions, the spatial reconstructions are approximate and may fail to account for certain effects.

8.2.6 Particle and Void Volume Fractions

The particle and void volume fractions are primary parameters of interest in damage-based constitutive models such as the Gurson (1977) model. While often assumed to be a material constant, it is not uncommon to see significant variation in the experimental volume fraction measurements. It is therefore of interest to generate particle fields of different sizes to observe the predicted variation of the particle and volume fractions. A total of 20 different AA5182 particle fields were generated within cubic elements with side lengths of 70, 100, 150 and 200 μm. The resulting confidence intervals for the voids and Mg_2Si particles are presented in Fig. 8.14a and the values for the Fe-rich particles in Fig. 8.14b. The general trend is that the variation in the volume fractions of the constituent objects decreases with increasing element size. As the element size increases, a larger number of objects are generated and the object dimensions consolidate about a mean value, giving rise

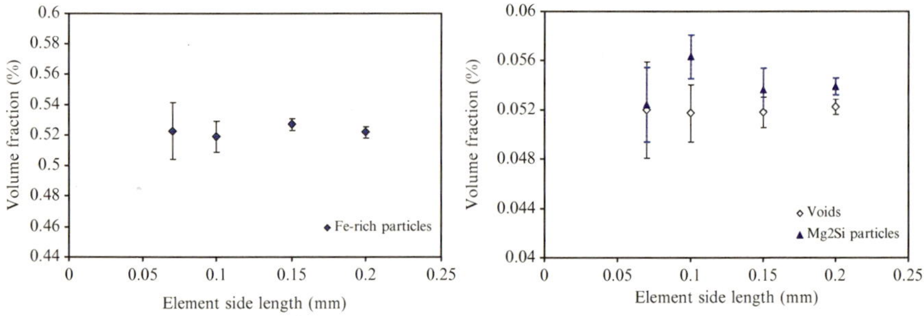

Fig. 8.14 Ninety-five percent Confidence intervals for the predicted volume fractions of the Fe-rich particles (*left*) and the voids and Mg2Si particles (*right*) in AA5182 sheet. 20 random particle fields were generated for each element volume

to a consistent prediction of the volume fraction. However, despite the decreased variation, there is no statistically significant difference because the confidence intervals overlap and we can conclude that the size of the RVE is not a significant factor when predicting the object volume fractions.

This is an interesting result since the size distributions of the objects is determined using rejection-sampling and therefore related to the element volume through the object density. This result is encouraging since the volume fraction of the voids in the element of 70 μm^3 (19 voids) is virtually the same as in the element of 200 μm^3 (447 voids). Additionally, the volume fractions for the voids, Fe-rich and Mg_2Si particles reported by Orlov (2006) for a volume of 70 μm^3 were 0.0529, 0.483 and 0.0485 %. The experimental values for the voids and Mg_2Si particles fall within the predicted range for that element size while the predictions for the Fe-rich particles are above the experimental value at about 0.525 %, which is still reasonable. The generally good agreement with the experimental volume fractions is encouraging for the performance of the particle field generator. It is also important to note that the experimental object densities were taken as constant values and future work should determine a tolerance for these values to better capture the variation in the volume fractions.

8.3 Summary

A particle field generation algorithm was developed and implemented into the percolation model to populate the elements with second-phase particles and initial voids. The particle field generator was designed for a general material and uses the probability density functions (assumed or experimental) to generate a representative particle field within the element. The particle field generator can accept any number of different particle types and accounts for the shape, orientation angle, cluster type, nearest neighbour information and particle properties.

The proposed algorithm has been implemented into the percolation model as a pre-processor to automatically create the particle distributions in the percolation elements at the start of a simulation. By generating different representative particle fields of an alloy, a series of finite-element simulations can be performed to capture the variation in the fracture strains and predicted material properties of a material of interest. The particle field generator algorithm was validated for an aluminum-magnesium alloy using the microtomography results of Orlov (2006). The void and particle size distributions, global volume fractions and inter-particle and void spacings were successfully generated by the model and in good agreement with the experimental observations.

Chapter 9
Estimation of the Stress State Within Particles and Inclusions and a Nucleation Model for Particle Cracking

Despite great strides in developing physically motivated models for void growth, shape evolution and coalescence, a suitable treatment for void nucleation remains an open question. Accurate modeling of void nucleation is difficult within a Gurson-based framework due to the intrinsic assumption that the material does not contain any second-phase particles. Consequently, the nucleation models employed in these constitutive models are overly simplistic as the particle shape, composition, stress state and load-sharing are neglected, lumped into a single calibration parameter (Beremin 1981) or indirectly accounted for in a phenomenological manner (Chu and Needlman 1980). The lack of progress in developing physically sound nucleation models has not been for lack of effort but a result of the inherently complex nature of the nucleation process. Void nucleation is very difficult to capture experimentally since it is a relatively random and instantaneous event that cannot be captured in-situ without the aid of high resolution x-ray tomography. Additionally, the local stress state near a particle of interest is typically unknown, as well as the particle composition and mechanical properties. The nucleation mechanism can occur by debonding or particle cracking and is influenced by the particle size, shape, composition, distribution, strain rate and temperature. From an engineering perspective, one can clearly see the attraction in adopting a phenomenological nucleation model whose parameters can be adjusted to give good agreement with the experiment data. Nevertheless, there is ample opportunity to improve the physical foundation of the current nucleation models, especially in regards to percolation modeling.

A promising procedure to account for the stress state in the particles has been proposed by Butcher (2011) by integrating a secant-based homogenization technique for particle-reinforced plasticity into an existing damage-based material model. This model has been developed to predict ductile fracture of industrial alloys during sheet metal forming operations where the loading is proportional and the particle content is small. The subsequent sections will introduce the particle-based homogenization model and its integration into a general damage-based constitutive model. The ability to determine the stress state within the particles will be used to model nucleation in Chaps. 10 and 11.

Z. Chen and C. Butcher, *Micromechanics Modelling of Ductile Fracture*, Solid Mechanics and Its Applications 195, DOI 10.1007/978-94-007-6098-1_9,
© Springer Science+Business Media Dordrecht 2013

9.1 Particle-Based Homogenization Theories

The development of particle-based homogenization theories to predict the bulk behavior of a material from its constituents encompasses a large branch of materials mechanics. As such, only a very brief review of the development of homogenization techniques is presented here with a focus on methods amenable for implementation into a damage-based framework. Excellent reviews on the efforts to develop particle-based homogenization techniques in the plastic regime can be found in Ponte-Casteneda and Suquet (1998) and Chaboche et al. (2005).

The framework for particle-based homogenization problems was pioneered by Eshelby (1957) when he obtained closed-form solutions for the stress field within ellipsoidal inclusions embedded in a matrix material by theorizing a stress-free transformation strain (eigen-strain) between the matrix and inclusions. Eshelby's work provided the analytical techniques required to estimate the average stress within the inclusions and matrix material using his so-called fourth-order **S** tensor. Hill (1965) later developed a rigorous solution for composite materials using the incremental theory of anisotropic elasticity. Incremental approaches to the homogenization process are based upon the tangent stiffness tensors of the constituents. It was soon recognized that this approach led to overestimations in the flow stress of the material due to the anisotropic nature of the tangent stiffness tensor in the plastic regime.

To obtain more realistic predictions for the composite flow stress, Berveiller and Zaoui (1979) modified Hill's incremental solution to obtain a total-strain formulation for proportional loading using the secant moduli of the matrix material. This model was further improved by Weng (1984, 1990) and Tandon and Weng (1986) using the mean field methods of Mori and Tanaka (1973) to account for inclusion interactions. Although the experimental material behaviour was better described using these models, errors arose when the inclusions remained elastic, resulting in an overly stiff response and an overestimation of the composite flow stress. The reason for this error was attributed to the fact that the plastic strain in the matrix was determined from a reference equivalent stress using the volumetric average of the matrix stress tensor. This equivalent stress can be significantly lower than the phase-average of the equivalent stress due to the severe stress gradients that develop in the plastic regime, resulting in the composite stress to be overestimated (Pierard et al. 2007).

This limitation led to the development of a modified secant method that defines the stress field in the non-linear matrix phase using the second-order moment of the volumetric stress tensor by Suquet (1995). The modified secant approach coincides with the variational approach of Ponte Castenada (1996) and gives very good agreement with the finite-element solution for a ductile matrix embedded with spherical elastic inclusions (Segurado, J., & Llorca, J. 2002). Despite these improvements, a fundamental limitation of secant-based methods is that they cannot be applied for non-proportional loadings, resulting in a renewed interest in recent years to improve the incremental formulations. Significant improvements have been obtained in the incremental models by using only the isotropic component of the anisotropic tangent stiffness tensors (Gonzalez and Llorca 2000; Doghri and Ouaar 2003).

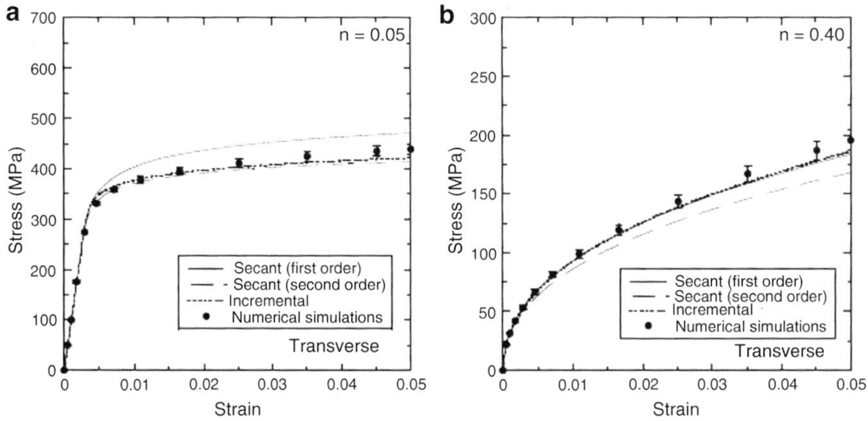

Fig. 9.1 Predictions of the tensile stress–strain curves in the longitudinal direction for the composite reinforced with ellipsoidal inclusions with a volume fraction of 25 %. (**a**) n = 0.05; (**b**) n = 0.40. The error bars represent the standard deviation in the numerical simulations (Reprinted with permission from Pierard et al. (2007). Copyright 2007 Elsevier)

Today, it is generally agreed that both secant and tangent-based homogenization models can provide acceptable approximations to the material behavior of composite materials in the elastic and plastic regimes (Pierard et al. 2007). As each type of model has its advantages, the desired application of the model is perhaps the most important factor in its selection. For materials with dilute concentrations of inclusions (less than 10 %), Mueller and Mortensen (2006) found no discernible difference in the predictions for the bulk and shear moduli of the composite when they evaluated five different homogenization schemes in the plastic regime (secant, self-consistent, generalized self-consistent, differential effective-medium, identical hard spheres approximation). This result suggests that a simple homogenization model can be suitable for a large range of engineering alloys.

A recent work by Pierard et al. (2007) provides additional insight into the importance of selecting the appropriate homogenization scheme. In this work, a finite-element study was conducted for a material containing aligned ellipsoidal inclusions to obtain an 'exact' solution to the homogenization problem. A classical secant (Mori and Tanaka 1973), modified secant (Suquet 1997) and a modified incremental model (Doghri and Ouaar 2003) were then evaluated to measure their predictive abilities. It was observed that the modified secant method gave the best agreement with the numerical results at the onset of plastic deformation while the classical method gave the best prediction for the composite hardening rate in the fully plastic regime. The incremental model of Doghri and Ouaar (2003) provided very good results in all of the cases considered. A comparison of the results with the 'exact solution' obtained using finite-element simulations is shown in Fig. 9.1.

What is most interesting from the work of Pierard et al. (2007) with respect to void nucleation is that while the incremental and modified secant methods gave the best estimates for the composite stress, the stress within the particles was best predicted using the secant method shown in Fig. 9.2. This is an important result and

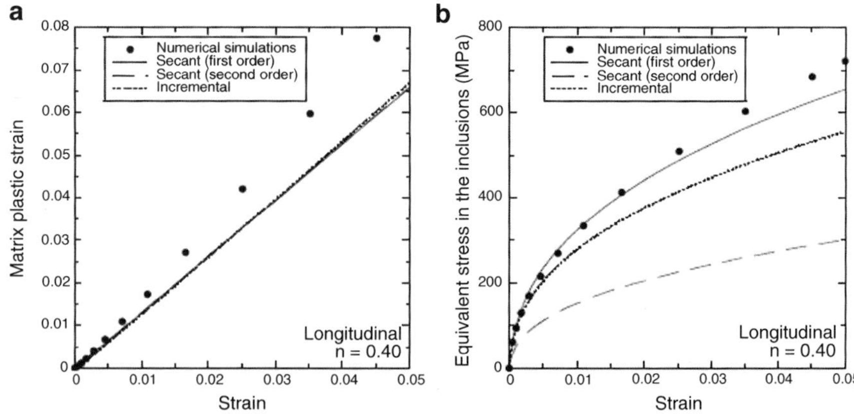

Fig. 9.2 (**a**) Evolution of the plastic strain in the matrix as a function of the applied strain. (**b**) Evolution of the von Mises equivalent stress in the ellipsoids as a function of the applied strain. The composite was loaded in the longitudinal direction and the matrix hardening coefficient was 0.40 (Reprinted with permission from Pierard et al. (2007). Copyright 2007 Elsevier)

it emphasizes the fact that that the ability of a model to predict the overall response of a material from its constituents does not mean that the predicted stress within the constituents is accurate as well. It is also important to note that all of the models underestimated the plastic strain in the matrix as the strong strain-gradients that develop between inclusions are not captured. These results serve to highlight the approximate nature of the homogenization models and one can see that in many cases, there may not be a distinct advantage in selecting one model over the other. The specific application and degree of accuracy in the metric of interest should be used in selecting an appropriate model.

9.2 Selection of a Homogenization Theory for Modelling Void Nucleation

For many ductile industrial alloys the volume fraction of second-phase particles is on the order of a few percent. As such, the reinforcement of these particles on the stress–strain curve is not of prime concern since they are generally not considered at all in Gurson-based material models. The influence of the particles in Gurson-based models is implicitly captured by using the experimental flow stress relation and assuming it describes the behaviour of the 'virgin' matrix. Consequently, our objective is to augment existing damage-based constitutive models by improving their ability to predict void nucleation through knowledge of the stress within the inclusions, not to predict the macroscopic response of the material from homogenization theory. The initiation and evolution of voids is the principal concern. If the material of interest contains a large volume fraction of reinforcing particles, a sophisticated homogenization scheme may be required to predict the overall response of the material to capture the load loss as particles nucleate voids.

For the purpose of modeling void nucleation, the secant method of Tandon and Weng (1986) will be used to estimate the stress within the inclusions. The advantages of adopting a secant-based method are that they are computationally efficient, easily implemented into numerical codes and can provide reasonable estimations for the stress in the matrix and inclusions. The issue of computational efficiency is of paramount importance in a damage percolation model because the homogenization procedure must be repeated for each of the large number of particles within the microstructure at each loading step. The adoption of a more sophisticated homogenization model is left for future work.

9.3 A Particle-Based Homogenization Model for a Dual-Phase Composite Subjected to a Prescribed Traction

The particle-based homogenization scheme considered in this work was developed by Tandon and Weng (1986) for a dual-phase composite subjected to a prescribed traction. The composite is composed of three-dimensional, randomly oriented elastic particles embedded within a ductile matrix. The particles are spheroidal and characterized by their elastic properties, volume fraction and aspect ratio. The homogenization theory is valid for inclusion shapes ranging from flat discs, to ellipsoids, to elongated fibres.

The homogenization model of Tandon and Weng (1986) is a secant-based approach that utilizes Berveiller and Zaoui's (1979) modification to the solution of Hill (1965) for proportional loading and incorporates the mean-field method of Mori and Tanaka (1973) to account for particle interactions. Additional details of the derivation of this model can be found in Tandon and Weng (1986). The following section will present the details of this theory relevant for integration with a damage-based constitutive model.

9.4 Effective Moduli of a Randomly-Oriented Composite

The ductile matrix material is treated as phase 0 and the embedded elastic particles are defined as phase 1 with a volume fraction, f_1, and aspect ratio, W_1. The respective isotropic Poisson's ratio, and bulk, shear and elastic moduli of the r-th phase are denoted by ν_r, κ_r, μ_r and E_r with a superscript, s, used to denote a secant quantity such as the secant shear modulus of the matrix, μ_0^s. The stress and strain tensors for the r-th phase are denoted by $\sigma_{ij}^{(r)}$ and $\varepsilon_{ij}^{(r)}$ which can be decomposed into their respective deviatoric and hydrostatic components as $\sigma_{ij}^{(r)} = \sigma_{ij}^{('r)} + \delta_{ij}\sigma_{\text{hyd}}^{(r)}$ with

$\sigma_{\text{hyd}}^{(r)} = \sigma_{kk}^{(r)}/3$ and $\varepsilon_{ij}^{(r)} = \varepsilon_{ij}^{\prime(r)} + \delta_{ij}\varepsilon_{\text{hyd}}^{(r)}$ with $\varepsilon_{\text{hyd}}^{(r)} = \varepsilon_{kk}^{(r)}/3$ where δ_{ij} is the Kronecker delta. Any property or quantity associated with the composite is denoted using an overbar symbol such as $\bar{\sigma}_{ij}$ for the composite stress tensor.

In a composite material the constituents are generally in a triaxial state of stress that should be characterized using the equivalent measures for the stress, strain and plastic strain as

$$\sigma_{\text{eq}}^{(r)} = \sqrt{(3/2)\sigma_{ij}^{(\prime r)}\sigma_{ij}^{(\prime r)}} \qquad \varepsilon_{\text{eq}}^{(r)} = \sqrt{(2/3)\varepsilon_{ij}^{(\prime r)}\varepsilon_{ij}^{(\prime r)}}$$
$$\bar{\varepsilon}_{\text{eq}}^{\text{p}} = \sqrt{(2/3)\bar{\varepsilon}_{ij}^{\text{p}}\bar{\varepsilon}_{ij}^{\text{p}}} \qquad (9.1\text{a, b, c})$$

The components of the stress and strain in the composite are related through the effective secant shear and bulk moduli as

$$\bar{\sigma}_{ij}^{\prime} = 2\bar{\mu}^{s}\bar{\varepsilon}_{ij}^{\prime} \qquad \bar{\sigma}_{\text{hyd}} = 3\bar{\kappa}^{s}\bar{\varepsilon}_{\text{hyd}} \qquad (9.2\text{a, b})$$

To determine the stress and strain in the composite and its constituents during plastic deformation, the effective secant elastic moduli of the matrix and composite must first be determined as functions of the matrix plastic strain, $\varepsilon_{\text{eq}}^{\text{p}(0)}$. The secant elastic modulus and Poisson's ratio of the matrix are expressed as

$$\text{E}_0^s = \frac{1}{\frac{1}{\text{E}_0} + \frac{\varepsilon_{\text{eq}}^{\text{p}(0)}}{\sigma_{\text{eq}}^{(0)}}} = \frac{3\text{E}_0\mu_0^s}{\text{E}_0 + \mu_0^s(1 - 2\nu_0)} \qquad \nu_0^s = \frac{1}{2} - \left(\frac{1}{2} - \nu_0\right)\frac{E_0^s}{E_0} \qquad (9.3, 9.4)$$

The secant bulk and shear moduli are obtained using the standard isotropic relations $\mu_0^s = \text{E}_0^s/2(1 + \nu_0^s)$ and $\kappa_0^s = \text{E}_0^s/3(1 - 2\nu_0^s)$. The matrix material is assumed to be plastically incompressible and thus the secant bulk modulus remains constant at $\kappa_0^s = \kappa_0$. The effective secant moduli of the composite material can be determined as

$$\bar{\kappa}^s = \frac{\kappa_0}{(1 + f_1 p_{2s}/p_{1s})} \qquad \bar{\mu}^s = \frac{\mu_0^s}{(1 + f_1 q_{2s}/q_{1s})} \qquad (9.5, 9.6)$$

where p_{is} and q_{is} are functions of the particle shape, volume fraction, elastic moduli of the constituents and the fourth-order Eshelby (1957) **S** tensor. The expressions for p_{is} and q_{is} are rather lengthy and are not presented here for brevity but can be found in Tandon and Weng (1986).

9.5 Average Stress in the Composite and Its Constituents

The respective hydrostatic and deviatoric stresses in each constituent are determined from the composite stress tensor using the stress concentration factors, $a^{(r)}$ and $b^{(r)}$ as

$$\sigma_{\text{hyd}}^{(r)} = a^{(r)} \bar{\sigma}_{\text{hyd}} \qquad \sigma_{ij}^{('r)} = b^{(r)} \bar{\sigma}'_{ij} \qquad \sigma_{eq}^{(r)} = b^{(r)} \bar{\sigma}_{eq} \tag{9.7a, b, c}$$

that are defined for the matrix material and particles as

$$a^{(0)} = 1/p_{1s} \qquad a^{(1)} = \left(1 - (1 - f_1)a^{(0)}\right)/f_1 \tag{9.8a, b}$$

$$b^{(0)} = 1/q_{1s} \qquad b^{(1)} = \left(1 - (1 - f_1)b^{(0)}\right)/f_1 \tag{9.9a, b}$$

The stress in the composite and its constituents must be in equilibrium and thus

$$\bar{\sigma}_{ij} = (1 - f_1)\sigma_{ij}^{(0)} + f_1 \sigma_{ij}^{(1)} \tag{9.10}$$

The onset of yielding of the matrix occurs when $\bar{\sigma}_{\text{eq}} \geq \sigma_{\text{y}}^{(0)}/b^{(0)}$ with $b^{(0)}$ determined using the elastic moduli of the matrix in Eq. (9.3) and (9.4).

9.6 Average Strain in the Composite and Its Constituents

Due to the presence of the elastic inclusions, the composite is not plastically incompressible and the composite plastic strain must be determined from the unloading process as

$$\bar{\varepsilon}_{ij}^{\text{p}} = \left(\frac{1}{2\bar{\mu}_s} - \frac{1}{2\bar{\mu}}\right)\bar{\sigma}'_{ij} + \delta_{ij}\left(\frac{1}{3\bar{\kappa}_s} - \frac{1}{3\bar{\kappa}}\right)\bar{\sigma}_{\text{hyd}} \tag{9.11}$$

The matrix is assumed to be isotropic and obey J_2 plasticity (von Mises material) from which the plastic strain components can be readily determined by integrating the J_2 flow rule for proportional loading to yield

$$\varepsilon_{ij}^{\text{p}(0)} = \frac{3}{2}\frac{\varepsilon_{\text{eq}}^{\text{p}(0)}}{\bar{\sigma}}\bar{\sigma}'_{ij} \tag{9.12}$$

The particles are assumed to remain elastic during deformation and the strain in the particles can be expressed as

$$\varepsilon_{ij}^{(1)} = \frac{b^{(1)}\bar{\sigma}_{ij}'}{2\mu_1} + \delta_{ij}\frac{a^{(1)}\bar{\sigma}_{\text{hyd}}}{3\kappa_1} \tag{9.13}$$

The solution for the secant moduli of the composite and subsequent stress and strain in the constituents is nonlinear and an iterative solution is required in the plastic regime. The required material parameters to determine the behaviour of the composite and stress in the constituents are E_0, ν_0, E_1, ν_1, f_1, and W_1. A simple fixed-point algorithm can be easily implemented to determine the required value for $\varepsilon_{\text{eq}}^{\text{p}(0)}$ and is described in Tandon and Weng (1986). However, this solution method must be modified in the present work because the homogenization method is to be coupled with a damage-based constitutive model to account for the influence of voids on the subsequent stress and strain in the constituents.

9.7 Procedure for Integrating a Particle-Based Homogenization Theory into an Existing Damage-Based Constitutive Model

To integrate a particle-based homogenization theory into a damage-based constitutive model, we assume that the bulk material can be idealized as a three-phase composite composed of a matrix material with embedded particles and voids. It is assumed that this idealized three-phase composite can be decomposed into its constituents by applying two successive homogenization schemes: (i) a Gurson-based constitutive model for the voids embedded in a ductile 'composite matrix' which is composed of the particles and matrix material and (ii) separation of the composite matrix into its constituents to determine the stress within the matrix and particles using a secant-based homogenization scheme.

Let us consider a bulk material which contains both voids and hard elastic particles/inclusions within a ductile matrix. To mitigate the influence of the voids, the experimental flow stress relation for the bulk material can be obtained from a torsion or compression-type test or from a tensile test if the initial porosity is negligible (Pardoen 2006). This flow stress relation is essentially that of a two-phase composite composed of the matrix and particles.

Now, the bulk material is subjected to a deformation process such as a sheet metal forming operation where the pre-existing voids will grow and additional voids will be nucleated from particle cracking and/or debonding from the matrix. The presence of the voids results in material softening which further promotes void evolution resulting in ductile fracture as the voids coalesce and link-up throughout the material. The influence of the voids on the response of the bulk material can be described using a damage-based constitutive model such as the Gurson (1977) model. Gurson-based models are the result of a homogenization procedure for a material composed of voids embedded within a virgin matrix. Therefore, a Gurson-based model can be applied to the bulk material to account for void damage by modeling the bulk material as a material which contains voids embedded within a so-called composite matrix. The composite matrix is composed of the virgin matrix and particles whose composite behaviour is described using the experimental flow-stress relation for a two-phase composite as described above.

The typical procedure for the integration of the Gurson-based material model is now applied to the bulk material. The material has a void volume fraction or porosity, f, and is subjected to a monotonic proportional loading with a macroscopic strain, $\bar{E}_{ij}$. The stress state is then integrated using the damage-based yield surface to determine the macroscopic stress, Σ_{ij}. The equivalent plastic strain and flow stress within the composite matrix are, $\bar{E}^{\mathrm{p}}_{\mathrm{eq}}$ and $\bar{\sigma} = \bar{\sigma}(\bar{E}^{\mathrm{p}}_{\mathrm{eq}})$. Void evolution is very sensitive to the stress state which is characterized by the stress triaxiality ratio defined as $T = \bar{\Sigma}_{\mathrm{hyd}}/\bar{\Sigma}_{\mathrm{eq}}$.

Due to the presence of the voids, the bulk material is softer and thus the composite matrix must work-harden to a greater extent to reach the applied strain of $\bar{E}_{ij}$ than if no voids were present. Since the voids do not contribute to load-sharing, the entire stress must be borne by the matrix and particles. Therefore, from the perspective of the constituents, it is equivalent to subjecting the composite matrix to a larger applied strain denoted $\bar{\varepsilon}_{ij}$, that results in the same equivalent stress and plastic strain as when softening was considered.

The situation for the composite matrix now resembles that of a particle-based homogenization problem for a prescribed traction. The equivalent stress state within the composite matrix is known and the stress state in the constituents must be determined that is in equilibrium with this prescribed stress. The secant-based homogenization method of Tandon and Weng (1986) can now be applied to determine the effective secant moduli of the composite matrix which satisfy the stress state defined by the damage-based stress integration.

The general integration procedure is presented in Fig. 9.3 and can be summarized in the following steps:

(a) A bulk material containing voids and particles within a ductile matrix is subjected to a macroscopic loading.
(b) The voids within the material are homogenized into an equivalent void embedded in a composite matrix.
(c) The material is now that of a Gurson-based material with the composite matrix taking the place of the virgin matrix
(d) A damage-based constitutive model is applied to determine the stress in the composite matrix by accounting for the influence of the voids. This decouples the void from the composite matrix.
(e) With the voids removed from the composite matrix, a particle-based homogenization theory for a prescribed traction is applied to the composite matrix to determine the stress within the matrix and particles.
(f) The particles can now be tested for nucleation using an appropriate model for the material.
(g) The evolution of the voids must obey the plasticity of the matrix. Combine the isolated void and the virgin matrix to obtain the traditional voided unit cell.

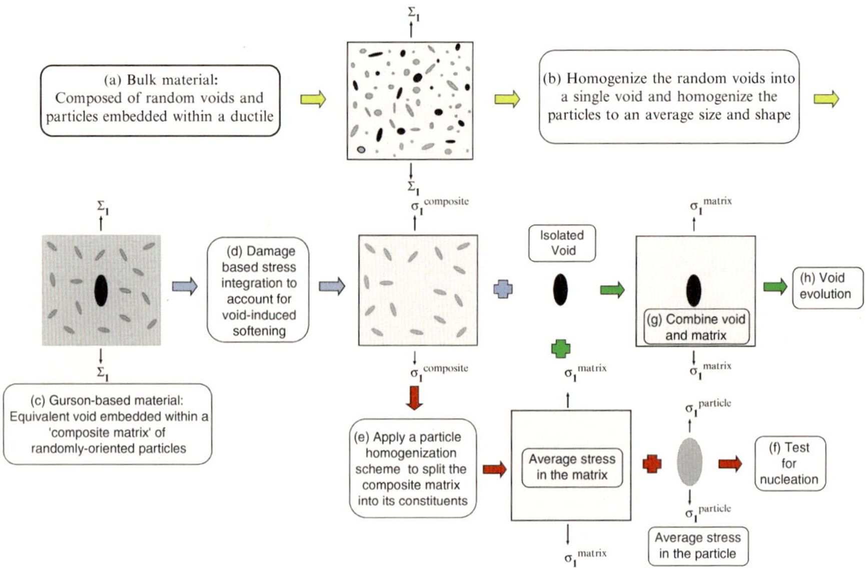

Fig. 9.3 Schematic of the integration process of a particle-based homogenization theory into a damage-based constitutive model (Butcher 2011)

(h) The standard models for void evolution are evaluated using the stress state and plastic strain of the matrix.

To create a stress state in the composite matrix equivalent to that of the voided bulk material, the applied strain, $\bar{\varepsilon}_{ij}$, must be determined to produce a stress, $\bar{\sigma}_{ij}$, subject to the constraints that $\bar{\sigma}_{eq} = \bar{\sigma}$, $\bar{\varepsilon}^{p}_{eq} = \bar{E}^{p}_{eq}$ and $\bar{\sigma}_{hyd}/\bar{\sigma}_{eq} = T$. It is important to mention that the equivalent stress, Σ_{eq}, identified in the damage-based integration is not equal to the flow stress within the composite matrix due to the presence of the voids. The procedure to determine the applied strain, $\bar{\varepsilon}_{ij}$, and stress, $\bar{\sigma}_{ij}$, within the composite matrix using a secant-based homogenization scheme for monotonic, proportional loading is described in the following section.

First, the hydrostatic stress of the composite matrix can be computed directly from the requirement of an equivalent triaxiality as

$$\bar{\sigma}_{hyd} = \bar{\sigma}(\bar{\Sigma}_{hyd}/\bar{\Sigma}_{eq}) \tag{9.14}$$

From the definition of the effective secant moduli in Eq. (9.2) and the equivalent stress and strain in Eq. (9.1a, b), the equivalent strain in the composite matrix is

$$\bar{\varepsilon}_{eq} = \bar{\sigma}/3\bar{\mu}^{s} \tag{9.15}$$

Only the deviatoric stress and strain of the composite matrix remain unknown and only one is required to be pre-determined as they are related through the secant shear modulus. The deviatoric strain in the composite is assumed to be proportional to the deviatoric strain in the damage-based model or three-phase composite as

$$\bar{\varepsilon}'_{ij} = \xi \bar{E}'_{ij} \tag{9.16}$$

so that $\xi = \bar{\varepsilon}_{eq}/\bar{E}_{eq}$, and the stress and applied strain in the composite matrix can be derived as

$$\bar{\sigma}_{ij} = \left(\frac{2}{3}\frac{\bar{\sigma}}{\bar{E}_{eq}}\right)\bar{E}'_{ij} + \delta_{ij}\bar{\sigma}_{hyd} \qquad \bar{\varepsilon}_{ij} = \left(\frac{1}{3\bar{\mu}^s}\frac{\bar{\sigma}}{\bar{E}^p_{eq}}\right)\bar{E}'_{ij} + \delta_{ij}\frac{\bar{\sigma}_{hyd}}{3\bar{\kappa}^s} \tag{9.17, 9.18}$$

which satisfies the requirements that $\bar{\sigma}_{hyd}/\bar{\sigma}_{eq} = T$ and $\bar{\sigma}_{eq} = \bar{\sigma}$. The stress within the composite matrix can be evaluated immediately following the stress integration of the damage-based constitutive model while $\bar{\varepsilon}_{ij}$ requires knowledge of the effective secant moduli which have yet to be determined.

9.8 Iterative Solution for the Effective Secant Moduli

An iterative procedure is required to determine the effective secant moduli of the composite matrix from which the stress and strain within the constituents can be determined. The iterative solution is developed through the final constraint that the secant elastic moduli must result in a plastic strain in the composite matrix of $\bar{\varepsilon}^p_{eq} = \bar{E}^p_{eq}$. The plastic strain of the composite matrix must account for the elastic heterogeneity of the particles and from substituting Eq. (9.11) into Eq. (9.2) and utilizing Eq. (9.1c), the required secant shear modulus of the composite matrix is

$$\bar{\mu}^{s*} = \bar{\mu}\bar{\sigma}\left(\bar{\sigma} + \bar{\mu}\sqrt{\left(3\bar{E}^p_{eq}\right)^2 - 2\left(1/\bar{\kappa}^s - 1/\bar{\kappa}\right)^2\bar{\sigma}^2_{hyd}}\right)^{-1} \tag{9.19}$$

All of the parameters in Eq. (9.19) are constant during the iteration loop except for the secant bulk modulus, $\bar{\kappa}^s$, which varies with the plastic strain of the matrix.

A fixed-point algorithm is used to obtain the solution to the non-linear equations for the secant moduli by iterating upon the secant elastic modulus of the matrix, E^s_0. A trial value for E^s_0 is assumed from which the subsequent trial values for the secant shear, bulk and Poisson's ratios of the matrix can be determined from Eqs. (9.3) and (9.4) and used to evaluate the expressions for p_{is} and q_{is}, as well as $\bar{\kappa}_s$ in Eq. (9.5) and $\bar{\mu}^{s*}$ in Eq. (9.19). A new estimate for μ^s_0 can be determined by setting $\bar{\mu}^s = \bar{\mu}^{s*}$ in Eq. (9.6) and evaluating Eq. (9.3) to obtain a new estimate for E^s_0. If this new value of E^s_0 is equal to the trial value, the solution has been obtained, otherwise the algorithm is repeated using the current E_0^s as the new trial value. Upon convergence, the stress concentration factors to determine the stress and strain tensors of the matrix and particles can be evaluated using Eqs. (9.7a–c), (9.12) and (9.13). This algorithm generally achieves convergence within three iterations to a tolerance of 0.001 and is straightforward in its implementation.

9.9 Application of the Particle-Based Homogenization Scheme into a Gurson-Based Constitutive Model for Ductile Fracture

The integration of a homogenization scheme into a Gurson-based damage framework improves all aspects of the damage model and couples the various damage mechanisms that are often considered independently. The advantages of the proposed fully-coupled damage-based constitutive model are:

- Modeling of void nucleation is improved through knowledge of the stress state within the particles as a function of their shape, content and composition.
- As particles nucleate voids, the stress state within the matrix becomes more severe due to increased material softening (higher porosity) as fewer particles are available for load sharing.
- The stress state within the matrix and particles becomes progressively more severe as the particle content decreases due to nucleation. This promotes additional nucleation, void growth and material softening.
- The plastic strain within the matrix material is higher than in the composite matrix and thus promotes void evolution and coalescence.
- The model reverts to its original formulation if the material does not contain any second-phase particles or inclusions.
- The flow stress relation of the matrix material does not have to be predetermined.
- An absolute minimum number of new parameters have been introduced into the damage-based framework. The elastic properties of the constituents as well as the average particle shape and content are typically known from standard material characterization techniques used in damage-based modeling.
- The particle-based homogenization scheme is computationally efficient and easily implemented into any existing damage-based constitutive model.

The proposed integration procedure has been implemented into the well-known GT constitutive model in Butcher (2011) and is also used in the percolation model by applying the homogenization procedure to each particle in the material. The variation in the principal stress state in the particles with the macroscopic stress state is shown in Fig. 9.4.

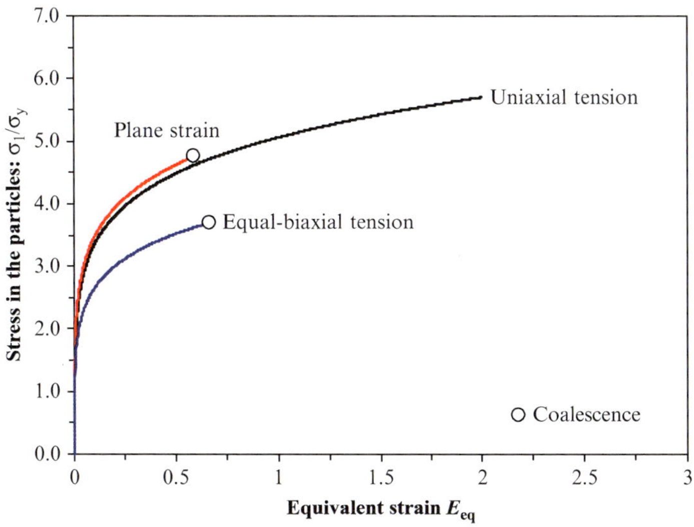

Fig. 9.4 Normalized principal stress within the spherical elastic particles/inclusions with a volume fraction of 10 % for three loading conditions in a model material based upon AA5083 (Butcher 2011). The initial voids in the material have an aspect ratio of 1/6 and a porosity of 0.1 %. The bulk material ruptures due to void coalescence when loaded in plane strain and equal-biaxial tension

9.10 Continuum Nucleation

The treatment of void nucleation on the continuum scale in a damage-framework can be substantially improved using the the proposed integration scheme to obtain the stress in the particles. For example, the nucleation stress in the Chu and Needleman (1980) nucleation model can now be predicted as a function of the particle content, shape and elastic moduli as

$$\dot{f}_{\text{nucleation}}(f_p, W_p, G_p, \kappa_p, \Sigma_{ij}) = \frac{W_{\text{N}}\xi_{\text{n}} f_{\text{p}}}{s_{\text{N}}\sigma_{\text{y}}\sqrt{2\pi}} \exp\left[-\frac{1}{2}\left(\frac{\sigma^p - \sigma_{\text{N}}}{s_{\text{N}}\sigma_{\text{y}}}\right)^2\right]\dot{\sigma}^p \qquad (9.20)$$

where W_{N} is the aspect ratio of the nucleated void and ξ_{n} is the fraction of particles that will nucleate voids. The particle stress term, σ_{p}, can be either the equivalent stress in the particles for general nucleation modelling or the maximum principal stress in the particles to better represent particle cracking. The yield stress of the material, σ_{y}, is simply a normalizing factor and could be changed to the particle yield stress if it happens to be known. Unlike the traditional Chu and Needleman nucleation model (1980), the stress in the particles will evolve as particles nucleate voids, increasing the stress in the remaining particles and promoting nucleation. Furthermore, the determination of the mean nucleation stress, σ_{N}, through calibration with the experiment will have a stronger physical foundation since the stress in the particles was determined as a function of the global stress state and the particle morphology.

9.11 Void Nucleation in a Particle Field

The dominant void nucleation mechanisms in ductile materials are particle cracking and interface fracture (debonding). Particle cracking occurs primarily in strong, rigid particles in a brittle-type fracture in the direction transverse to the maximum principal stress. Void nucleation due to interface separation is strongly related to the particle geometry and the interface strength between the inclusion and the surrounding matrix. Similar to particle cracking, debonding occurs preferentially in the principal loading direction unless the stress triaxiality is high.

From a modeling perspective, particle cracking and debonding will occur simultaneously in a material and thus two nucleation models and their parameters must be determined for each type of inclusion present. The identification of the appropriate material parameters for even a single nucleation model is a complicated process as nucleation is exceptionally difficult to measure experimentally. Consequently, most treatments of nucleation rely on a phenomenological model such as a normal-distribution (Chu and Needleman 1980) with either assumed or calibrated parameters obtained from a combination of finite-element simulations and experiment data (Butcher and Chen 2011).

In a damage percolation model, an approximate treatment for nucleation is required to reduce the number of required nucleation parameters while retaining a physically sound description of the nucleation mechanism. A proper accounting of nucleation requires both a nucleation model and a treatment for determining the initial dimensions and orientation of the nucleated void. For practical considerations, an emphasis is placed upon developing an accurate representation of particle cracking as this is the dominant nucleation mode in many advanced automotive alloys. Additionally, the nucleated void by particle cracking gives rise to a penny-shaped void whose growth and evolution is relatively well understood from finite-element simulations in Chap. 4. The implementation of a sophisticated particle debonding criterion into the percolation framework is left for future work.

9.12 Modeling Void Nucleation Using Penny-Shaped Voids

In particle cracking, the void is assumed to nucleate transversely to the maximum principal direction as illustrated in Fig. 9.5. This cracking behaviour is similar for both oblate and prolate ellipsoidal inclusions with prolate inclusions particularly prone to cracking. At the instant of nucleation, the particle is in two pieces that have yet to appreciably separate and the void is said to be penny-shaped. From unit cell simulations of penny-shaped voids, it has been shown that the growth of the void and response of the material converges for aspect ratios lower than 1/100. This is an extremely beneficial result that allows us to assume an aspect ratio for the nucleated void with the confidence that is a physically sound approximation.

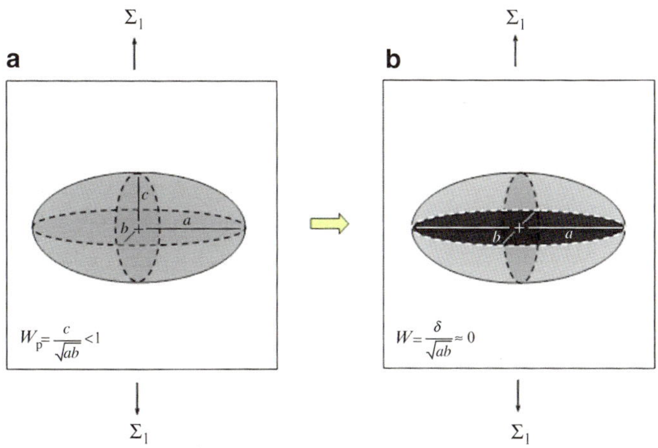

Fig. 9.5 Void nucleation by the cracking of an oblate ellipsoidal inclusion. Nucleation occurs in the same manner for a prolate ellipsoid. (**a**) Oblate ellipsoidal particle or inclusion and (**b**) particle cracks to form a penny shaped void

For particle debonding, we will first consider an oblate ellipsoidal particle as viewed from the principal loading direction (the particle would appear as a prolate ellipsoidal if viewed from the transverse direction). The interface may fracture at the top and/or bottom surfaces of the inclusion. At the onset of interface fracture, the vertical height of the debonded void, δ, is negligible and we assume nucleation occurs over the entire top/bottom surface. Since it is not possible to estimate whether the top and/or the bottom interface will debond first, we can take the average of the three possible cases and simply place the void at the center of the particle. As shown in Fig. 9.6, this approximation yields a penny-shaped void and can receive the same treatment for particle cracking to provide a good estimate for the initial void dimensions.

For the debonding of a prolate ellipsoidal inclusion, interface separation occurs at the top and/or bottom poles of the inclusion as shown in Fig. 9.7. Similar to the oblate inclusion, the debonded regions can be approximated by a penny-shaped void at the centroid of the inclusion.

This type of debonding is not as well represented using this treatment compared to debonding of the oblate inclusion. However, this approximation is not entirely unreasonable if it is assumed that sufficient material debonds at the pole so that the cross-sectional area of the inclusion is a good estimate for the debonded area. Nevertheless, the approximation of all nucleated voids as penny-shaped voids sufficiently covers the range of nucleation modes to be an effective and straightforward treatment as only one nucleation model is required per inclusion type. This nucleation treatment can be considered as 'cracking-centric' since it best represents particle cracking and is only approximate for debonding.

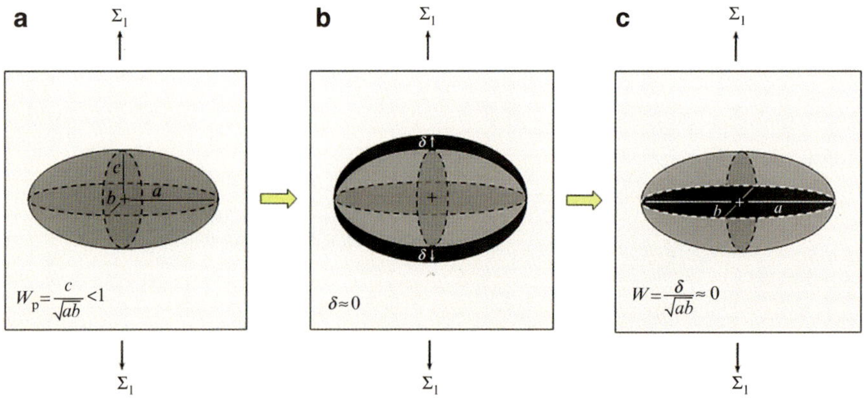

Fig. 9.6 Void nucleation by the debonding of an oblate ellipsoidal inclusion and its approximation as a penny-shaped void. (**a**) Oblate ellipsoidal particle or inclusion, (**b**) debonding may occur at the *top* and/or *bottom* surfaces of the particle, and (**c**) debonded regions approximated as a penny-shaped void at the particle centroid

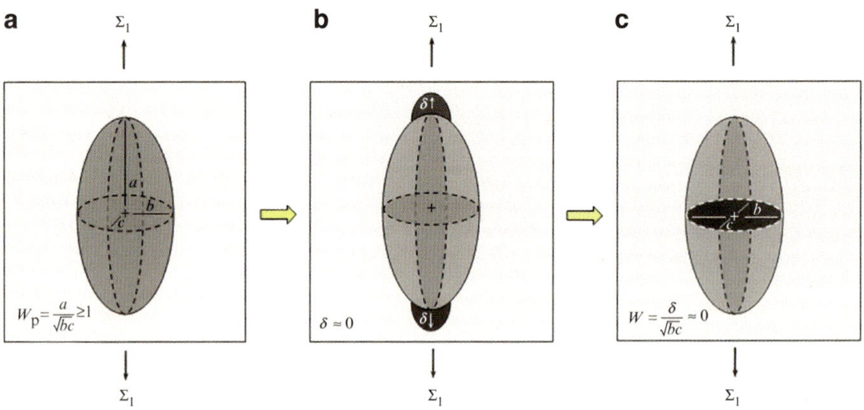

Fig. 9.7 Void nucleation by the debonding of a prolate ellipsoidal inclusion and its approximation as a penny-shaped void. (**a**) Prolate ellipsoidal particle or inclusion. (**b**) Debonding may occur at the *top* and/or *bottom* poles of the particle. (**c**) Debonded regions approximated as a penny-shaped void at the particle centroid

9.13 A Nucleation Model for Particle Cracking

A void nucleation model for particle cracking is fundamentally a model of brittle fracture and must account for the stress within the particle, particle size, shape and composition. It is well known that hard large particles are predisposed to cracking since they are irregularly shaped and experience higher stresses via load sharing with the matrix compared to their smaller neighbours. From a statistical perspective, a large particle has a higher probability of containing internal defects than a smaller particle. This effect can be clearly seen in the SEM micrographs in Fig. 9.8 that show the presence of surface cracks on the particles in an as cast AA5182 alloy.

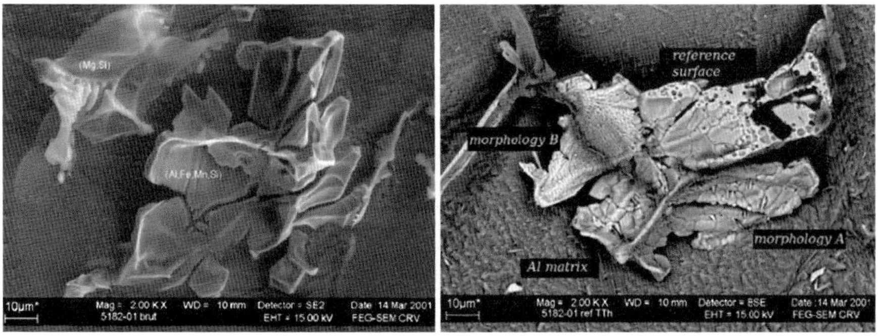

Fig. 9.8 SEM observation from Moulin et al. (2009) showing the intermetallic particles in as-cast AA5182 (*left*) and the typical morphology of the large Fe-rich particles (*right*) (Reprinted with permission from Moulin et al. (2009). Copyright 2009 John Wiley & Sons)

These cracks are a result of the solidification process and act as preferential sites for particle break-up during the hot and cold rolling processes. In a metal forming operation, these particles are smaller but still contain cracks that can nucleate voids. Since brittle fracture is associated with negligible plastic deformation, the particles can be assumed to remain elastic during the deformation process, enabling the use of linear elastic fracture mechanics to relate the stress state with the distribution of internal cracks within the inclusions.

To model the break-up of the large intermetallic particles in AA5182 during the rolling process, Moulin et al. (2009) performed an extensive finite-element and experimental study of irregularly shaped inclusions. It was observed that the stress distribution within these particles is not homogeneous and that only certain regions of the particle will achieve the stress required for particle break-up. These high stress regions were termed the 'active volume' of the inclusion, V_a. Since the particles are brittle and contain cracks, the fracture mechanism is one of mode I failure and is a function of the critical material toughness, K_{1C}, and the distribution of the cracks. Assuming a homogeneous distribution of cracks with an effective length, a_{eff}, the crack will propagate according to Griffith's criterion for mode I fracture when the principal stress in the particle satisfies the inequality

$$\sigma_1^P \geq \frac{K_{1C}}{\sqrt{\pi a_{eff}}} \tag{9.21}$$

This equation originates from the energy requirement that the stored elastic energy within the inclusion must satisfy the energy released by the creation of the two fracture surfaces. By assuming a homogenous distribution of cracks, the inclusion is expected to contain a crack that is perpendicular to the principal loading direction that can propagate to cause fracture. The critical length of this crack, a_c, that must be present within the active volume of the inclusion can be estimated as

$$a_c \approx \frac{\alpha}{\pi}\left(\frac{K_{1C}}{\sigma_{th}}\right)^2 \qquad \alpha \approx 1 \tag{9.22}$$

where α is a geometry parameter to account for various effects such as crack blunting and interactions and σ_{th} is a threshold stress for the maximum principal stress within the active volume. The critical volume that contains a crack transverse to the loading direction is defined as $V_c^3 = a_c^3$ and particle fracture will occur when the critical volume is equal to the active volume (Moulin et al. 2009)

$$V_a \geq V_c \geq \left(\frac{\alpha}{\pi}\right)^3 \left(\frac{K_{1C}}{\sigma_{th}}\right)^6 \tag{9.23}$$

9.13.1 Stress State and Nucleation

The criterion of Moulin et al. (2009) in Eq. (9.23) is ideally suited for the modeling of void nucleation since it is a volume-based metric that accounts for the stress within the inclusion, the material toughness and has a strong physical foundation based upon the energy required for crack propagation via the Griffith criterion. However, the model cannot be readily implemented into the percolation model because the threshold stress and active volume of the particles are functions of the particle size, shape, composition and loading condition. In Moulin et al. (2009), they were determined by using finite-element techniques.

Fortunately, since the inclusions in the percolation model are assumed to be ellipsoidal, the stress and strain distributions within the inclusion are uniform as proven by Eshelby (1957). This favourable result indicates that the activated volume in Eq. (9.23) must be equal to the volume of the inclusion and that the threshold stress must be equal to the principal stress in the inclusion. The homogenization scheme for particle-reinforced plasticity is now used to determine the stress within the particle as a function of the particle size, shape and composition. The fracture criterion can now be expressed in terms of a nucleation stress

$$\sigma_1^{\mathrm{p}} \geq \sigma_N \tag{9.24}$$

$$\sigma_N = \frac{1}{\sqrt{\pi}} \frac{K_{1C}^*}{V_p^{1/6}} \qquad K_{1C}^* = K_{1C}\sqrt{\alpha} \qquad \alpha \approx 1$$

where K_{1C}^* is the effective critical toughness of the particle material and is the sole parameter in the nucleation model. The K_{1c} value can be estimated if the composition of the particle is known or else it can be identified by calibrating the effective K_{1C}^* with the experimental nucleation data. A realistic range for the value of K_{1c} for

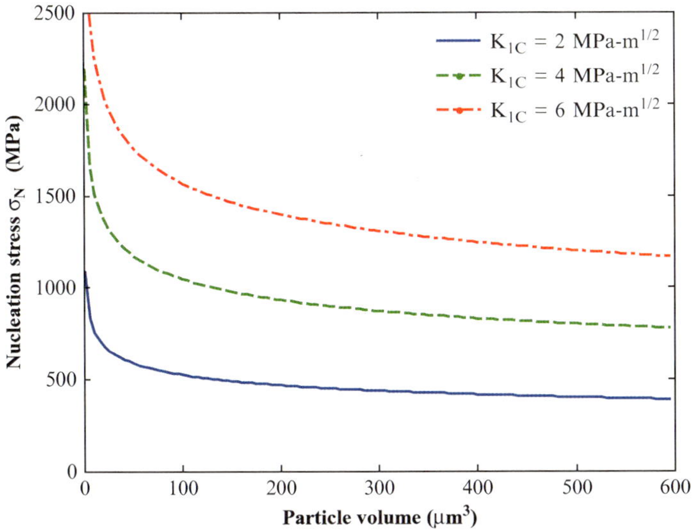

Fig. 9.9 Variation of the nucleation stress with particle size and toughness

brittle materials is 1–10 MPa · $m^{1/2}$ (Moulin et al.). The geometry factor, α, does not have to be determined and can be lumped into the K^*_{1C} value.

The variation of the particle nucleation stress with the volume is presented in Fig. 9.9. The criterion captures the particle size-effect where small particles nucleate at high strains while large particles nucleate at low stresses and are roughly independent of the particle size. The nucleation model also predicts that brittle phases are more likely to crack than more ductile phases. It should be emphasized that the nucleation stress is deeply coupled with the particle size, shape, composition, fracture toughness and the applied stress state as

$$\sigma_N = \sigma_N\left(f_p, W_p, G_p, \kappa_p, \Sigma_{ij}, K^p_{1c}\right) \tag{9.25}$$

and that successful prediction of nucleation relies upon the interrelationships among variables.

The principal contribution of this nucleation treatment is that it is physically realistic and does not contain any calibration parameters. The percolation model is now completely deterministic where fracture is a natural consequence of void nucleation and evolution and no adjustable parameters are employed. The particle distribution, stress state and material properties are solely responsible for the fracture process. All parameters are intrinsic material properties that can be quantified or estimated such as the K_{1c} value.

9.14 Determination of the Initial Dimensions for a Nucleated Void

A three-dimensional ellipsoidal inclusion with semi-axes (a, b, c) and orientation vectors ($\boldsymbol{n}_1$, $\boldsymbol{n}_2$, $\boldsymbol{n}_3$) is embedded within a ductile matrix as shown in Fig. 9.10. The matrix and inclusion are subjected to an external loading that gives rise to a maximum principal stress, Σ_1, along the direction, $\boldsymbol{u}$. The loading is severe enough to induce the particle to fracture or debond from the matrix to form a void. It is assumed that the particle cracks through its center in the direction transverse to the principal loading direction. The dimensions and orientation of the void in this plane are obtained by sectioning the inclusion to form an ellipse on the $\boldsymbol{p}$-$\boldsymbol{n}$ plane where the vectors $\boldsymbol{p}$ and $\boldsymbol{n}$ correspond to the semi-axes of the ellipse denoted a_s and b_s. The void height, c_W, is determined from the assumed aspect ratio for the void at nucleation, W_N. For penny-shaped voids, any value for $W_N < 0.01$ can accurately represent the void. The height of the void can then be calculated as

$$c_W = W_N R_{\text{axi}} = W_N \sqrt{a_s b_s} \tag{9.26}$$

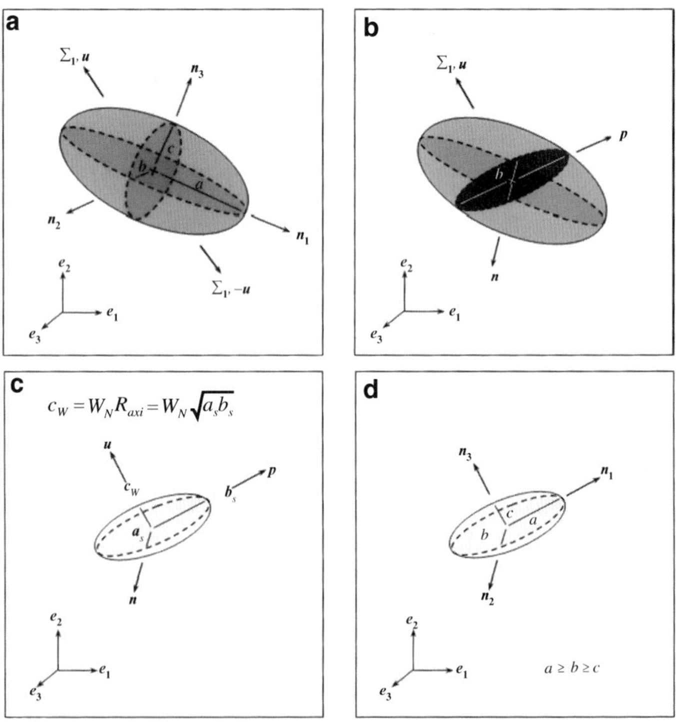

Fig. 9.10 Procedure for modeling void nucleation. (**a**) Ellipsoidal inclusion in a ductile matrix, (**b**) cross-sectional area of the nucleated void, (**c**) penny-shaped void geometry, and (**d**) ordered void geometry and orientation

Finally, the void semi-axes, (c_W,a_s,b_s), and their respective directions, ($\boldsymbol{u}$, $\boldsymbol{p}$, $\boldsymbol{n}$), are ordered so that the semi-axes are now defined as (a, b, c) with directions ($\boldsymbol{n}_1$, $\boldsymbol{n}_2$, $\boldsymbol{n}_3$) and $a \geq b \geq c$. The inclusion that nucleated the void is then removed from further analysis as it is assumed that the particle can only crack once, has a negligible ability to reinforce the material (true for particle cracking) and that any pieces of the broken particle do not significantly impact the subsequent growth of the void (Lassance et al. 2006).

When the principal stresses are comparable as in equal-biaxial tension, the particle may crack transverse to either direction. In this case, the cracking direction is selected as the direction that has the largest particle aspect ratio because prolate particles are more prone to cracking than oblate particles.

9.15 Summary

An approximate integration scheme has been presented to implement a secant-based homogenization theory for particlere inforced plasticity into an existing damage-based constitutive model for ductile fracture. The resulting model can account for the influence of the second-phase particles on void growth, shape evolution, coalescence and material softening. The stress state within particles can also be determined as a function of the particle content, shape and elastic properties. Additionally, the stress state within the matrix material can be estimated with no prior knowledge of its hardening profile. The results for the local stress states will be used to predict void nucleation through particle particle fracture and decohesion in the percolation modelling of ductile fracture in actual particle fields. The present model is best suited for application to sheet metal forming of damage-sensitive industrial alloys where the loading is proportional and the volume fraction of the second-phase particles is low, which is true for most metal forming processes.

Chapter 10
Modelling Void Growth to Coalescence in a 3-D Particle Field

10.1 Void Growth and Shape Evolution

Modeling of the growth and shape evolution of the voids and cracks in the percolation model requires certain assumptions since the evolution models are designed for spheroids and not for general 3-D ellipsoids. The main challenge in adapting these models to the general case is their implicit dependence upon the loading direction. By assuming a periodic distribution of axisymmetric voids, the void aspect ratio can be defined as a state variable with a definitive initial value that can evolve during deformation. However, in the general case, the loading direction is not constrained to a specific direction and the aspect ratio is not an independent variable, but a function of the loading direction. This point is best illustrated if we consider a penny-shaped void that has just nucleated from a cracked particle. If the principal loading direction happens to be aligned with the opening direction of the void, it will appear as a penny-shaped void as viewed from the loading direction and there is no issue. If the loading direction happens to be transverse to the void opening direction, the penny-shaped void appears as an extremely prolate or needle-shaped void that will experience negligible growth and shape evolution. Certainly the void growth and shape evolution rules for a penny-shaped void do not apply in this case.

With this in mind, all void/crack growth and shape evolution will be governed by the principal loading direction and how the geometry of the void appears from this direction. The development of general three-dimensional void evolution models that explicitly accounts for the viewing direction would be a welcome contribution to the percolation model.

Z. Chen and C. Butcher, *Micromechanics Modelling of Ductile Fracture*, Solid Mechanics and Its Applications 195, DOI 10.1007/978-94-007-6098-1_10,

© Springer Science+Business Media Dordrecht 2013

10.2 Procedure for Modeling Void Growth and Shape Evolution

Eliminating the void aspect ratio as a state variable requires the determination of the void dimensions and shape relative to the principal loading direction. Consider an arbitrary void or crack with the semi-axes (a, b, c) corresponding to the vectors $\mathbf{n}_1$, $\mathbf{n}_2$, $\mathbf{n}_3$. The principal loading direction is defined by the vector $\boldsymbol{u}$, as seen in Fig. 10.1. The distance from the center of the void to its surface along the direction $\boldsymbol{u}$, is denoted R_1. A line-projection of the void is then taken with $\boldsymbol{u}$ as the viewing direction to obtain an ellipse with semi-axes R_2 and R_3 that coincide with the $\boldsymbol{p}$ and $\boldsymbol{n}$ directions. The reconstructed geometry of the void as viewed from the principal loading direction is an ellipsoid with semi-axes, R_1, R_2, R_3 and the equivalent aspect ratio is defined as

$$W = \frac{R_1}{\sqrt{R_2 R_3}} \tag{10.1}$$

This aspect ratio is used to evaluate the void growth and shape evolution models but the porosity of the actual void is used in these models and not the porosity of the equivalent void.

The void growth and shape evolution rates can now be determined from knowledge of this effective aspect ratio. The void growth rate, $\dot{f}/f$, is determined using the calibrated void growth rule of Ragab (2004b) using the unit cell correlations obtained in Chap. 4. Similarly, the evolution rate of the effective void aspect ratio, $\dot{W}/W$, is also determined using the unit cell correlations in Chap. 4. These evolution rates were obtained for the equivalent void and are not truly representative of the actual void since these growth rates are not aligned with the semi-axes of the actual void. A clever application of the unit cell geometric relationships from Chap. 4

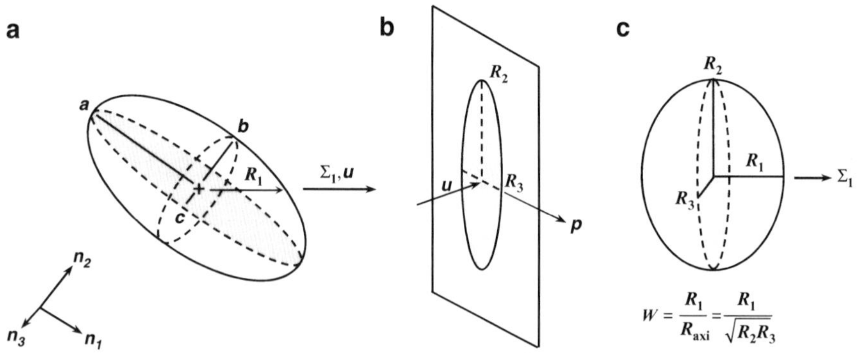

Fig. 10.1 Procedure for modeling an arbitrary ellipsoidal void as an axisymmetric void as viewed by the principal loading direction. (**a**) Ellipsoidal void, (**b**) void projection and (**c**) model ellipsoidal void

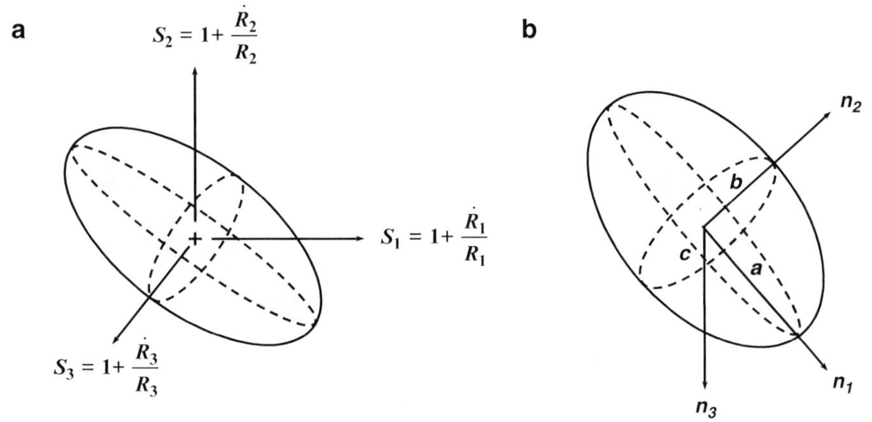

Fig. 10.2 Stretching of a void in three arbitrary directions resulting in a new void size and orientation. (**a**) Stretching of the void in the u, p, n directions and (**b**) new void size and orientation

enable the calculation of the radial growth rates of the equivalent axisymmetric void as a function of the principal strain rates as

$$\frac{\dot{R}_1}{R_1} = \dot{E}_{\text{hyd}} + \frac{1}{3}\left(\frac{\dot{f}}{f} + 2\frac{\dot{W}}{W}\right) \tag{10.2}$$

$$\frac{\dot{R}_2}{R_2} = \frac{\dot{R}_3}{R_3} = \frac{\dot{R}_1}{R_1} - \frac{\dot{W}}{W} \tag{10.3}$$

The evolution rates of the equivalent void semi-axes are taken as the growth rates of the actual void but in the directions $\boldsymbol{u}, \boldsymbol{p}, \boldsymbol{n}$. The problem can now be treated as the stretching of an ellipsoid along three arbitrary directions as shown in Fig. 10.2. The stretch values of the void in the $\boldsymbol{u}, \boldsymbol{p}, \boldsymbol{n}$ directions are

$$S_1 = 1 + \frac{\dot{R}_1}{R_1} \qquad S_2 = 1 + \frac{\dot{R}_2}{R_2} \qquad S_3 = S_2 \tag{10.4a, b, c}$$

The solution for the arbitrary stretching of an ellipsoid requires an eigenvalue solution where the eigenvalues are related to the new void semi-axes with their orientation defined by the eigenvectors. The solutions for the line projection of an ellipsoid, the distance from an ellipsoid center to its surface and for the arbitrary stretching of an ellipsoid are presented in Butcher (2011).

The novelty of this modeling procedure is that the evolution of the void orientation is naturally accounted for in the model, and this is not captured using the standard growth and shape models in the literature. The void will naturally grow and rotate itself to be aligned with the preferential loading direction. If the

The in-plane stresses can be determined by selecting an arbitrary coordinate system using the vectors ***u***, ***v***, and ***w***. Within this plane exists an angle, θ, that will rotate ***u*** and ***v*** about ***w*** to be aligned with the in-plane stress vectors. The solution for these vectors has been framed as an optimization problem to enable its extension in future work to minimize the plastic limit-load by accounting for the variation in the dimensions of the voids as a function of θ. For an assumed vector, $\boldsymbol{v} = [v_1,v_2,v_3]$, and center-to-center vector, $\boldsymbol{w} = [w_1,w_2,w_3]$, the trial in-plane stress is defined using a vector, $\boldsymbol{s} = [s_1,s_2,s_3]$, as a function of θ by

$$\Sigma_1^{ip}(\theta) = s^T \Sigma_{ij}^c s \tag{10.6}$$

$$\begin{aligned}
s_1 &= \left(v_1^2 + (1 - v_1^2)\cos\theta\right)w_1 + (v_1 v_2(1 - \cos\theta) - v_3 \sin\theta)w_2 \\
&\quad + (v_1 v_3(1 - \cos\theta) + v_2 \sin\theta)w_3 \\
s_2 &= (v_1 v_2(1 - \cos\theta) + v_3 \sin\theta)w_1 + \left(v_2^2 + (1 - v_2^2)\cos\theta\right)w_2 \\
&\quad + (v_2 v_3(1 - \cos\theta) - v_1 \sin\theta)w_3 \\
s_3 &= (v_1 v_3(1 - \cos\theta) - v_2 \sin\theta)w_1 + (v_2 v_3(1 - \cos\theta) - v_1 \sin\theta)w_2 \\
&\quad + \left(v_3^2 + (1 - v_3^2)\cos\theta\right)w_3
\end{aligned}$$

An iterative Newton–Raphson search quickly converges to the solution by iterating with respect to θ. The value of θ in the *n-th* + 1 iteration is

$$\theta_{n+1} = \theta_n - \frac{d\Sigma_1^{ip}(\theta)/d\theta}{d^2\Sigma_1^{ip}(\theta)/d\theta^2} \tag{10.7}$$

Once θ has been determined, the vectors ***u*** and ***v*** can be rotated to the optimal directions and the in-plane stress values can be determined using the standard stress transformations

$$\Sigma_1^{ip} = u^T \Sigma_{ij}^c u \qquad \Sigma_2^{ip} = v^T \Sigma_{ij}^c v \tag{10.8}$$

In many cases, void coalescence will be triggered by the maximum in-plane tensile stress but coalescence can also occur in the second in-plane stress direction depending on the void alignment and stress state. It is cautioned that a preoccupation with the maximum tensile stress governing coalescence can lead to erroneous predictions in equal-biaxial stretching when $\Sigma_2^{ip} \approx \Sigma_1^{ip}$. In the interest of being conservative, coalescence will also be evaluated in the second in-plane stress direction whenever $\Sigma_2^{ip} > 0$. Knowledge of the tensile stress along the vector, ***w***, is not important even in triaxial loading since the void spacing ratio will be zero as the voids are aligned in this direction and the plastic limit-load to coalescence will be infinite. This type of coalescence is rare and is known as 'necklace' coalescence and cannot be predicted by the plastic limit-load model.

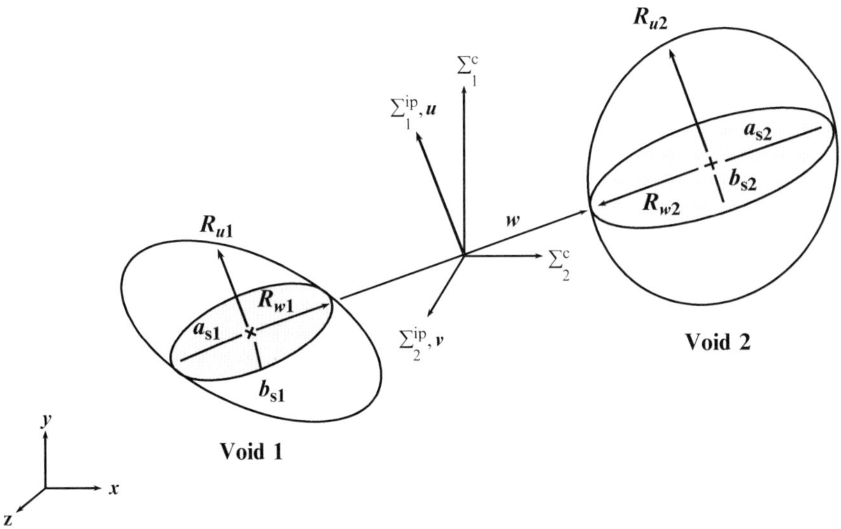

Fig. 10.4 Void coalescence geometry showing the identification of the relevant dimensions

10.3.3 Effective Geometry for the Evaluation of Coalescence

A procedure must be developed to effectively homogenize the two voids into an equivalent geometry of two identical voids to be amenable to the evaluation of the coalescence model. The plastic limit-load criterion requires knowledge of the void aspect ratio, W, spacing ratio, χ, and the maximum tensile stress transverse to the ligament. The maximum in-plane tensile stress has been determined in the optimization procedure above. The void spacing ratio is readily determined from the center-to-center distance, $d_{c2c} = \|x_{c1} - x_{c2}\|$, between the void centroids, $\boldsymbol{x}_{ci}$, as

$$\chi_c = \frac{R_{w1} + R_{w2}}{d_{c2c}} \tag{10.9}$$

where R_{w1} and R_{w2} are the distances from the voids, measured from their center to their surface along the center-to-center vector, $\boldsymbol{w}$, as seen in Fig. 10.4. The center-to-center distance is used for modeling coalescence so that the spacing ratio reduces to its original definition in the plastic limit-load formulation when the voids are aligned and have the same orientation. It is important to state that χ_c in Eq. (10.9) is not the absolute minimum spacing ratio between the two voids. The minimum spacing ratio is not employed because it is computationally expensive and requires an iterative solution for every void-neighbour pair in every element at each time step. Overall, the center-to-center spacing ratio is a good approximation to the minimum spacing ratio provided the voids are not very close together and have similar orientations.

The principal approximations and uncertainties introduced into the coalescence model are related to the determination of the equivalent aspect ratio. The voids have arbitrary sizes and orientations and are far from the idealized geometry of two identical spheroidal voids. The aspect ratio of the equivalent axisymmetric void is defined as

$$W_{eq} = \frac{R_u}{R_{axi}} \tag{10.10}$$

where R_u is the radius along the in-plane stress direction and R_{axi} is the equivalent axisymmetric radius. The height of the equivalent spheroidal void, R_u, can be estimated by averaging the distance from the void centers to their surfaces along the in-plane stress direction as

$$R_u = \frac{R_{u1} + R_{u1}}{2} \tag{10.11}$$

To determine the equivalent axisymmetric radius, each void is first sectioned transverse to the in-plane stress direction to obtain an ellipse with semi-axes (a_{si}, b_{si}) as shown in Fig. 10.4. For clarity, R_{w1} and R_{w2} are shown to be aligned with the section of the ellipse in Fig. 10.4 but the semi-axes of the sectioned ellipsoid are generally not equal to the R_w distance due to the void orientation.

Since the locations of the semi-axes in the section planes may be different for the two voids, these dimensions cannot be directly averaged. Instead an equivalent axisymmetric radius is defined for each void as

$$R_{1-axi} = \sqrt{a_{s1} b_{s1}} \quad R_{2-axi} = \sqrt{a_{s2} b_{s2}} \tag{10.12}$$

and the axisymmetric radius of the equivalent void and its aspect ratio are computed as

$$R_{axi} = \frac{R_{1-axi} + R_{2-axi}}{2} \qquad W_{eq} = \frac{R_{u1} + R_{u1}}{R_{1-axi} + R_{2-axi}} \tag{10.13, 10.14}$$

The coalescence criterion in Eq. (1.15) can now be evaluated using W_{eq}, χ_c, Σ_1^{ip}, and the matrix flow stress, $\bar{\sigma}$. The geometry of the voids used in the coalescence model is shown in Fig. 10.5.

10.3.4 Creation of a New Crack: Merging Operation

When the coalescence criterion is satisfied for two voids/cracks, a new crack must be created by merging the two voids together. The merging operation plays a significant role in the fracture process since the size of the resulting crack will

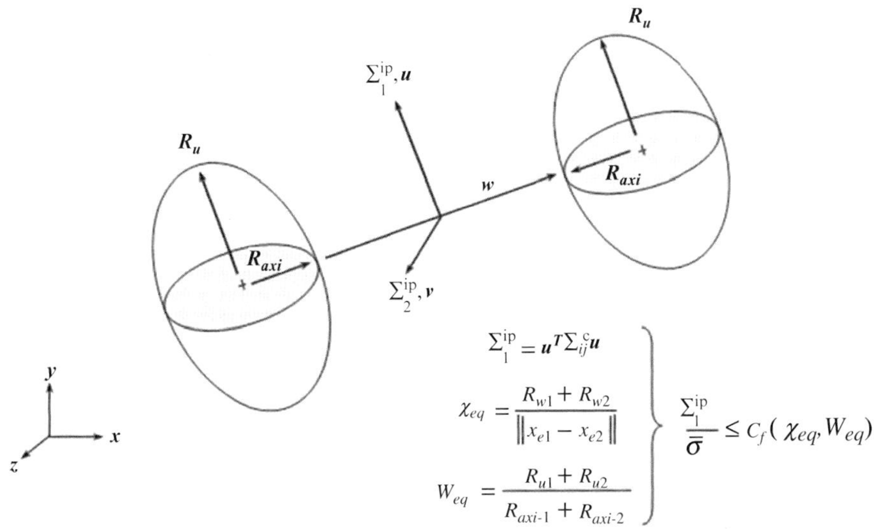

Fig. 10.5 Equivalent void geometry for evaluation of the plastic limit-load coalescence model

influence coalescence with other neighbouring voids. The simplest merging operation is to use the maximum void dimensions of the two voids to make a large crack that encompasses both voids as in Worswick et al. (2001). The disadvantage of this model is that the crack size is uncharacteristically large and coalescence quickly spreads throughout the particle field.

The bounding box method presented in Fig. 10.6 is used to define the new crack dimensions. The crack is assumed to be oriented with its semi-axes along the $\boldsymbol{u}$, $\boldsymbol{v}$, $\boldsymbol{w}$ vectors identified in the previous section for the in-plane stresses. The dimensions and points that will define the bounding box for the new crack must be identified in the three planes defined by $\boldsymbol{u}$, $\boldsymbol{v}$, $\boldsymbol{w}$. Each void bounding box will have four points that lie above and four points that lie below the plane and the distances from these points to the plane are calculated as D_{a1i}, D_{b1i} for void 1 and D_{a2i}, D_{b2i} for void 2. The semi-axis of the crack in this plane is then computed as

$$R = \frac{\max(D_{a1i}, D_{a2i}) + \max(D_{b1i}, D_{b2i})}{2} \qquad i = 1 \ldots 4 \tag{10.15}$$

This process is repeated for each plane to obtain the three semi-axes of the crack. The center of the crack can be determined by constructing the bounding box for the crack by using one void centroid as the reference point. The centroid is the midpoint of the bounding box coordinates.

The maximum distances above and below the planes are employed to ensure that the crack will progressively enlarge as it coalesces with other voids and cracks. The average void dimensions could be used but only for void-void coalescence and not for void-crack or crack-crack coalescence as the amplification effect of the crack

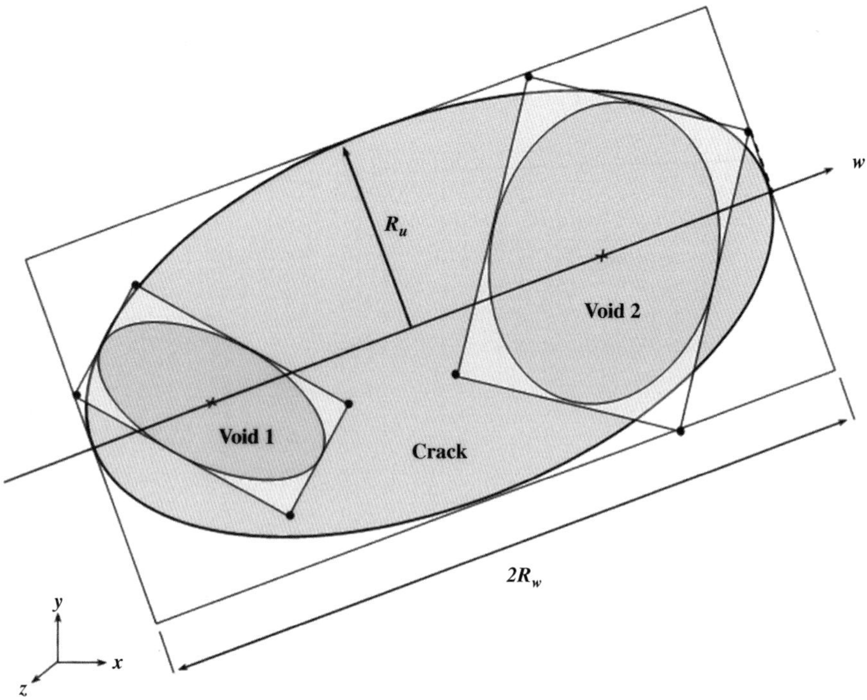

Fig. 10.6 Two-dimensional schematic of coalescence between two voids showing their bounding boxes and the creation of the new crack

size will be lost. If the voids that compose the crack belong to different parent elements, the parent element of the crack is identified as the element with the minimum distance from the element center to the crack centroid.

Overall, this merging procedure is about as good as one can achieve without resorting to an optimization algorithm to determine the minimum ellipsoid that envelops both voids. This merge procedure works very well for most void geometries except for coalescence between a large and small void where the crack size is overestimated as shown in Fig. 10.7. Fortunately, this does not become significant until large cracks have first formed which typically does not occur until prior to fracture when coalescence is sweeping throughout the element and failure is imminent. Nevertheless, future work could extend the merge operation to account for the relative sizes of the voids in determining the size of the new crack.

10.3.5 Mandatory Coalescence: Object Intersections

It is not uncommon for voids and cracks to overlap as they evolve with deformation or coalesce with their neighbours. This type of void/crack impingement is not realistic since the material between the voids will fail before the voids come into

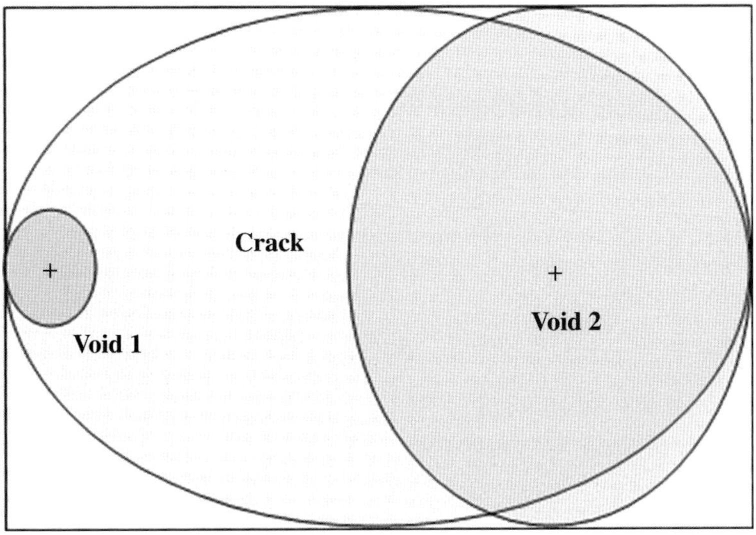

Fig. 10.7 Schematic of coalescence between a large and small void and the resulting crack geometry

contact (Thomason 1990). As a result, when a void or crack overlaps with another neighbouring void, the coalescence process is enforced using the merging process described above. The minimum distance algorithm of Lin and Han (2002) is evaluated prior to testing for coalescence at each time step. There are no restrictions placed upon particles coming into contact with voids, cracks or other particles for the sole reason of reducing the number of intersection tests. Each percolation element may contain thousands of particles and their overlapping is of secondary importance compared to the voids. A worthwhile extension would be to treat the particles as rigid bodies that cannot intersect to model the creation of deformation-induced particle clusters as the particles pile up with their neighbours. The proximity of a particle next to an existing crack would also promote premature void nucleation.

10.4 Development of the Percolation Model

In this chapter, the void growth and evolution models developed in the previous chapters are amalgamated into a complete framework to model damage evolution in a material with a heterogeneous particle distribution. The void evolution models are adapted to a general three-dimensional particle field and stress state by removing the periodic assumption used in their development in Chap. 5. The percolation model was then integrated into a commercial finite-element code, LS-DYNA (Hallquist 2006) to create so-called 'percolation elements' by mapping the particle distributions to the elements. This enables the percolation model to be directly coupled to the macro-scale as damage evolution at the micron-scale will control the stress state in the elements, and ultimately, that of the entire structure via the FE method. This represents a significant contribution in the modeling of ductile fracture as the particle distribution can be accounted for in finite-element simulations of metal forming

The development of a sophisticated constitutive model of this scale requires careful consideration and justification of each step in the modeling process to ensure that the model has a strong physical foundation. The development of the percolation model will be separated into two main categories: macroscopic models and microscopic models. The macroscopic models are used to transition from the micro-scale to the macro-scale and vice versa.

10.4.1 Principal Assumptions

The principal advantage of the damage percolation model is that the microstructure no longer is assumed to be periodic and a measured particle/void distribution can be used to model ductile fracture. A concerted effort has been made in this work to improve the physical foundation of the percolation model and significant progress has been achieved in reducing the number of assumptions inherent in the modeling of ductile fracture. Nevertheless, extending the micromechanical models to a general three-dimensional case requires numerous approximations and assumptions. The principal assumptions used in the development of the percolation model are:

- The stress state within the element is uniform and homogeneous. Local stress and strain heterogeneity within void and particle clusters is neglected.
- The particles, voids and cracks do not interact. Voids/cracks grow as isolated voids.
- Void evolution models obtained using axisymmetric unit cells can provide a reasonable approximation to the growth of voids in a real material if the shear stress is small compared to the tensile stress.
- The particles, voids and cracks remain ellipsoidal during deformation.
- Nucleation via debonding is not explicitly modeled. An emphasis is placed upon void nucleation via particle cracking.
- Void coalescence occurs by necking failure of the inter-void ligament. Ligament shearing is not considered.
- Size effects are negligible at the length scales considered and continuum-based models for void initiation and evolution are valid at the micron-scale and higher. Deformation of voids and particles that are smaller than 1 μm would require considering dislocation dynamics.
- The grains and texture-related effects are not accounted for in the model. It is assumed that the bulk properties of the particle field are isotropic and rate-independent although deformation of the voids at the local-scale is dependent upon the loading direction.

Each of the above assumptions will introduce a degree of uncertainty into the model with the major limitations being the assumption of a uniform stress distribution, no shear coalescence, negligible size effects and neglecting anisotropy and the grains in the material. The rationale for these assumptions is discussed in the following sections.

10.4.1.1 Assumption of a Homogeneous Stress State – No Void and Particle Interactions

Although the stress state is assumed to be homogeneous within the percolation element, the stress state is not homogeneous because severe local strain-gradients develop within particle and void clusters. As a result, the voids and particles will interact and not evolve as isolated objects. This assumption is unlikely to be overcome without explicitly modeling every void and particle using finite-element techniques; however, it can be improved by using a sufficient number of percolation elements. In the limit, each percolation element would contain a single void/particle and the isolated assumption becomes realistic.

10.4.1.2 Assumption of Internal Necking Coalescence as the Dominant Coalescence Mode

The omission of a shear-based coalescence model is an unfortunate limitation of the present work and is attributed to the lack of a robust model akin to the plastic limit-load criterion for internal necking coalescence. Butcher and Chen (2009a) proposed a shear-extension to the plastic limit-load for combined tension and shear based upon the work of Xue (2008). However, this model is better suited for a general constitutive model and not for individual voids. A recent work by Schyvearts et al. (2011) has found that the plastic limit-load model can give good predictions for necking coalescence in combined tension and shear by accounting for the void orientation with shear. The percolation model is also expected to perform well in this situation as it accounts for the orientation of the voids and cracks as well as their rotation when they evolve and change shape in shear loading.

10.4.1.3 Assumption of Size-Effects Being Negligible Above the Micron-Scale

All multi-scale models are only valid for a range of length scales. The percolation model has been designed to model void initiation and evolution from the micron-scale to the macro-scale. The typical size of a second-phase particle in many alloys is on the order of several microns or larger in diameter. It is assumed that continuum-scale models can be used to describe the deformation of the particles and voids at this scale. Size effects are unavoidable and it is well known that sub-micron sized voids tend to grow slower than their larger counterparts (Liu et al.

2003; Tvergaard and Niordson 2004; Wen et al. 2005). At the sub-micron scale, dislocation dynamics must be considered using a strain-gradient plasticity model (Fleck and Hutchinson 1997) that includes a length scale parameter for the size effects. Fortunately, a numerical study by Wen et al. (2005) has found that the void size effect is limited in uniaxial tension and for a small volume fraction of voids. Since most metal forming operations occur in the low triaxiality regime ($T < 1$) and have failure porosities on the order of several percent at most, the size effect can be neglected as a first-order approximation of the material behavior. An extension of the percolation model that accounts for dislocation dynamics on void nucleation and evolution would improve the physical foundation of the model and enable it to incorporate events from the nano-scale to the macro-scale.

10.4.1.4 Neglecting the Influence of Texture Effects and Grains

The percolation model does not account for influence of material texture or for the presence of grains in the material. This assumption implies that it is the presence of the micro-voids in the material that are responsible for the promotion of localization and fracture. For example, in a uniaxial tensile test, geometric softening causes the material to form a localized neck. The stress state in the center of the necked region becomes severe and spurs void nucleation and growth, softening the material in the neck and causing additional necking. Ultimately, the sample fails as the voids coalesce throughout the necked region. This type of fracture mechanism can be well described using a damage-based approach such as the percolation model.

This type of fracture process is not always the case since anisotropy and the grain structure of the material can be responsible for the initial localization by the formation of shear bands. These shear bands provide the high local plastic strains required to drive void nucleation and growth, softening the material in the shear band and promoting additional localization until fracture occurs. In this situation, the isotropic percolation model would overestimate the strain and porosity at fracture since higher levels of porosity would be required to trigger the localization process without the presence of the shear bands.

Overall, the percolation model provides an excellent tool for the modeling of ductile fracture and can describe a large portion of the fracture process. A future extension of the percolation model should account for material anisotropy in the yield criterion and void evolution models. The influence of the grains could be accounted for by coupling the percolation model with a crystal plasticity model. The coupled crystal plasticity – percolation model would provide insight into the complete fracture process by accounting for the entire microstructure.

10.4.1.5 Terminology Used in Development of the Percolation Model

The particle field is assumed to be composed of particles/inclusions, voids and cracks and referred to as ‘objects’ if the modeling treatment is the same for each constituent. Particles are modeled as a single class but their composition and type may be variable to accommodate different particle types in the material. Cracks are subjected to the same modeling treatments as voids but are defined as a crack to discriminate between the primary and nucleated voids with the voids that form due to coalescence. A crack is formed by the coalescence between a void and a crack, two voids or two cracks.

10.4.2 Macroscopic Models

Quantities at the element scale are defined as macroscopic because the stress state within the element is obtained by homogenizing the voids within the microstructure using the GT yield criterion. In this definition, the element is analogous to the typical unit cell except the cell contains a particle field instead of a single void. The purpose of the macroscopic modeling process is to accept the nodal displacements from the finite-element code and determine the stress and strain according to the adopted constitutive model for the element. The failure criterion for the element is then evaluated and the element is deleted or the stress returned to the finite-element solver. This is the typical procedure for any finite-element program with the only notable difference being that the constitutive model is rather complex. An overview of the macroscopic modeling process is presented in Fig. 10.8.

The elements provide the link to the relevant length scales for engineering (mm and higher) because the stress state within the elements controls the deformation of the global structure. Therefore, a brief review of the relevant kinematics of the finite-elements is required as they play a major role in the percolation model.

10.4.3 Relevant Finite-Element Kinematics

Finite-element discretization is achieved in the initial (reference) configuration by using either four or eight node isoparametric elements to interpolate the position of the element nodes, X_a, during deformation as

$$X = \sum_{a=1}^{n} N_a(\xi, \eta, \zeta) X_a \tag{10.16}$$

where $N_a(\xi, \eta, \zeta)$ are the standard shape functions defined in dimensionless element coordinates and n is the number of nodes (Fig. 10.9). During deformation the current position of the nodes, $x_a(x, y, z)$ as well as the nodal velocities, v_a, can be expressed in terms of the shape functions as

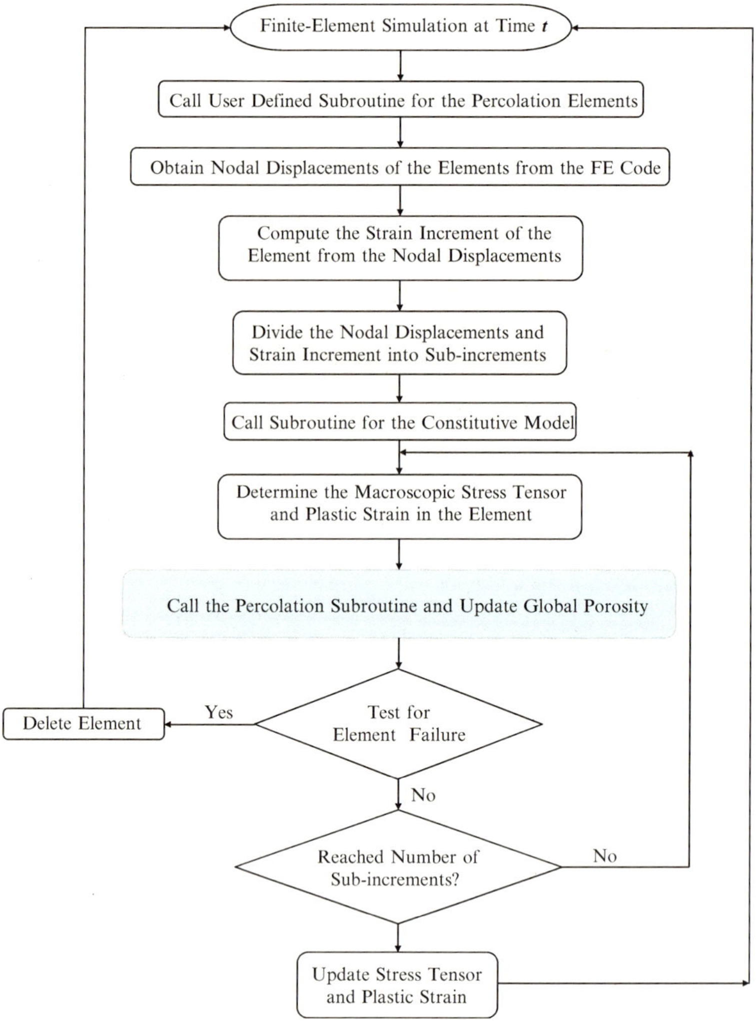

Fig. 10.8 Flow-chart of the macroscopic modeling process

$$x = \sum_{a=1}^{n} N_a x_a(t) \qquad v = \sum_{a=1}^{n} N_a v_a \tag{10.17, 10.18}$$

By enforcing the displacement of the nodes by an arbitrary increment, $\boldsymbol{u}_{\mathrm{a}}$, the displacement can be interpolated as

$$u = \sum_{a=1}^{n} N_a u_a \tag{10.19}$$

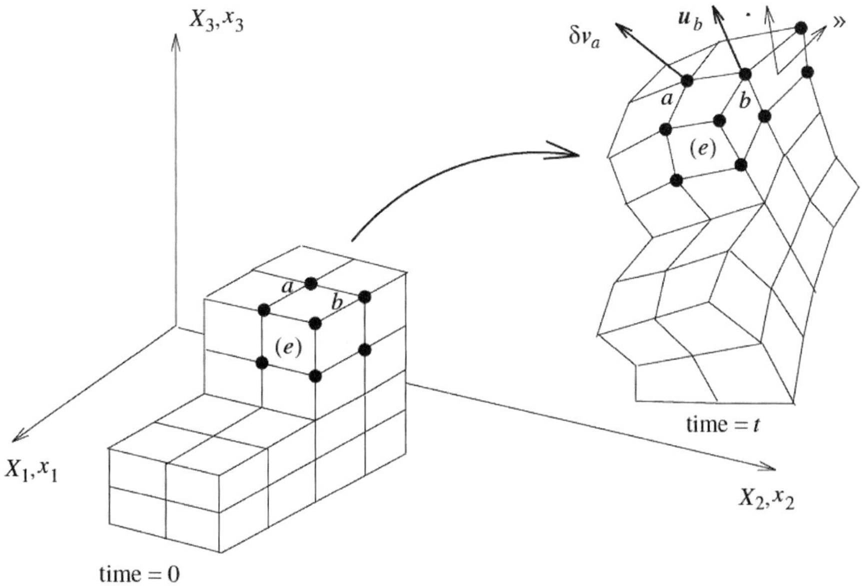

Fig. 10.9 Finite-element discretization (Reprinted with permission from Bonet and Wood (1997). Copyright 1997 Cambridge University Press)

The deformation gradient tensor, F, maps the element from its initial configuration to its deformed configuration and can be interpolated within the element using the relations

$$F = \sum_{a=1}^{n} x_a \otimes \frac{\partial N}{\partial x} \qquad \frac{\partial N}{\partial x} = J^{-1}\frac{\partial N}{\partial \xi} \qquad J = \sum_{a=1}^{n} x_a \otimes \frac{\partial N_a}{\partial \xi} \tag{10.20a, b, c}$$

where J is the Jacobian matrix. The deformation gradient can be used to determine the strain measure of interest such as the Cauchy-Green strain. The strain matrix for specific elements will be discussed in a subsequent section.

The velocity gradient tensor, L, is related to the rate of change of the deformation gradient as

$$L = \dot{F}F^{-1} \tag{10.21}$$

The deformation gradient tensor can be decomposed into a stretch and rotation tensor. The rotational rate of deformation can be measured using the antisymmetric spin tensor, Ω, as

$$\Omega = \frac{1}{2}(L - L^T) \tag{10.22}$$

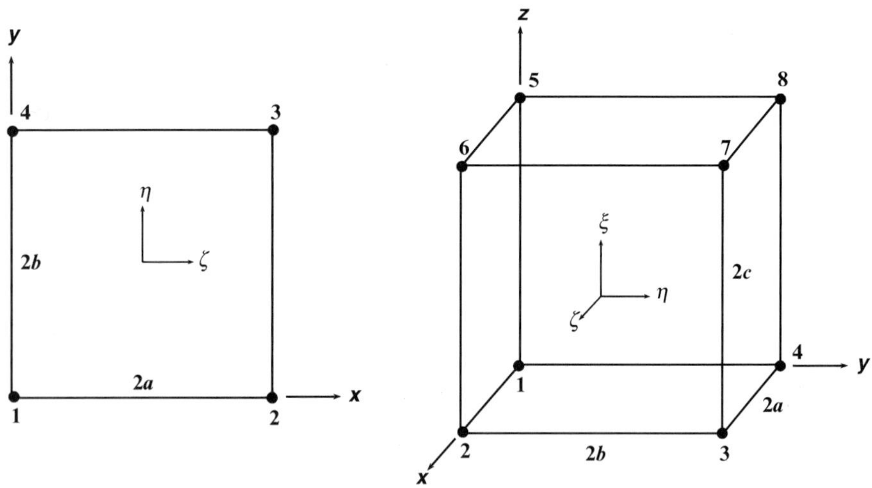

Fig. 10.10 Local node numbering and coordinate systems for a 4-node (*left*) and 8-node (*right*) isoparametric element

10.4.4 Percolation Element Types

The types of percolation elements considered are eight-node isoparametric brick elements and four-node isoparametric plane elements. The type of element integration such as reduced-point integration is controlled by the finite-element code. The geometry and local node numbering of the elements are shown in Fig. 10.10.

10.4.4.1 Eight-Node Isoparametric Brick Element

The geometry and local node numbering of the eight-node brick element is shown in Fig. 10.10. The center of the element is computed from the element nodal points as

$$x_c = \frac{1}{n}\sum_{a=1}^{n} x_a \tag{10.23}$$

where n is the number of element nodes. The half lengths of the element sides (a, b, c) are

$$a = \frac{1}{8}[(x_2 - x_1) + (x_3 - x_4) + (x_6 - x_5) + (x_7 - x_8)] \tag{10.24a}$$

$$b = \frac{1}{8}[(y_4 - y_1) + (y_3 - y_2) + (y_7 - y_6) + (y_8 - y_5)] \tag{10.24b}$$

$$c = \frac{1}{8}[(z_5 - z_1) + (z_6 - z_2) + (z_7 - z_3) + (z_8 - z_4)] \tag{10.24c}$$

Any point of interest within the element can be transformed to dimensionless element coordinates using the relations

$$\xi = \frac{x - x_c}{2a} \qquad \eta = \frac{y - y_c}{2b} \qquad \zeta = \frac{z - z_c}{2c} \qquad (10.25\text{a, b, c})$$

The shape functions for an isoparametric eight-node brick element at a node, a, are

$$N_a = \frac{1}{8}(1 + \xi_a \xi)(1 + \eta_a \eta)(1 + \zeta_a \zeta) \qquad (10.26)$$

with derivatives

$$\frac{\partial N_a}{\partial \xi} = \left(\frac{\xi_a}{1 + \xi_a \xi} \right) N_a \qquad (10.27)$$

The strain in the element can be interpolated using the shape functions. For an eight-node brick element, the macroscopic strain rate tensor is defined as

$$\dot{E}_{ij} = \begin{Bmatrix} \dot{E}_x \\ \dot{E}_y \\ \dot{E}_z \\ 2\dot{E}_{xy} \\ 2\dot{E}_{yz} \\ 2\dot{E}_{xz} \end{Bmatrix} = \sum_{a=1}^{n} \begin{bmatrix} \frac{\partial N_a}{\partial x} & 0 & 0 \\ 0 & \frac{\partial N_a}{\partial y} & 0 \\ 0 & 0 & \frac{\partial N_a}{\partial z} \\ \frac{\partial N_a}{\partial y} & \frac{\partial N_a}{\partial x} & 0 \\ 0 & \frac{\partial N_a}{\partial z} & \frac{\partial N_a}{\partial y} \\ \frac{\partial N_a}{\partial z} & 0 & \frac{\partial N_a}{\partial x} \end{bmatrix} \begin{Bmatrix} u_a \\ v_a \\ w_a \end{Bmatrix} = \sum_{a=1}^{n} \mathrm{B}_a u_a \qquad (10.28)$$

10.4.4.2 Four-Node Isoparametric Plane Element

The relevant equations for a four-node isoparametric element can be readily determined from the equations for the eight-node element above by replacing the factor of 1/8 with 1/4 and setting any z-related quantities to zero. The strain rate tensor for the plane elements is slightly different and is written as

$$\dot{E}_{ij} = \begin{Bmatrix} \dot{E}_x \\ \dot{E}_y \\ \dot{E}_z \\ 2\dot{E}_{xy} \end{Bmatrix} = \sum_{a=1}^{n} \begin{bmatrix} \frac{\partial N_a}{\partial x} & 0 \\ 0 & \frac{\partial N_a}{\partial y} \\ 0 & 0 \\ \frac{\partial N_a}{\partial y} & \frac{\partial N_a}{\partial x} \end{bmatrix} \begin{Bmatrix} u_a \\ v_a \end{Bmatrix} + \begin{Bmatrix} 0 \\ 0 \\ \dot{\varepsilon}_z \end{Bmatrix} = \sum_{a=1}^{n} \mathrm{B}_a u_a + \dot{E}_z \qquad (10.29)$$

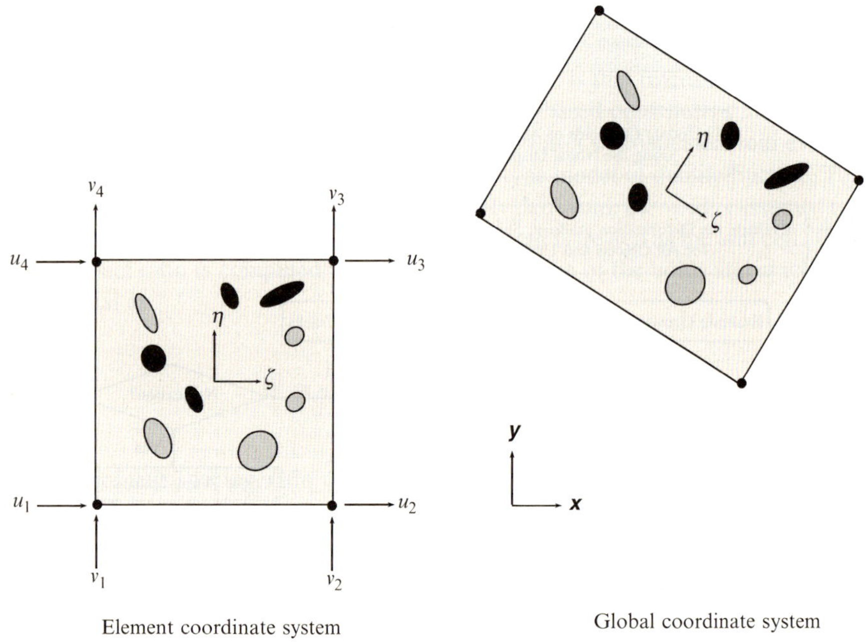

Fig. 10.12 Percolation element in the dimensionless element coordinate system (*left*) and in the global coordinate system (*right*) where the voids and particles have translated and rotated with the deformation of the element

10.5.1.2 Object Rotation

The rigid body rotation of the objects is computed from the material spin tensor of the element in Eq. (10.22). However, this rotation is not sufficient to describe the complete motion of the objects within the material as they may rotate due to plastic deformation within the element. In uniaxial tension, the material spin is zero but the objects will still rotate to align themselves with the principal loading direction. This effect should be considered because of its implications on the fracture mechanism because the objects will orient themselves to create favourable conditions for nucleation, growth and coalescence. Voids are more likely to coalesce when they are aligned transversely to the principal loading direction and particles will crack transversely to the loading direction. Particle and void rotation is especially important in torsion as shown in Fig. 10.13 where the voids and particles will rotate to align themselves at 45° to the loading direction.

The influence of plasticity-induced rotation has largely been neglected in the modeling of damage-induced ductile fracture. Typically, the voids have been assumed to remain stationary or to rotate with the material spin (Benzerga 2002; Pardoen 2006). More recent models by Keralavarma and Benzerga (2010) and Schyvaerts et al. (2011) have employed the nonlinear homogenization solution of Kailasam and Ponte Casteneda (1998) in its reduced form for voids as used by

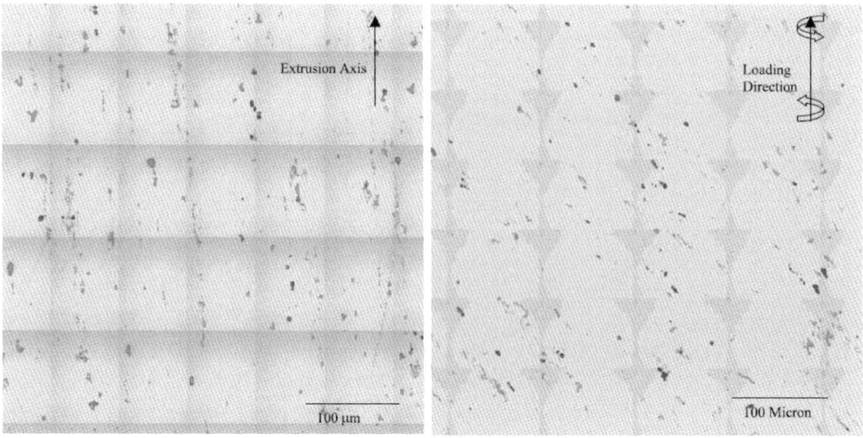

Fig. 10.13 Digitally compressed montage of 500 × 500 μm² area of an undeformed AA6061 alloy (*left*) and a montage after being subjected to a strain of 98 % in torsion (*right*) (Reprinted with permission from Agrawal et al. (2002). Copyright 2002 Elsevier)

Aravas and Ponte Casteneda (2004). These solutions are based upon a rigorous homogenization scheme to obtain expressions for the plastic spin rate in a composite material. This model has been shown to give very good agreement with the void rotation as validated by Schyvaerts et al. (2011) using finite-element simulations (Fig. 10.14).

The model of Kailasam and Ponte Casteneda (1998) will also be adopted to account for particle and void/crack rotations within the percolation element, albeit in a slightly reduced version. The solution of Kailasam and Casteneda (1998) can account for any number of different particle types in addition to the voids as well as their distributions. However, the computation of the spin rates requires many intermediate calculations of fourth-order tensors and matrix operations. With thousands of particles and voids within the percolation element and different particle types to be evaluated, this can be a computationally expensive process. Therefore, the rotation rate of each object type will be evaluated by neglecting interactions with the other object types, considerably reducing the expression for the rotation rate.

By also neglecting the distributional effects of the ellipsoidal objects within the material, the average deformation rate, D^{o}, and the average spin, Ω^{o}, of the local representative ellipsoidal object can be expressed as

$$D^{o} = A^{o} : \dot{E}^{p} \qquad \Omega^{o} = \Omega - C^{o} : \dot{E}^{p} \qquad (10.37, 10.38)$$

Where $\boldsymbol{\Omega}$ is the material spin rate of the element and $\mathbf{A}^{o}$ and $\mathbf{C}^{o}$ are fourth-order “concentration tensors” for the objects and are defined as

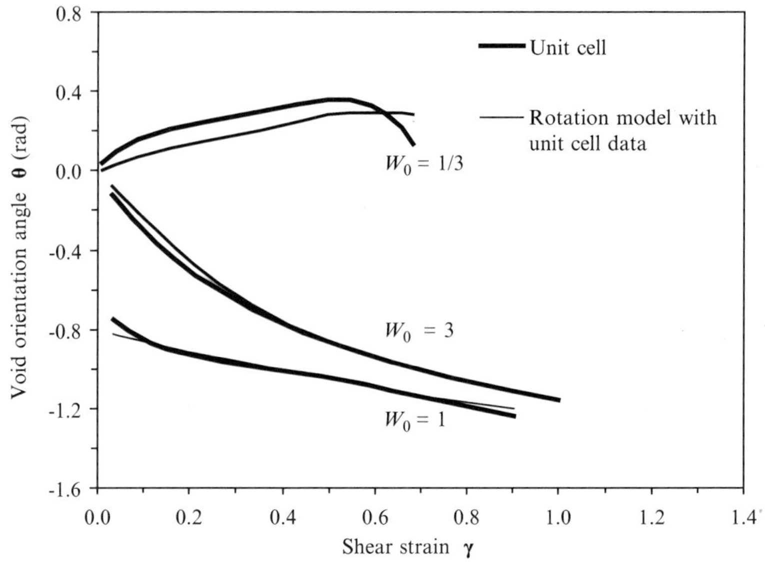

Fig. 10.14 Comparison of the analytical rotation model of Kailasam and Ponte Casteneda (1998) with the void rotation obtained from unit cell simulations (Reprinted with permission from Schyvaerts et al. (2010). Copyright 2010 Elsevier)

$$\mathbf{A}^{\mathrm{o}} = \left[\mathbf{I} - (1 - f^o)\mathbf{S}(\mathbf{I} - [\mathbf{L}^{\mathrm{M}}]^{-1}\mathbf{L}^o)\right]^{-1} \tag{10.39}$$

$$C^{\mathrm{o}} = (1 - f^o)\Pi^o\left[\left(\mathbf{L}^o - \mathbf{L}^{\mathrm{M}}\right)^{-1}\mathbf{L}^{\mathrm{M}} + (1 - f^o)\mathbf{S}\right]^{-1} \tag{10.40}$$

where f^o is the object volume fraction; $\mathbf{L}^{\mathrm{M}}$ and $\mathbf{L}^{\mathrm{o}}$ are the fourth-order viscosity (elastic) tensors of the matrix and object that are defined as a function of their shear and bulk moduli as $\mathbf{L} = (3\kappa, 2\,\mu)$. The fourth order tensors, $\mathbf{S}$ and $\mathbf{\Pi}$, are Eshelby tensors (1957) whose expressions are given in Butcher (2011). For the rotation of the voids and cracks, the expressions in Eq. (10.39) and (10.40) are computed by setting $L^o = 0$. In the limit that the object volume fraction approaches zero, the solutions reduce to those of Eshelby (1957).

The ellipsoidal object is defined as having semi-axes (a, b, c) where $a > b > c$ are corresponding vectors, $\mathbf{n}_1$, $\mathbf{n}_2$, $\mathbf{n}_3$. The microstructural spin, ω, in the global frame is defined as

$$\omega = \Omega^o + \frac{1}{2}\sum_{\substack{i,j=1\\ i\neq j\\ w_i\neq w_j}}^{3} \frac{w_i^2 + w_j^2}{w_i^2 - w_j^2}\left[\left(\boldsymbol{n}_i\boldsymbol{n}_j + \boldsymbol{n}_j\boldsymbol{n}_i\right) : D^o\right]\boldsymbol{n}_i\boldsymbol{n}_j \quad (w_3 = 1) \tag{10.41}$$

where w_i are the aspect ratios of the object and are

$$w_1 = \frac{c}{a} \qquad w_2 = \frac{c}{b} \qquad w_3 = 1 \tag{10.42a, b, c}$$

Finally, the plastic spin of the object is expressed as

$$\Omega^p = \Omega - w \tag{10.43}$$

and the rotation rates of the semi-axes of the ellipsoid relative to the rotation of the continuum are

$$\dot{\boldsymbol{n}}_i = \Omega^p \boldsymbol{n}_i \tag{10.44}$$

The plastic spin will be zero if the material spin is zero and the object is aligned in the loading direction. Otherwise, the object will rotate until it is aligned with the principal loading direction. In the case that two of the aspect ratios are equal such as $w_1 = w_2$, Ω_{12}^p in Eq. (10.43) becomes indeterminate and is set to zero (Aravas and Ponte Casteneda 2004). If the object is spherical, the material is locally isotropic the plastic spin vanishes since $\mathbf{C}^o = 0$.

10.5.2 Implications of the Rotation Model on the Percolation Model

The adoption of the above rotation model significantly improves the physical foundation of the percolation model and has implications on void nucleation, growth and coalescence as these are all related to the direction of the maximum principal stress. During plastic deformation, the predictions of the rotation model are improved by computing the elastic tensor of the matrix, $\mathbf{L}^M$, using the secant shear modulus, $\bar{\mu}_s$, determined as part of the homogenization process to determine the stress within the particles. In this manner, another level of integration is achieved in the percolation model.

The rotation model can be evaluated for the bulk values of each object type and applied uniformly to rotate each object within the particle field. However, this method neglects the individual object dimensions and any strain-gradients that exist within the element. This has been accounted for in the implementation of the model by evaluating the material deformation gradient, spin tensor and strain increment at the centroid of each object in the element using the equations in the previous section. The equations for the local plastic spin can then be determined using the above procedure by assuming that the local strain rate is a close estimate to the local plastic strain rate. This implementation provides an immense contribution to the model when shear loading is present as the shear-induced rotational distribution is captured by the model. Objects located in the center of the material will experience less rotation than the objects near the surface of the material where the shear traction is applied. Overall, the treatment of particle and void rotation in the percolation model is physically sound and superior to the existing implementations of this model.

10.5.3 Neighbours – Element and Object

Since fracture is a local phenomenon originating within specific regions of the material, the spatial distribution of the neighbours of each object must be accounted for. This is accomplished by creating neighbour lists on the macro- and the micro-scale through the use of characteristic length parameters. If the distance from the centroid of an object to the centroid of a neighbour object is less than the characteristic length, it is appended to a neighbour list of that object. The size of the characteristic length controls the sphere of influence of the object. In the percolation model, up to four distinct characteristic lengths can be input to determine element neighbours, particle neighbours and void/crack neighbours.

At the global level, a characteristic length can be defined for the elements, L_e, to locate the neighbouring percolation elements and identify the objects within them for potential interactions (Fig. 10.15). This characteristic length is not a significant parameter except for crack propagation prior to final fracture as it enables the crack front to progress through neighbouring elements. For the most part, this parameter is used to reduce the number of interaction searches for each time step by limiting the size of the global neighbourhood. The simplest choice of selecting element neighbours is by using common nodes so that only the surrounding elements are considered. If more elements are required, the center-to-center distance between elements is used to identify the neighbours.

Within the global neighbourhood defined by the element length, up to three additional characteristic lengths can be adopted for:

Particle-particle neighbours: L_{pp}
Particle-void/crack neighbours: L_{pv}
Void-void, void-crack, or crack-crack neighbours: L_{vc}

It is important to have different characteristic lengths to be able to easily test for different interactions. For example, in modeling void nucleation within particle clusters, the particle-particle characteristic length will play a central role which is not likely to be the same as for void-crack interactions. The particle-void characteristic length could be used in modeling nucleation since the proximity of a particle to a crack will increase its propensity to nucleate a void. In the simplest case where no interactions are considered (isolated voids and particles) a single length can be used to define the local neighbourhood. In this case, any reasonable choice for this value such as five average diameters will be sufficient to identify the neighbouring particles, voids and cracks for nucleation and coalescence.

In the event of void coalescence, the new void inherits the neighbours of its parent particle and the new void is added to the lists of its neighbours. Similarly,

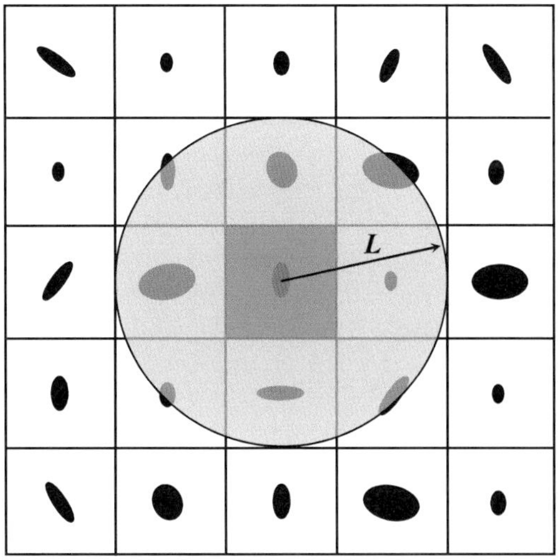

Fig. 10.15 Schematic of the identification process for neighbouring elements in an assemply of series of percolation elements at the continuum-scale where the voids within each percolation element have been homogenized into a single void located in the *center* of the element. This process is a natural analog to the non-local treatments of void damage

when two voids or cracks coalesce, the new crack will inherit the neighbour lists both parents and is added to these neighbours' lists as well. In this manner, coalescence will naturally sweep through the particle field, inheriting neighbours and coalescing with them as cracks propagate through the elements. The advantage of this method is that it is computationally efficient, physically sound, and relegates the characteristic length to a third-tier parameter. Of course, if the computational resources are available, every object within every element can be compiled into a neighbour list and no characteristic lengths are required.

10.6 Summary

Damage percolation modeling provides a unique tool to study and understand ductile fracture in heterogeneous materials. The present model has made significant in-roads into developing a physically sound framework to model fracture from first principles. The significant features of the present percolation model are:

- The model accepts a three-dimensional particle distribution of arbitrary ellipsoidal particles and voids.
- The particle fields are mapped to finite-elements to capture the development of the complex stress- and strain-gradients that develop in the microstructure.
- The initiation and evolution of damage at the micro-scale controls the bulk material behavior of the element and ultimately the structure in the finite-element model
- The Gurson-based yield criterion has been calibrated through the development of a library of correlations to model void evolution.

- The stress state within the particles is determined based upon a homogenization scheme and is a function of the particle shape, size and composition.
- A sophisticated void nucleation model for particle cracking is used to predict nucleation as a function of the particle stress (from homogenization theory), composition, fracture toughness, size and shape.
- Physically sound treatments for void growth, shape and coalescence have been developed that place no restriction on the void shape or orientation.
- The void growth, shape evolution and coalescence models have been calibrated and validated from an extensive study of voided unit cells.
- An advanced rotation law based upon homogenization theory has been adopted to account for the plastic spin of the particles, voids and cracks within the material.
- No material calibration parameters are required by the model or have been introduced. All parameters are directly related to the material such as the particle composition, flow stress relation of the matrix, etc. The only parameter that could be calibrated with experiment would be the fracture toughness of the particles if it cannot be predetermined. Otherwise fracture is a natural consequence of microstructure evolution.

The theoretical development of the percolation model concludes with this chapter. The subsequent chapters will develop a particle field generator to populate the percolation elements with statistically representative particle fields. Finally, the complete finite-element percolation model will be evaluated in Chap. 11 to predict damage initiation and failure in an aluminum-magnesium alloy.

Chapter 11
Application of the Complete Percolation Model

The performance of the complete percolation model is evaluated by applying the model to a notched tensile test specimen of AA5182 sheet. All of the fundamentals established in the previous chapters are used in this application. The calibrated void evolution models in Chap. 4 are used to model the individual voids in the particle field. The homogenization scheme of Chap. 9 is used to obtain the stress state in the particles as a function of their composition, size and shape to predict nucleation using the particle cracking-based model of Chap. 9. The particle field generator of Chap. 8 is used to generate five representative particle fields of AA5182 to capture the experimentally observed variation in the fracture strains. The percolation elements developed in Chap. 10 are placed at the notch root of the sample to capture the initiation of a macro-crack as observed experimentally. Finally, the predicted fracture strains, porosity and nucleation trends are compared and validated with the experiment data and as well as the data available in the literature.

11.1 Void Nucleation by the Constituent Particles

The porosity and number of nucleated voids in AA5182 were measured using x-ray microtomography by Orlov (2006) in a tensile sample at five strain levels. It was observed that all of the nucleated voids were attributed to the Fe-rich particles and that the Mg_2Si particles did not appreciably nucleate voids. A similar result has been reported by Agarwal et al. (2002) for an AA5083 alloy. From these results we can safely assume that void nucleation by the Mg_2Si particles is of minor importance. Therefore, all of these particles are assumed to remain intact in the percolation model. The Fe-rich particles are hard and brittle. It is assumed that they nucleate via particle cracking using the nucleation model described in Chap. 9. It should be noted that even if a percentage of the Mg_2Si particles did nucleate voids, the volume fraction of this phase is only one-tenth that of the Fe-rich particles and their contribution would be marginal by comparison.

Z. Chen and C. Butcher, *Micromechanics Modelling of Ductile Fracture*, Solid Mechanics and Its Applications 195, DOI 10.1007/978-94-007-6098-1_11,
© Springer Science+Business Media Dordrecht 2013

11.1.1 Material Properties of the Constituent Particles

The properties of the constituent intermetallic particles are not often known and must be estimated from their composition. Fortunately, the composition of the Mg_2Si particles is well documented since this intermetallic has gained interest as a possible structural material due to its low density of 1.99 g/cm^3, high melting point, and compressive strength of 1,640 MPa (Xiong et al. 2007). The elastic modulus and Poisson's ratio of these particles are taken as 133.5 GPa and 0.161, respectively (Tani and Kido 2008). The fracture toughness of the Mg_2Si particles is not required since they are not considered for nucleation but have a fracture toughness ranging from 0.6 to 1.2 MPa-m$^{1/2}$.

The composition of the Fe-rich particles depends upon the casting process such as casting method, cooling time and temperature. The general composition of the Fe-rich particles are Al_3(Fe, Mn), Al_6(Fe, Mn), Al_8Mg_5 and others such as Al_2Fe_5 (Li and Arnberg 2004; Rathod and Kutsuna 2004). Li and Arnberg (2004) observed that the majority of the Fe-rich particles in DC-cast AA5182 were $A1_3Fe$ and Al_mFe where $m \approx 4$ in a DC-cast AA5182 alloy. The Fe-rich particles are perhaps best treated as a single class for the purposes of modeling nucleation and estimating their composition. Experimental data for the Fe-rich particles is scarce. Yamagiwa et al. (2003) estimated that the Fe-rich intermetallics have the elastic modulus of steel, 200 GPa, but with half the density. The Poisson ratio can be taken as about 0.30 as suggested by Somekawa et al. (2004).

The fracture toughness of the Fe-rich particles is required to predict particle cracking and is the controlling parameter in the percolation model. Since the percolation model only contains this sole damage-related parameter, it can be determined by calibration with experiment. In the interest of evaluating the true predictive ability of the model, it should be first evaluated using only the material data. The fracture toughness for two types of Fe-rich intermetallics in AA5182, Al3Fe and Al_2Fe_5, have been reported as 2.15 and 2.30 MPa-m$^{1/2}$ (Rathod and Kutsuna 2004). Since the fracture toughness values for these two different particles are so similar, we can assume that these values are representative of the other Fe-rich particles. This is a reasonable assumption in light of the absence of material data and the validity of this assumption will be evaluated by comparing the predictions of percolation model with experiment data. The material parameters are presented in Table 11.1.

11.1.2 Finite-Element Model of the Notched Tensile Tests

The details of the finite element models for the notched tensile specimens are discussed in Chap. 2. For computational efficiency, only percolation elements are placed at the notch root (Fig. 11.1) to capture the initiation of the macro-crack while the remaining elements obey J_2 plasticity using the hardening rule in Eq. (9.12).

Table 11.1 Material parameters of AA5182 required by the percolation model

Constituent	E (GPa)	ν	σ_y (MPa)	K_{Ic} (MPa-m$^{1/2}$)
Matrix	65.33	0.33	122.7	–
Fe-rich particles	200	0.30	–	2.15–2.30
Mg_2Si particles	113.5	0.161	–	–

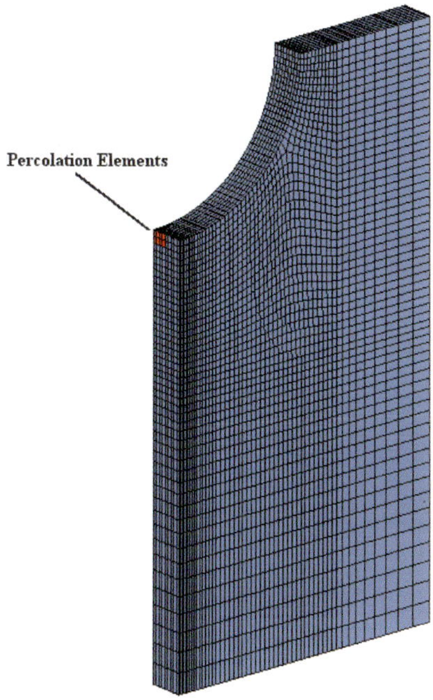

Fig. 11.1 Finite-element mesh showing the location of the percolation elements. The remaining elements obey J2 plasticity

At present, the percolation model is computationally expensive and the size of the global particle field in this specific model is limited to a size of 4,000 particles split into four elements. The placement of only several percolation elements at the notch root is acceptable for this specific notch geometry because deformation is highly localized at the notch root. A previous study by Butcher and Chen (2011) using a Gurson-based constitutive model described in Chap. 2 has shown that the only appreciable damage occurs within this region (Fig. 2.43).

11.1.3 Identification of the Fracture Parameters

The finite-element simulation is terminated at the onset of element deletion signifying the initiation of a macro-crack at the notch root (Butcher and Chen 2011). The resulting ligament and axial strains are then obtained from the finite-element model and compared with the experiment values. The data in each of the percolation elements is combined into a single data set and used for

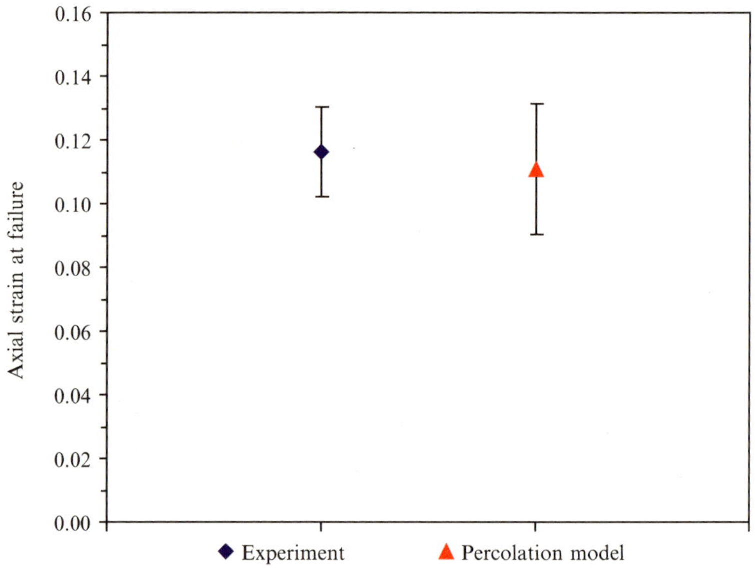

Fig. 11.2 Comparison of the experimental and predicted 95 % confidence intervals for the axial strain at failure

post-processing. The simulation is repeated for five particle fields, $P1$–$P5$, generated using the distributions and procedures in Chap. 8.

11.1.3.1 Ligament and Axial Strain at Failure

The predictions of the percolation model for the axial and ligament strain with the experimental values are shown in Figs. 11.2 and 11.3, respectively. The fracture strains are presented as 95 % confidence intervals due to the stochastic nature of the percolation model predictions and because of experiment data exhibits variation as well. It is remarkable that the predictions of the percolation model are in excellent agreement for the axial strain at failure and are reasonable for the ligament strain. Although only five particle fields were considered, the variation in the axial strain measurements in the percolation model is comparable to the experimental variation. It is important to state that this agreement is not a result of a calibration and is a testament to the physical foundation and predictive ability of the model.

Since the confidence intervals overlap, there is no significant difference between the numerical and experimental predictions but the evidence is not overwhelming for the ligament strain. Overall, the axial strain is the more reliable measure in this case because it is directly measured by the extensometer and is located away from the fracture surface. The ligament strain is difficult to measure with a micrometer because it is distorted by the fracture process. The use of a digital-image-correlation system to map the strain field within the ligament during the experiment and stop the test at the formation of a crack at the notch root would be preferable in future studies.

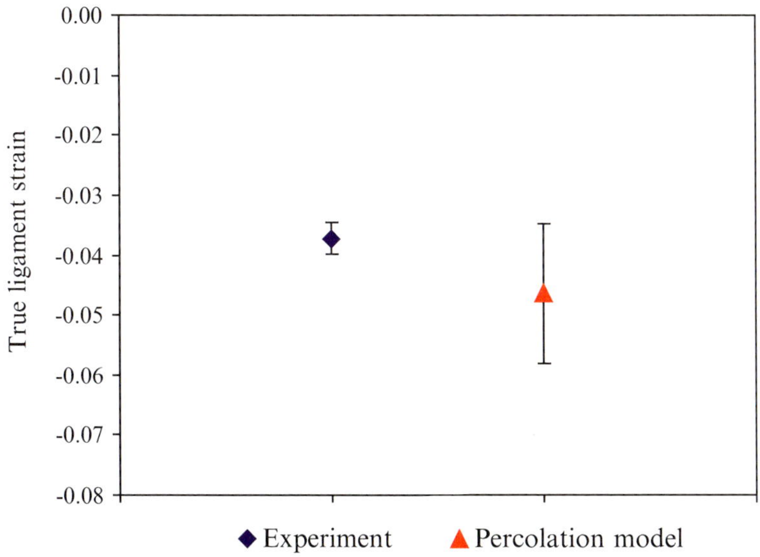

Fig. 11.3 Comparison of the experimental and predicted 95 % confidence intervals for the ligament strain at failure

11.2 Porosity and Void Nucleation

The stress state at the notch root is essentially that of pure uniaxial tension and the stress triaxiality is approximately 1/3 throughout the deformation process. The evolution of the plastic strain in the notch root with the sample elongation is shown in Fig. 11.4. Since the stress state is nearly identical to uniaxial tension, the predicted porosity and nucleation trends can be compared with the x-ray tomography results of Orlov (2006) from a standard uniaxial tension test without a notch. The predicted porosity distributions of the five particle fields considered are shown Fig. 11.5. All of the particle fields are in excellent agreement with the experimental porosity data at a plastic strain up to 0.10 and show generally good agreement at the higher strain levels.

All of the particle fields considered, P1–P5, exhibit the same behaviour where deformation is relatively homogenous until the commencement of void nucleation at higher strains. The start of nucleation is shortly followed by localized coalescence that quickly becomes unstable and sweeps throughout the particle field causing failure. The fracture porosities are in good agreement with the metallographic observations of Smerd et al. (2005) who reported failure porosities on the order of 0.3 %.

The number of voids in the particle field is representative of void nucleation and the experimental and numerical predicted values are presented in Fig. 11.6. The predictions of all of the particle fields are in very good agreement with the results of Orlov (2006) with little variation. The convergence of the nucleation predictions

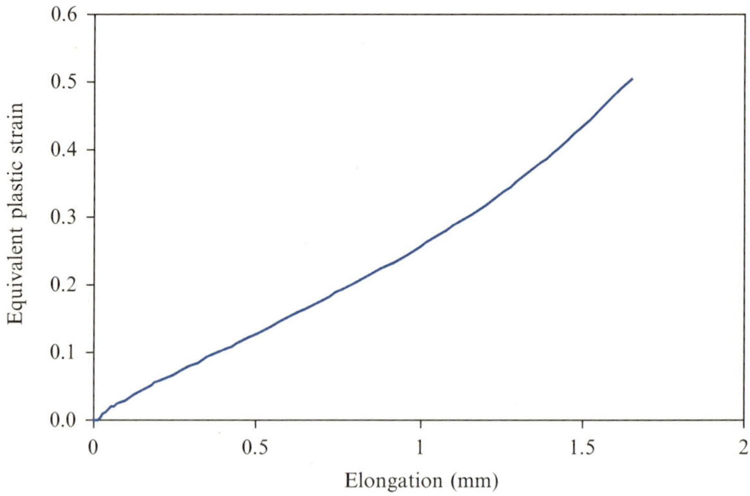

Fig. 11.4 Evolution of the equivalent plastic strain at the center of the notch root

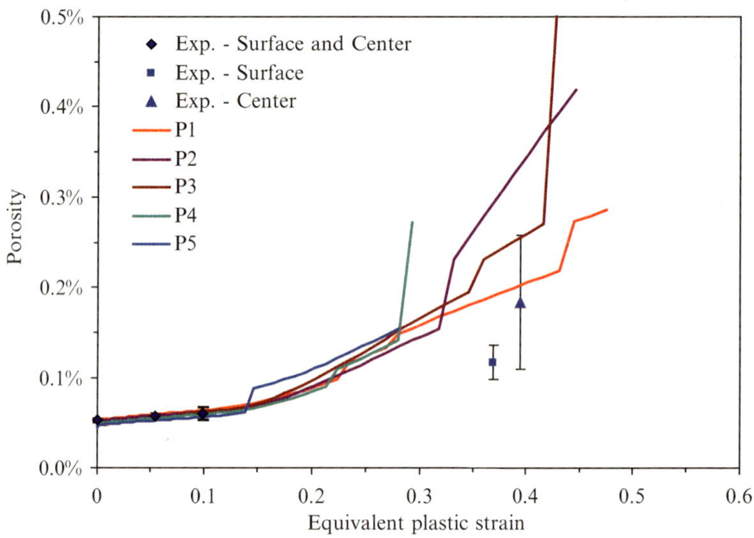

Fig. 11.5 Comparison of the predicted porosity in the *center* of the notch root with the experiment data of Orlov (2006) for a standard tensile specimen

demonstrates that only several particle fields are required to obtain the general results. The porosity and fracture strains in the previous figures are expected to exhibit the most variation because they are related to coalescence which is strongly dependent upon the local microstructure. This agreement with the experimental nucleation trends is very encouraging for the physical foundation of both the percolation and the nucleation models. The void nucleation model relies upon

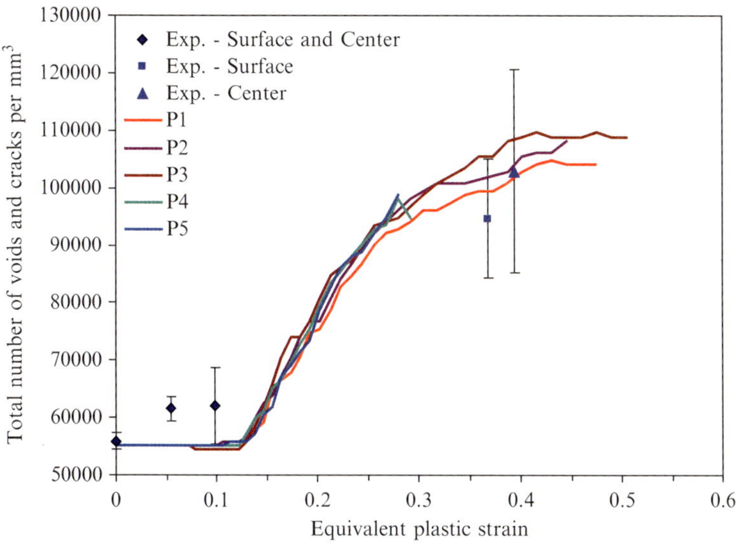

Fig. 11.6 Comparison of the total number of voids and cracks predicted by the percolation model with the experimental results of Orlov (2006)

estimation of the stress state within the particle, its elastic properties, shape, size and fracture toughness as well as the intrinsic physics related to the assumption of particle cracking. It is no small feat to have all of these factors to align in a convincing fashion without the use of heuristics and calibrations.

11.2.1 Nucleation Stress and Strain

Traditional void nucleation models rely upon strain-based measurements to describe nucleation using the plastic strain of the material. This is a convenient method because the particles do not have to be explicitly modeled or the stress estimated within them using a homogenization method. The bulk deformation is used to quantify the initiation of voids and this can be used to recreate the expected trend in the simulation. The problem with this methodology is that the nucleation model cannot be applied to any other situation other than the one it was determined in. Strain measurements are path-dependent while stress measurements are not. The plastic strain obtained in a tensile test is not the same as the plastic strain in tube bending and hydroforming. It is of great interest to adopt a stress-based nucleation model so that it can be transferred to different stress states and used to predict fracture in metal forming operations.

The average particle principal stress at nucleation is shown in Fig. 11.7. Large particles crack at low stresses early in the deformation process while the smaller particles do not nucleate until the later in the deformation process. The average

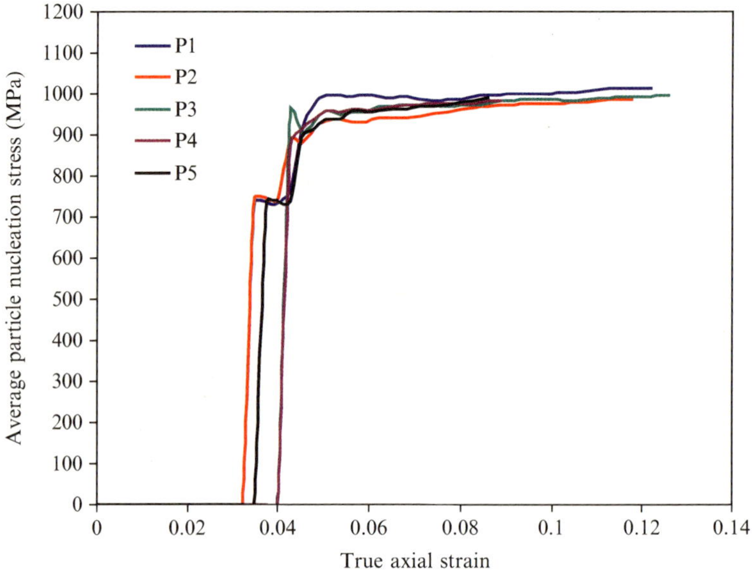

Fig. 11.7 Comparison of the average maximum principal stress in the particles at void nucleation for each particle field considered

nucleation stress is approximately constant because the early nucleation of the large particles leaves the population of roughly the same size. The predicted nucleation stress for the Fe-rich particles in AA5182 is about 950 MPa and is a reasonable value for a hard and brittle particle. It is expected that the Fe-rich particles would have a lower nucleation stress than the yield stress of the Mg2Si particles of 1,640 MPa which do not appreciably nucleate voids. The predicted nucleation stress of the Fe-rich particles could be validated using a micro-indentation method. The fracture toughness can also be determined using this procedure to verify if the value of 2.30 MPa-m$^{1/2}$ is truly representative of the particles.

A strain-based nucleation criterion can also be developed by comparing the average volume of a broken particle to the global plastic strain at which it nucleated as shown in Fig. 11.8. All of the particle fields are in good agreement with each other and display the well-known behaviour that small voids will nucleate only at high strains while larger particles display a negligible size effect and have a near constant nucleation stress or strain. Orlov (2006) observed that no particles that had a volume smaller than 17.8 μm^3 nucleated a void and this is in accordance with the predictions of the model.

It is interesting to mention that the strain-based nucleation trend in Fig. 11.9 is quite similar to the phenomenological nucleation model of Butcher and Chen (2009a) that was used in a previous 2-D percolation model in Fig. 7.1. The advantage of the present model is that the trends in Fig. 11.9 were obtained from first-principles, do not contain calibration parameters, and use the stress in the particles instead of the matrix strain. Nevertheless, it is encouraging that the

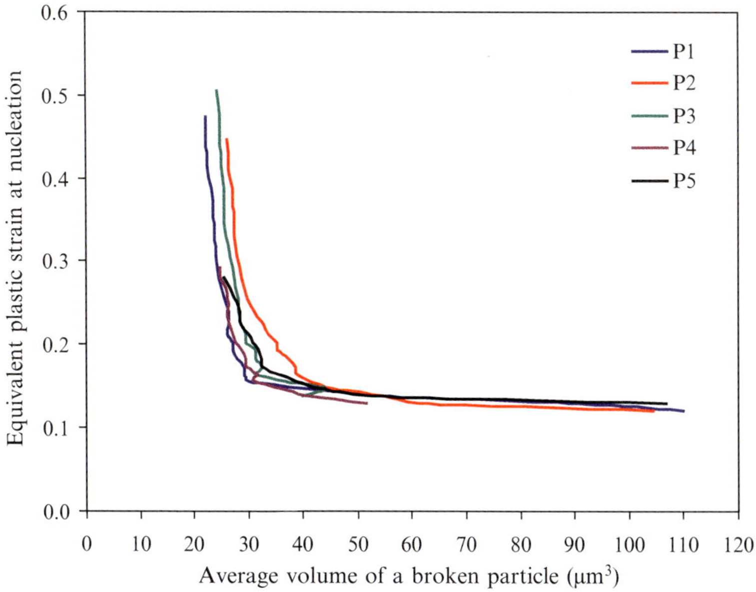

Fig. 11.8 Predicted average plastic strain of the particles upon nucleation via cracking

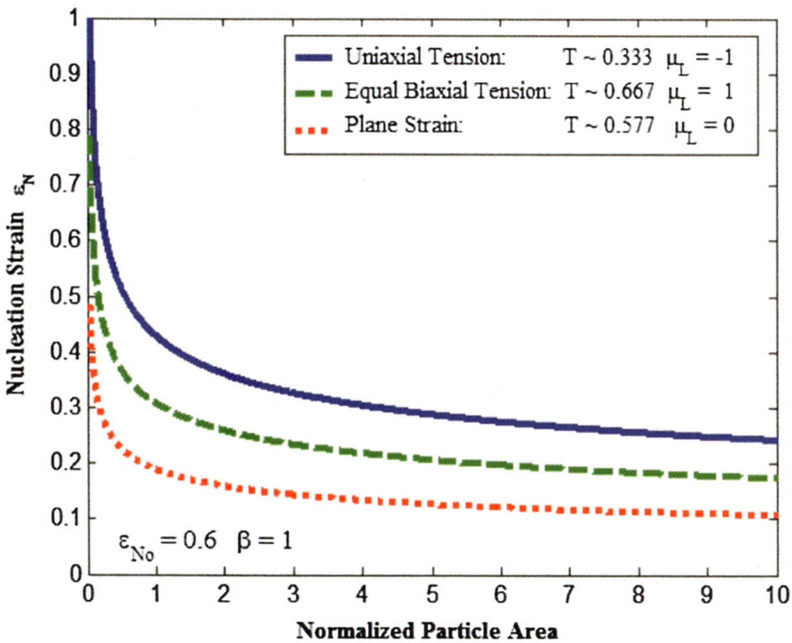

Fig. 11.9 Variation of the nucleation strain with the particle size and stress state using the phenomenological model of Butcher and Chen (2009a). This nucleation model was used to predict nucleation in a previous version of the percolation model that was restricted to 2-D voids

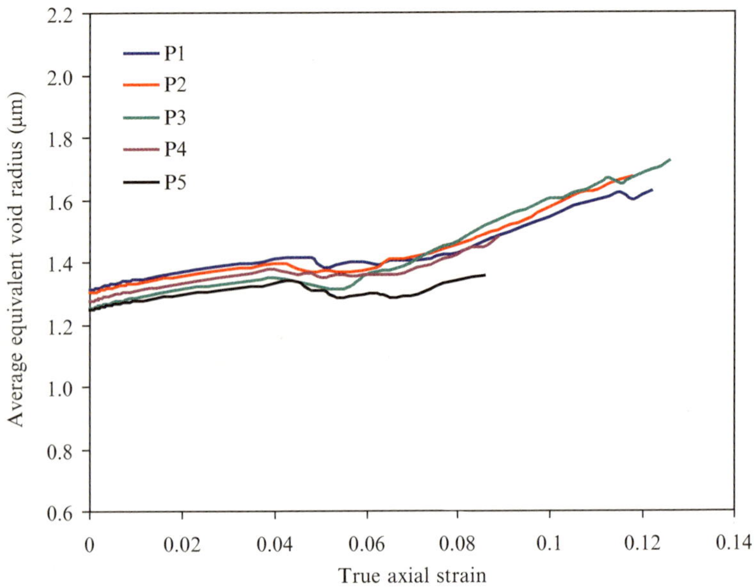

Fig. 11.10 Predicted average equivalent radius of the voids in the various particle fields

phenomenological model of Butcher and Chen (2009a) described in Chap. 7 can provide similar trends and could certainly be calibrated using the percolation model results.

The particle-cracking void nucleation model has given predictions that are in excellent agreement with the experimental data, especially considering that no calibration parameters were employed. The agreement could no doubt be improved by introducing calibration parameters such as a critical particle volume or stress. The physics of the nucleation process are well represented with large particles nucleating at low strains compared to smaller particles. It should be emphasized that this nucleation rule does not account for particle clusters and this will play a role on the nucleation process. A non-local averaging scheme could be employed to account for the influence of neighbour particles when determining the stress in the particle. This process will introduce a length-scale that will require calibration and was not considered at present to avoid introducing parameters with unknown values. Future work will address the issue of particle nucleation within clusters.

11.2.1.1 Void and Crack Size

Figure 11.10 shows the variation of the average equivalent radius of the voids during deformation. The average size of the voids steadily increases as the primary voids evolve with the matrix. The average void size then dips as the nucleation begins and creates penny-shaped voids of almost negligible size, increasing the

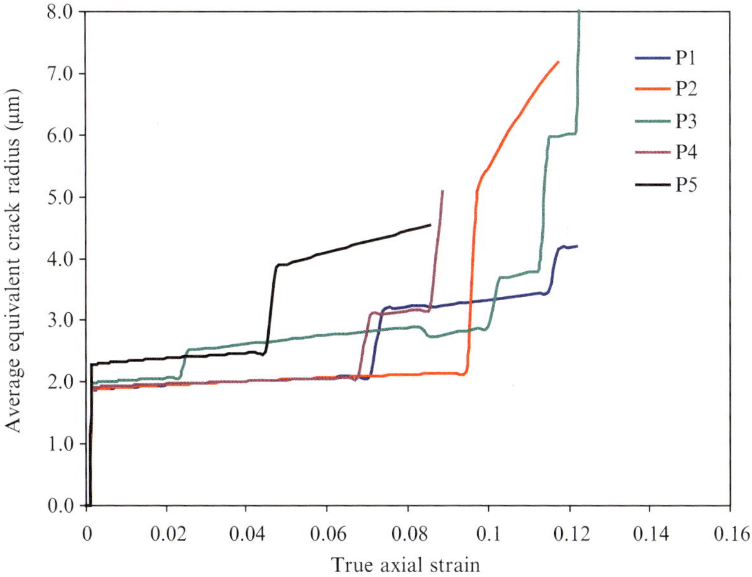

Fig. 11.11 Predicted average equivalent radius of the cracks in the various particle fields

number of voids and reducing the average size. The average void size quickly recovers from this dip and grows at a faster rate since penny-shaped voids grow at a high rate in the opening direction. The trends are similar for each of the particle fields considered and the average void radius at fracture is approximately 1.3–1.7 μm.

The average equivalent crack radius exhibits significant variation between particle fields as shown in Fig. 11.11. This is not unexpected since coalescence is primarily a function of geometry and each particle field will contain different clusters that will coalesce in different ways and directions. What can be gleaned from this figure is that the general trend is the same for each particle field in that there are few coalescence events before fracture. The crack size early in the deformation process is similar to that of the voids, which is a result of coalescence between two voids that are in close proximity and belong to the same particle cluster. Crack propagation begins and ends abruptly as seen by the sharp vertical jumps in the radius in a small applied strain increment. The average crack radius ranges from about 4 to 8 μm for the field considered.

The predicted ranges for the equivalent void and crack radii are in qualitative agreement with the sizes of the voids obtained from an SEM micrograph of the fracture surface at the notch root in Fig. 11.12. The SEM micrograph was obtained for a similar notch ratio of 0.125 that exhibits the same fracture behaviour. The dimpled fracture surface corresponds to the size of the voids and the large dimples correspond to 'cracks' as defined in the percolation model. The presence of a large primary void near the top of the image and the relatively uniform size of the smaller remaining voids is consistent with a large void instigating the unstable coalescence process.

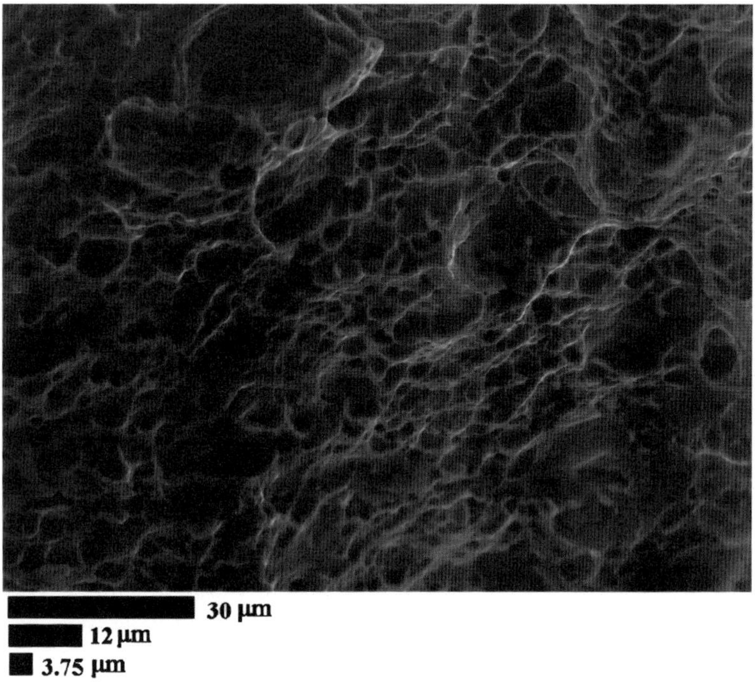

Fig. 11.12 SEM micrograph of the fracture surface at the notch root for an AA5182 tensile specimen with a notch ratio of 0.125

11.2.1.2 Shape of the Voids and Cracks

The variation of the average aspect ratio of the voids and cracks is presented in Fig. 11.13. The initial voids are approximately spherical and elongate in the loading direction, increasing the aspect ratio. The onset of void nucleation quickly reduces the average aspect ratio since the nucleated voids are penny-shaped and have an aspect ratio of 0.01. However, once nucleated, these penny-shaped voids exhibit remarkable growth and shape evolution and the average aspect ratio of the voids recovers and slightly surpasses its pre-nucleation value at fracture.

Overall, the average void aspect ratio is roughly spherical for the bulk of the deformation process. This result has several implications for constitutive modeling. The first being that the standard Gurson (1977) yield criterion could be used to account for material softening of the bulk particle field. This is beneficial because the Gurson (1977) model is widely available and simpler to implement than other more advanced models that account for void shape. It should be emphasized that this does not mean that void shape can be neglected in the constitutive model since the void growth predictions for a constant spherical void would not at all be in accordance with penny-shaped voids. The assumption of a spherical void is only valid for evaluating the yield criterion to account for material softening and not the evolution

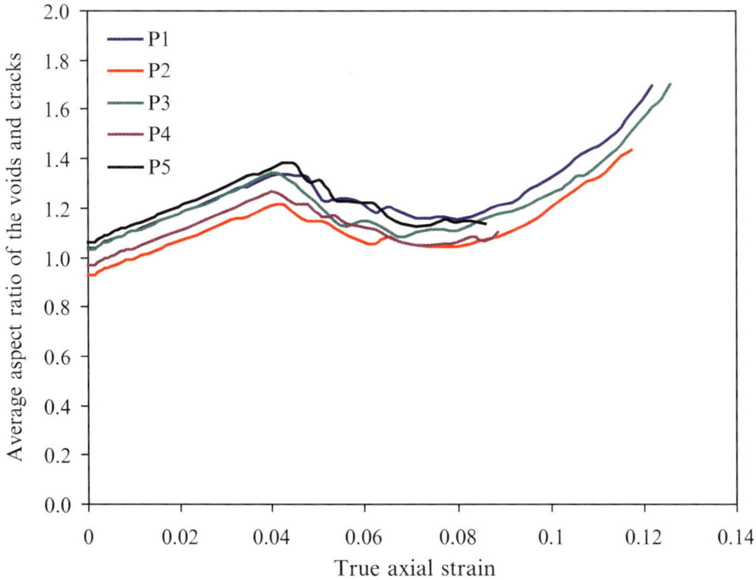

Fig. 11.13 Evolution of the global aspect ratio (voids and cracks) in the various particle fields. The aspect ratio is determined relative to the principal loading direction

of the individual or equivalent void. If only a single void is considered as in the traditional damage-based constitutive models, the void shape must be considered.

11.2.2 Comparison of the Percolation Model with a Traditional Damage-Based Model

It is of great interest to compare the predictions of the percolation model with that of a typical damage-based constitutive model. Butcher and Chen (2011) employed the classical Gurson-Tvergaard (GT) model for spherical voids to describe void growth and material softening in AA5182 notched tensile specimens. Void nucleation was represented using the stress- and strain-based continuum models of Chu and Needleman (1980) in Eqs. (1.5) and (1.7). The unit cell geometry used in the coalescence model was determined based upon the average inter-particle spacings from 2-D images of the material. The initial porosity and volume fraction of second-phase particles of Orlov (2006) were employed to ensure that all of the parameters were based upon experimental measurements. The mean nucleation stress and strain values were then identified by calibrating the model to give fracture strains in accordance with the experiment data. In short, considerable effort went into ensuring that the modeling procedure was sound and that the parameters were physically realistic.

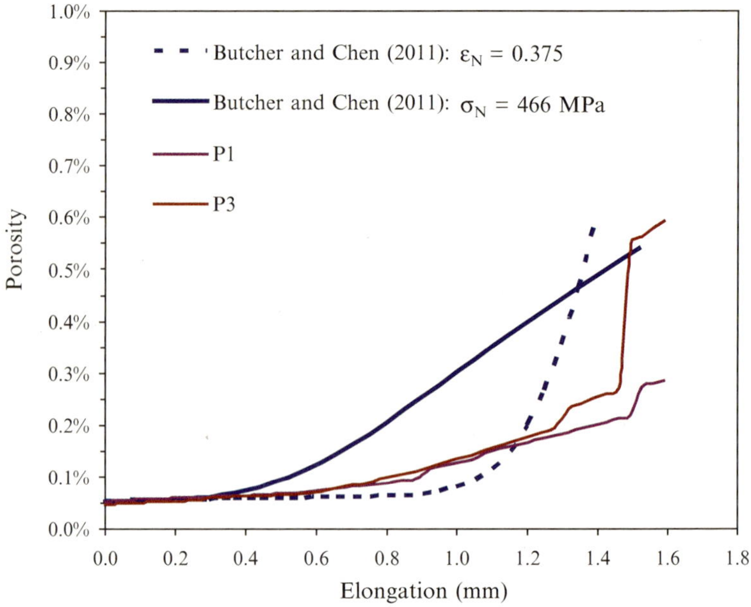

Fig. 11.14 Predicted porosity in the notch root using the percolation model and the Gurson-based model of Butcher and Chen (2011) who considered both stress- and strain-based nucleation models

The comparison of the porosity in the notch root as predicted by Butcher and Chen (2011) and the percolation model is presented in Fig. 11.14. Only two of the particle fields are presented for clarity. The predictions of the percolation model are in pretty good agreement with that of the strain-controlled nucleation model. Porosity is low for the majority of deformation prior to the onset of rapid nucleation causing fracture. The porosity at failure is similar for both the continuum and percolation models. The ability of the Gurson-based model to predict fracture at this realistic porosity is because the void spacing was accounted for in defining the representative unit cell in the coalescence model and that the assumption of a global spherical void is reasonable (Fig. 11.13). As established in Chap. 4, if the void geometry and spacing ratio are well predicted, the coalescence model will give good predictions for failure.

From these results, the performance of the traditional model with proper material characterization appears to be quite reasonable. Although the nucleation model was calibrated to provide sufficient porosity to cause coalescence at the experimental fracture strain, the resulting trend for the porosity and its value at coalescence were not calibrated, so their agreement with the percolation model is encouraging. However, if we dig deeper into the Gurson-based model, its predictions for nucleation expose the flaws in the model as shown in Fig. 11.15.

The porosity created by nucleated voids is extremely large compared to the percolation model whose trend is too low to be clearly shown in the figure.

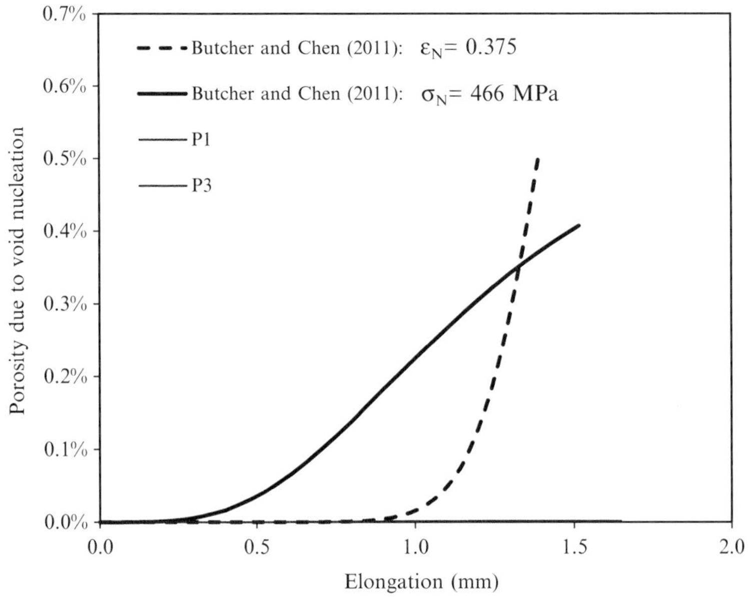

Fig. 11.15 Comparison of the predicted porosity due to nucleated voids of the percolation model with the Gurson-based model used by Butcher and Chen (2011). The constitutive model of Butcher and Chen (2011) considered continuum-based stress and strain-controlled nucleation models for this alloy

The contribution of the porosity due to nucleated voids in percolation model is negligible because the voids are initially penny-shaped. The calibration of the nucleation parameter results in the model predicting a massive amount of void nucleation in order to trigger fracture at the desired experimental strain. As a result, the porosity due to nucleation is 0.4–0.5 % which means that 80–100 % of the second phase-particles have become voids because the total particle fraction in this material is only about 0.5 %. In reality, the experimental percentage of particles that nucleate voids is a paltry 6.7 % by comparison. The calibration process of the nucleation model has given the model a decent prediction of the porosity history and fracture strains but at the cost of the validity of the nucleation predictions.

The purpose of this comparison was to emphasize the limitations of the traditional damage-based constitutive models and the difficulties in calibrating them to offset their limitations while giving physically realistic predictions. Recent extensions to these models have improved their physical foundations, especially for void growth and coalescence, but the nucleation treatments remain inadequate. The homogenization techniques developed in Chap. 9 along with the nucleation model in Chap. 7 can be employed in future studies to improve the physical foundation of these models.

11.3 Summary

The complete damage percolation model was used to predict fracture and damage evolution in a notched tensile sheet specimen of AA5182 sheet. Representative particle distributions were created and mapped to the percolation elements located at the notch root where fracture initiates in the sample. The fracture strain, porosity, and nucleation predictions of the model are in very good agreement with the experiment data of Orlov (2006).

No calibration or adjustable parameters were employed in the model and its good predictions of the experiment data attest to the strong physical foundation of the model. Fracture is a sole consequence of the stress state, material composition and the particle distribution. The present percolation model represents a significant contribution to the field of damage-induced ductile fracture by enabling a designer to optimize a material microstructure by simulating its performance in a metal forming operation. Unlike traditional damage-based models, the physical foundation of the model is sound and the inter-relationships between the competing fracture mechanisms are well represented.

References

Agarwal, H., Gokhale, A. M., Graham, S., & Horsemeyer, M. F. (2002). Anisotropy of intermetallic particle cracking damage evolution in an Al-Mg-Si base wrought aluminum alloy under uniaxial compression. *Metallurgical and Materials Transactions A, 33*, 3443–3448.

Agrawal, H., Gokhale, A. M., Graham, S., Horstemeyer, M. F., & Bamman, D. J. (2002). Rotations of brittle particles during plastic deformation of ductile alloys. *Materials Science and Engineering A, 328*, 310–316.

Aravas, N., & Ponte Castaneda, P. (2004). Numerical methods for porous metals with deformation-induced anisotropy. *Computational Methods Applied Mechanical Engineering, 193*, 3767–3805.

Argon, A. S., Im, J., & Needleman, A. (1975a). Distribution of plastic strain and negative pressure in necked steel and copper bars. *Metallurgical Transactions, 6A*, 815–824.

Argon, A. S., Im, J., & Safoglu, R. (1975b). Cavity formation from inclusions in ductile fracture. *Metallurgical Transactions, 6A*, 825–837.

Argon, A. S., Im, J., & Safoglu, R. (1975c). Separation of inclusions in spheroidized 1045 steel, Cu-0.6% Cr alloy and maraging steel in plastic straining. *Metallurgical Transactions, 6A*, 839–851.

Asnafi, N., & Skogsgardh, A. (2000). Theoretical and experimental analysis of stroke-controlled tube hydroforming. *Materials Science and Engineering, A279*, 95–110.

Bandstra, J., & Koss, D. (2008). On the influence of void clusters on void growth and coalescence during ductile fracture. *Acta Materialia, 56*, 4429–4439.

Baradari, G. J. (2006). *Damage in hydroforming of pre-bent aluminum alloy tubes*. Ph.D. thesis, University of Waterloo, Waterloo, Canada.

Bardelcik, A. (2006). *Effect of pre-bending and hydroforming parameters on the formability of advanced high strength steel*. MASc thesis, University of Waterloo, Waterloo, Canada.

Barlat, F. (1987). Prediction of tricomponent plane stress yield surfaces and associated flow and failure behavior of strongly textured f.c.c. polycrystalline sheets. *Materials Science Engineering, 95*, 15–29.

Barlat, F., & Lian, J. (1989). Plastic behavior and stretchability of sheet metals (Part I): A yield function for orthotropic sheet under plane stress conditions. *International Journal of Plasticity, 5*, 51–56.

Barlat, F., Lege, D. J., & Brem, J. C. (1991). A six-component yield function for anisotropic materials. *International Journal of Plasticity, 7*, 693.

Barlat, F., Maeda, Y., Chung, K., Yanagawa, M., Brem, J. C., Hayashida, Y., Lege, D. J., Matsui, K., Murtha, S. J., Hattori, S., Becker, R. C., & Makosey, S. (1997). Yield function development for aluminum alloy sheets. *Journal of the Mechanics and Physics of Solids, 45*, 1727–1763.

Barsoum, I., & Faleskog, J. (2007a). Rupture mechanisms in combined tension and shear – Experiments. *International Journal of Solids and Structures, 44*, 1768–1786.

Z. Chen and C. Butcher, *Micromechanics Modelling of Ductile Fracture*, Solid Mechanics and Its Applications 195, DOI 10.1007/978-94-007-6098-1,

© Springer Science+Business Media Dordrecht 2013

Barsoum, I., & Faleskog, J. (2007b). Rupture mechanism in combined tension and shear – Micromechanics. *International Journal of Fracture, 44*, 5481–5498.

Bayoumi, S. E. A. (1999). *Engineering solid mechanics: Fundamentals and applications, chapt. 7* (p. 306). Boca Raton: CRC Press.

Becker, R. (1987). The effect of porosity distribution on ductile fracture. *Journal of Mechanics and Physics of Solids, 35*, 577–599.

Benson, D. J. (1995). The effects of void cluster size on ductile fracture. *International Journal of Plasticity, 11*, 571–582.

Bonet, J., & Wood, R. D. (1997). *Non-linear continuum mechanics for finite element analysis*. Cambridge: Cambridge University Press.

Benzerga, A. A. (2002). Micromechanics of coalescence in ductile fracture. *Journal of Mechanics and Physics of Solids, 50*, 1331–1362.

Benzerga, A. A., & Besson, J. (2001). Plastic potential for anisotropic porous solids. *European Journal of Mechanics – A/Solids, 20*, 397–434.

Benzerga, A. A., Besson, J., & Pineau, A. (2004). Anisotropic ductile fracture part II: Theory. *Acta Materialia, 52*, 4639–4650. 223.

Beremin, F. M. (1981). Cavity formation from inclusions in ductile fracture of A50S steel. *Metallurgical Transactions, 12A*, 723–731.

Berveiller, M., & Zaoui, A. (1979). An extension of the self-consistent scheme to plastically-flowing polycrystals. *Journal of the Mechanics and Physics of Solids, 26*, 325–344.

Besson, J., Steglich, D., & Brocks, W. (2003). Modelling of plane strain ductile rupture. *International Journal of Plasticity, 15*, 1517–1541.

Bourcier, R. J., Koss, D. A., Smelser, R. E., & Richmond, O. (1986). The influence of porosity on the deformation and fracture of alloys. *Acta Metallurgica, 34*, 2443–2453.

Brocks, W., Sun, D. Z., & Honig, A. (1996). Verification of micromechanical models for ductile fracture by cell model calculations. *Computational Materials Science, 7*, 235–241.

Brown, L. M., & Embury, J. D. (1973). The initiation and growth of voids at second phase particles. In *Proceedings of the 3rd international conference on the strength of metals and alloys* (pp. 164–169). London: Institute of Metals.

Brown, L. M., & Stobbs, W. M. (1976). Work-hardening of copper-silica-5. Equilibrium plastic relaxation by secondary dislocations. *Philosophical Magazine, 34*, 351–372.

Brunet, M., Morestin, F., & Walter-Leberre, H. (2005). Failure analysis of anisotropic sheet-metals using a non-local plastic damage model. *Journal of Materials Processing Technology, 170*, 457–470.

Budiansky, B., Hutchinson, J. W., & Slutsky, S. (1982). Void growth and collapse in viscous solids. In H. G. Hopkins, & M. J. Sewell (Eds.), *Mechanics of solids – The Rodney Hill 60th anniversary volume* (pp. 13–45). Oxford: Pergamon press.

Butcher, C. (2011). *A multi-scale damage percolation model of ductile fracture*. Ph.D. thesis, University of New Brunswick, Fredericton, NB, Canada.

Butcher, C., Chen, Z. T., & Worswick, M. J. (2013). Integration of a particle-based homogenization theory into a general damage-based constitutive model to improve the modelling of void nucleation to coalescence. *Acta Mechanica, 224*, 139–156.

Butcher, C., & Chen, Z. T. (2009a). A void coalescence model for combined tension and shear. *Modeling and Simulation in Materials Science and Engineering, 17*, 1–15.

Butcher, C., & Chen, Z. T. (2009b). Damage percolation modeling of void nucleation within heterogeneous particle distributions. *Modeling and Simulation in Materials Science and Engineering, 17*, 075003.

Butcher, C., & Chen, Z. T. (2009c). A continuum void nucleation model for an Al-Mg alloy sheet based on measured particle distribution. *Acta Mechanica Solida Sinica, 22*, 391–398.

Butcher, C., & Chen, Z. T. (2011). Characterizing void nucleation in a damage-based constitutive model using notched tensile sheet specimens. *Theoretical and Applied Fracture Mechanics, 55*, 140–147. 224.

Butcher, C., Chen, Z. T., & Worswick, M. J. (2006). A lower bound damage-based finite element simulation of stretch flange forming of Al-Mg alloys. *International Journal of Fracture, 142*, 289–298.

Butcher, C., Chen, Z. T., Bardelcik, A., & Worswick, M. (2009). Damage-based finite-element modeling of tube hydroforming. *International Journal of Fracture, 155*, 55–65.

Bilger, N., Auslender, F., Bornert, M., Michel, J. C., Moulinec, H., Suquet, P., & Zaoui, A. (2005). Effect of nonuniform distribution of voids on the plastic response of voided materials: A computational and statistical analysis. *International Journal of Solids and Structures, 42*, 517–538.

Cazacu, O., & Barlat, F. (2003). Application of representation theory to describe yielding of anisotropic aluminum alloys. *International Journal of Engineering Science, 41*, 1367–1385.

Cazacu, O., & Stewart, J. B. (2009). Analytic plastic potential for porous aggregates with matrix exhibiting tension–compression asymmetry. *Journal of the Mechanics and Physics of Solids, 57*, 325–341.

Cazacu, O., Plunkett, B., & Barlat, F. (2006). Orthotropic yield criterion for hexagonal closed packed metals. *International Journal of Plasticity, 22*, 1171–1194.

Chaboche, J. L., Kanoute, P., & Roos, A. (2005). On the capabilities of mean-field approaches for the description of plasticity in metal-matrix composites. *International Journal of Plasticity, 21*, 1409–1434.

Chakrabarty, J. (1987). *Theory of plasticity*. New York: McGraw-Hill.

Chein, W., Pan, J., & Tang, S. (2001). Modified anisotropic Gurson yield criterion for porous ductile sheet metals. *Journal of Engineering Materials and Technology, 132*, 409–416.

Chen, Z. T. (2004). *The role of heterogeneous particle distribution in the prediction of ductile fracture*. Ph.D. thesis, University of Waterloo, Canada.

Chen, Y., & Lambert, S. (2003). Analysis of ductile tearing of pipeline-steel in single edge notch tension specimens. *International Journal of Fracture, 124*, 179–199.

Chen, Z. T. & Worswick, M. J. (2008). Investigation of void nucleation in Al-Mg sheet. *Materials Science and Engineering A, 483–484*, 99–101.

Chen, Z. T., Worswick, M. J., Cinotti, N., Pilkey, A. K., & Lloyd, D. (2003). A linked FE-damage percolation model of aluminium alloy sheet forming. *International Journal of Plasticity, 19*, 2099–2120.

Chen, Z. T., Worswick, M. J., Pilkey, A. K., & Lloyd, D. J. (2005). Damage percolation during stretch flanging forming of aluminium alloy sheet. *Journal of the Mechanics and Physics of Solids, 53*, 2692–2717.

Chu, C. C., & Needleman, A. (1980). Void nucleation effects in biaxially stretched sheets. *Journal Engineering Materials Technology, 102*, 249–256.

Cinotti, N. (2003). *Stretch flange formability of aluminum alloys*. M.Sc., thesis, University of Waterloo, Ontario, Canada.

Cox, T. B., & Low, J. R., Jr. (1974). An investigation of the plastic fracture of AISI 4340 and 18 Nickel-200 grade maraging steels. *Metallurgical Transactions, 5*, 1457–1470.

Cramer, M., & Sevostianov, I. (2009). Effect of pore distribution on elastic stiffness and fracture toughness of porous materials. *International Journal of Fracture, 160*, 189–196.

Danas, K., & Aravas, N. (2012). Numerical modeling of elasto-plastic porous materials with void shape effects at finite deformations. *Composites Part B: Engineering, 6*, 2544–2559.

Danas, K., & Ponte Castaneda, P. (2009). A finite-strain model for anisotropic viscoplastic media: I – Theory. *European Journal of Mechanics/A Solids, 28*, 387–401.

Dighe, M. D., Gokhale, A. M., & Horstemeyer, M. F. (2002). Effect of loading condition and stress state on damage evolution of silicon particles in an Al-Si-Mg base cast alloy. *Metallurgical and Materials Transactions-A, 33A*, 555–565.

Doghri, I., & Ouaar, A. (2003). Homogenization of two-phase elasto-plastic composite materials and structures: Study of cyclic plasticity and numerical algorithms. *International Journal of Solids and Structures, 40*, 1681–1712.

Drabek, T., & Bohm, H. J. (2005). Damage models for studying ductile matrix failure in composites. *Computational Materials Science, 32*, 329–336.

Dubensky, E. M., & Koss, D. A. (1987). Void/pore distributions and ductile fracture. *Metallurgical Transactions A, 18A*, 1887–1895.

Embury, J. D. (1985). Plastic flow in dispersion hardened materials. *Metallurgical Transactions A, 16*, 2191–2200.
Eshelby, J. D. (1957). The determination of the elastic field of an ellipsoidal inclusion, and related problems. *Proceedings of the Royal Society of London A, 241*, 376–396.
Fabregue, D., & Pardoen, T. (2008). A constitutive model for elastoplastic solids containing primary and secondary voids. *Journal of the Mechanics and Physics of Solids, 56*, 719–741.
Faleskog, J., & Shih, C. (1997). Micromechanics of coalescence – I. Synergistic effects of elasticity, plastic yielding and multi-size-scale voids. *Journal of the Mechanics and Physics of Solids, 45*, 21–50.
Fisher, J. R., & Gurland, J. (1981a). Void nucleation in spheroidized carbon steels, Part 1: Experimental. *Metal Science, 15*, 185–192.
Fisher, J. R., & Gurland, J. (1981b). Void nucleation in spheroidized carbon steels, Part 2: Model. *Metal Science, 15*, 193–202. 225.
Finn, M. J. (1999). Private communications.
Fleck, N. A., & Hutchinson, J. W. (1997). Strain gradient plasticity. In J. U. W. Hutchinson & T. Y. Wu (Eds.), *Advances in applied mechanics* (Vol. 33, pp. 295–361). New York: Academic.
Francescato, P., Pastor, J., & Riveill-Reydet, B. (2004). Ductile failure of cylindrically porous materials. Part 1: Plane stress problem and experimental results. *European Journal of Mechanics A Solids, 23*, 181–190.
Gangalee, A., & Gurland, J. (1967). On the fracture of silicon particles in aluminum–silicon alloys. *Transactions of the Metallurgical Society of AIME, 239*, 269–272.
Gelin, J. C. (1998). Modelling of damage in metal forming processes. *Journal of Materials Processing Technology, 80–81*, 24–32.
Geltmacher, A. B., Koss, D. A., Matic, P., & Stout, M. G. (1996). A modeling study of the effect of stress state on void linking during ductile fracture. *Acta Materialia, 44*, 2201–2210.
Geltmacher, A. B., Koss, D. A., Stout, M. G., & Matic, P. (1998). Flow localization in sheet specimens with pairs of holes. *Metallurgical and Materials Transactions A, 29*, 775–780.
Gurson, A. L. (1975). *Plastic flow and fracture behaviour of ductile materials incorporating void nucleation, growth and interaction*. Ph.D. thesis, Brown University, Providence, R.I.
Gologanu, M., Leblond, J., & Devaux, J. (1993). Approximate models for ductile metals containing non-spherical voids – Case of axisymmetric prolate ellipsoidal cavities. *Journal of the Mechanics and Physics of Solids, 41*, 1723–1754.
Gologanu, M., Leblond, J., & Devaux, J. (1994). Approximate models for ductile metals containing non-spherical voids – case of axisymmetric oblate ellipsoidal cavities. *Journal of Engineering Materials and Technology, 116*, 290–297.
Gologanu, M., Leblond, J-B., Perrin, G., & Devaux, J. (1997). Recent extensions of Gurson's model for porous ductile metals. In P. Suquet (Ed.), *Continuum micromechanics* (CISM Courses and Lectures no. 377, pp. 61–130). New York: Springer.
Gonzalez, C., & LLorca, J. (2000). A self-consistent approach to the elasto-plastic behaviour of two-phase materials including damage. *Journal of the Mechanics and Physics of Solids, 48*, 675–692.
Goods, S., & Brown, L. (1979). The nucleation of cavities by plastic deformation. *Acta Metallurgica, 27*, 1–15.
Goto, D. M., & Koss, D. A. (1996). An experimental model of the growth of neighbouring voids during ductile fracture. *Scripta Materialia, 35*, 459–463.
Green, R. J. (1972). A plasticity theory for porous solids. *International Journal of Mechanical Sciences, 14*, 215–224.
Griffin, J., Butcher, C., & Chen, Z. T. (2011). On using a dual bound approach to characterize the yield behavior of porous ductile materials containing void clusters. *International Journal of Fracture, 169*, 97–104.
Gurland, J. (1972). Observations on the fracture of cementite particles in a spheroidized 1.05% steel deformed at room temperature. *Acta Metallurgica, 20*, 735–741.
Gurland, J., & Plateau, J. (1963). The mechanism of ductile rupture of metals containing inclusions. *Transactions of the ASME, 56*, 442–454.

Gurson, A. L. (1977). Continuum theory of ductile rupture by void nucleation and growth – Part I. Yield criteria and flow rules for porous ductile media. *Journal of Engineering Materials and Technology, 99*, 2–15.

Hadianfard, M. J., Smerd, R., Winkler, S., & Worswick, M. (2008). Effects of strain rate on mechanical properties and failure mechanism of structural Al-Mg alloys. *Materials Science and Engineering A, 492*, 283–292.

Hallquist, J. O. (2006). *LS-DYNA theory manual*. Livermore Software Technology Corporation.

Hill, R. (1948). A theory of the yielding and plastic flow of anisotropic metals. *Proceedings of the Royal Society London, A193*, 281–297.

Hill, R. (1965). Continuum micro-mechanics of elastoplastic polycrystals. *Journal of the Mechanics and Physics of Solids, 13*, 89–101.

Horstemeyer, M. F., & Gokhale, A. M. (1999). A void-crack nucleation model for ductile metals. *International Journal of Solids and Structures, 36*, 5029–5055.

Horstemeyer, M. F., Lathrop, J., Gokhale, A. M., & Dighe, M. (2000a). Modeling stress state dependent damage evolution in a cast Al-Si-Mg aluminum alloy. *Theoretical and Applied Fracture Mechanics, 33*, 31–47.

Horstemeyer, M. F., Matalanis, M. M., Sieber, A. M., & Botos, M. L. (2000b). Micromechanical finite element calculations of temperature and void configuration effects on void growth and coalescence. *International Journal of Plasticity, 16*, 979–1015.

Horstemeyer, M. F., Ramaswamy, S., & Negrete, M. (2003). Using a micromechanical finite element parametric study to motivate a phenomenological macroscale model for void/crack nucleation in aluminum with a hard second phase. *Mechanics of Materials, 35*, 675–687.

Hu, J. G., Ishikawa, T., & Jonas, J. J. (2000). Finite element analysis of damage evolution and the prediction of the limiting draw ratio in textured aluminum sheets. *Journal of Materials Processing Technology, 103*, 374–382.

Hu, X., Wilkinson, D., Jian, M., & Mishra, R. (2008). Modeling strain localization using a plane stress two-particle model and the influence of grain level matrix inhomogeneity. *Journal of Engineering Materials and Technology, 130*, 1–10.

Huang, Y. (1991). Accurate dilatation rates for spherical voids in triaxial stress fields. *Journal of Applied Mechanics, 58*, 1084–1086.

Jones, M. K., Horstemeyer, M. F., & Belvin, A. D. (2007). A multiscale analysis of void coalescence in nickel. *Journal of Engineering Materials and Technology, 129*, 94–104.

Kachanov, L. M. (1971). *Foundations of the theory of plasticity*. Amsterdam: North-Holland. 333.

Kachanov, L. M. (1974). *Fundamentals of the theory of plasticity*. Moscow: Mir Publishers.

Kachanov, M., Tsukrov, I., & Shafir, B. (1994). Effective moduli of solids with cavities of various shapes. *Applied Mechanics Review, 47*, 151–174.

Kailasam, K., & Ponte Castaneda, P. (1998). A general constitutive theory for linear and nonlinear particulate media with microstructure evolution. *Journal of the Mechanics and Physics of Solids, 46*, 427–465.

Kanetake, N., Nomura, M., & Choh, T. (1995). Continuous observation of microstructural degradation during tensile loading of particle reinforced aluminum composites. *Material Science and Technology, 11*, 1246–1252.

Keralavarma, S. M., & Benzerga, A. A. (2010). A constitutive model for plastically anisotropic solids with non-spherical voids. *Journal of Mechanics and Physics of Solids, 58*, 874–890.

Koplik, J., & Needleman, A. (1988). Void growth and coalescence in porous plastic solids. *International Journal of Solids and Structures, 24*, 835–853. 227.

Kuna, M., & Sun, D. Z. (1996). Three-dimensional cell model analyses of void growth in ductile materials. *International Journal of Fracture, 81*, 235–258.

Landry, M., & Chen, Z. T. (2011). An approximate lower bound damage-based yield criterion for porous ductile sheet metals. *Theoretical and Applied Fracture Mechanics, 55*, 76–81.

Lassance, D., Scheyvaerts, F., & Pardoen, T. (2006). Growth and coalescence of penny-shaped voids in metallic alloys. *Engineering Fracture Mechanics, 73*, 1009–1034.

Le Roy, G., Embury, J. D., Edward, G., & Ashby, M. F. (1981). A model of ductile fracture based on the nucleation and growth of voids. *Acta Metallurgica, 29*, 1509–1522.

Leblond, J., Perrin, G., & Devaux, J. (1994). Bifurcation effects in ductile metals with damage delocalization. *Journal of Applied Mechanics, 61*, 236–242.

Leblond, J.-B., Perrin, G., & Devaux, J. (1995). An improved Gurson-type model for hardenable ductile metals. *European Journal of Mechanics A/Solids, 14*(4), 499–527.

Lee, B. J., & Mear, M. E. (1992a). Axisymmetric deformation of power-law solids containing a dilute concentration of spheroidal voids. *Journal of Mechanics and Physics of Solids, 40*, 1805–1836.

Lee, B. J., & Mear, M. E. (1992b). Effect properties of power-law solids containing elliptical inhomogeneities – Part II: Voids. *Mechanics of Materials, 13*, 337.

Li, Y. J., & Arnberg, L. (2004). Solidification structures and phase selection of iron-bearing eutectic particles in a DC-cast AA5182 alloy. *Acta Materialia, 52*, 2673–2681.

Liao, K. C., Pan, J., & Tang, S. C. (1997). Approximate yield criterion for anisotropic porous ductile sheet metals. *Mechanics of Materials, 26*, 213–226.

Lievers, W. B., Pilkey, A. K., & Lloyd, D. J. (2004). Using incremental forming to calibrate a void nucleation model for automotive aluminum sheet alloys. *Acta Materialia, 52*, 3001–3007. doi:10.1016/j.actamat.2004.03.002.

Lin, A., & Han, S. P. (2002). On the distance between two ellipsoids. *Society for Industrial and Applied Mathematics, 13*, 298–308.

Liu, B., Qiu, X., Huang, Y., Hwang, K., Li, M., & Liu, C. (2003). The size effect on void growth in ductile materials. *Journal of Mechanics and Physics of Solids, 51*, 1171–1187.

Maire, E., Grenier, J. C., Daniel, D., Baldacci, A., Klocker, H., & Bigot, A. (2006). Quantitative 3D characterization of intermetallic phases in an Al-Mg industrial alloy by X-ray microtomography. *Scripta Materialia, 55*, 123–126.

Maire, E., Bouaziz, O., Di Michiel, M., & Verdu, C. (2008). Initiation and growth of damage in a dual-phase steel observed by X-ray microtomography. *Acta Materialia, 56*, 4954–4964.

Mangan, M. A., Spanos, G., & Kral, M. V. (1999). Correlation between the crystallography and morphology of proeutectoid cementite precipitates. *Acta Materialia, 47*, 4263–4274.

Marciniak, Z., & Kuczynski, K. (1967). Limit strains in the processes of stretch-forming sheet metal. *International Journal of Mechanical Science, 9*, 609–620.

Mazinani, M., and & Poole, W. J. (2007). Effect of plasticity on the deformation behavior of a low-carbon dual phase steel. *Metallurgical and Materials Transactions A*, 38, 328–339

McClintock, F. (1968). A criterion for ductile fracture by the growth of holes. *Journal of Applied Mechanics, 35*, 363–371.

Meyers, M. A., & Aimone, C. T. (1983). Dyanmic failure (spalling) of metals. *Progress in Materials Science, 28*, 1–96.

Monchiet, V., Cazacu, O., Charkaluk, E., & Kondo, D. (2008). Macroscopic yield criteria for plastic anisotropic materials containing spheroidal voids. *International Journal of Plasticity, 24*, 1158–1189.

Mueller, R., & Mortensen, A. (2006). Simplified prediction of the monotonic uniaxial stress–strain curve of non-linear particulate composites. *Acta Materialia, 54*, 2145–2155.

Mori, T., & Tanaka, K. (1973). Average stress in matrix and average elastic energy of materials with misfitting inclusions. *Acta Metallurgica, 21*, 571–574.

Moulin, N., Jeulin, D., & Klöcker, H. (2009). Stress concentrations in non-convex elastic particles embedded in a ductile matrix. *International Journal of Engineering Science, 47*, 170–191.

Needleman, A. (1972). Void growth in elastic–plastic medium. *Journal of Applied Mechanics, Trans. ASME, 39* (Ser. E, no. 4), 964–970.

Needleman, A. (1987). Continuum model for void nucleation by inclusion debonding. *Journal of Applied Mechanics, 54*, 525–531. 228.

Needleman, A., & Tvergaard, V. (1991). An analysis of dynamic, ductile crack growth in a double edge cracked specimen. *International Journal of Fracture, 49*, 41–67.

Nemat-Nasser, S., & Hori, M. (1993a). *Micromechanics: Overall properties of heterogeneous materials*. Amsterdam/New York: North-Holland.

Nemat-Nasser, S., & Hori, M. (1993b). *Micromechanics: Overall properties of heterogeneous solids*. Amsterdam: Elsevier Science Publishers.

Nielsen, K. L., & Tvergaard, V. (2011). Failure by void coalescence in metallic materials containing primary and secondary voids subject to intense shearing. *International Journal of Solids and Structures, 48*, 1255–1267.

Orlov, O. (2006). *A three-dimensional damage percolation model*. Ph.D. thesis, University of Waterloo, Waterloo, Ontario, Canada.

Orlov, O. (2003). Private communications.

Pardoen, T. (2006). Numerical simulation of low stress triaxiality ductile fracture. *Computers and Structures, 84*, 1641–1650.

Pardoen, T., & Delannay, F. (1998a). Assessment of void growth models from porosity measurements in cold drawn copper bars. *Metallurgical and Materials Transaction A, 29*, 1895–1909.

Pardoen, T., & Delannay, F. (1998b). The coalescence of voids in prestrained notched round copper bars. *Fatigue and Fracture of Engineering Materials and Structures, 21*, 1459–1472.

Pardoen, T., & Hutchinson, J. (2000). An extended model for void growth and coalescence. *Journal of the Mechanics and Physics of Solids, 48*, 2467–2512.

Perrin, G., & Leblond, J. (2000). Accelerated void growth in porous ductile solids containing two populations of cavities. *International Journal of Plasticity, 16*, 91–120.

Pierard, O., Gonzalez, C., Segurado, J., Llorca, J., & Doghri, I. (2007). Micromechanics of elasto-plastic materials reinforced with ellipsoidal inclusions. *International Journal of Solids and Structures, 44*, 6945–6962.

Pilkey, A. K. (2001). Private Communications.

Pilkey, A. K. (1997). *Effect of second phase particle clustering on aluminum-silicon alloy sheet formability*. Ph.D. thesis, Carleton University, Ottawa, Canada.

Pilkey, A. K., Fowler, J. P., Worswick, M. J., Burger, G., & Lloyd, D. J. (1995). Characterizing particle contributions in model aluminum alloy systems. *Microstructural Science, 22*, ASM.

Pilkey, A. K., Worswick, M. J., Thomson, C. I. A., Burger, G., & Lloyd, D. J. (1998). Effect of second phase particle distribution on the formability of Al-Si sheet. In D. S. Wilkinson, W. J. Poole, & A. Alpus (Eds.), *Advances in industrial materials*. Calgary: The Metallurgical Society of CIM.

Ponte-Castaneda, P., & Suquet, P. (1998). Nonlinear composites. *Advances in Applied Mechanics, 34*, 171–301.

Ponte Castañeda, P. (1996). Exact second-order estimates for the effective mechanical properties of nonlinear composites. *Journal of the Mechanics and Physics of Solids, 44*, 827–862.

Qiu, Y. P., & Weng, G. J. (1993). Plastic potential and yield function of porous materials with aligned randomly oriented spheroidal voids. *International Journal of Plasticity, 9*, 271–290.

Ragab, A. R. (2004a). A model for ductile fracture based on internal necking of spheroidal voids. *Acta Materialia, 52*, 3997–4005.

Ragab, A. R. (2004b). Application of an extended void growth model with strain hardening and void shape evolution to ductile fracture under axisymmetric tension. *Engineering Fracture Mechanics, 71*, 1515–1534.

Rathod, M. J., & Kutsuna, M. (2004, January). Joining of aluminum alloy 5052 and low-carbon steel by laser roll welding. *Welding Research, 83*, 16–26.

Reusch, F., Svendsen, B., & Klingbeil, D. (2003). Local and non-local Gurson-based ductile damage and failure modeling at large deformation. *European Journal of Mechanics and Solids, 22*, 779–792.

Rice, J. R., & Tracey, D. M. (1969). On the ductile enlargement of voids in triaxial stress fields. *Journal of the Mechanics and Physics of Solids, 17*, 201–217.

Richelsen, A. B., & Tvergaard, V. (1994). Dilatant plasticity on upper bound estimates for porous ductile solid. *Acta Metallurgica, 42*, 2561–2577. 229.

Richmond, O., & Smelser R. E. (1985). Alcoa technical center memorandum. In C. L. Hom & R. M. McMeeking (Eds.), *Void growth in elastic–plastic materials. Journal of Applied Mechanics, 56*, 309–317, 1989.

Riks, E. (1979). An incremental approach to the solution of snapping and buckling problems. *International Journal of Solids and Structures, 15*, 529–551.

Rousselier, G. (1987). Ductile fracture models and their potential in local approach of fracture. *Nuclear Engineering and Design, 105*, 97–111.

Saje, M., Pan, J., & Needleman, A. (1982). Void nucleation effects on shear localization in porous plastic solids. *International Journal of Fracture, 19*, 163–182.

Scheyvaerts, F., Pardoen, T., & Onck, P. R. (2010). A new model for void coalescence by internal necking. *International Journal of Damage Mechanics, 19*, 95–126.

Scheyvaerts, F., Onck, P. R., Tekoglu, C., & Pardoen, T. (2011). The growth and coalescence of ellipsoidal voids in plane strain under combined shear and tension. *Journal of Mechanics and Physics of Solids, 59*, 373–397.

Segurado, J., & Llorca, J. (2002). A numerical approximation to the elastic properties of sphere-reinforced composites. *Journal of the Mechanics and Physics of Solids, 50*, 2107–2121.

Shabrov, M., & Needleman, A. (2002). An analysis of inclusion morphology effects on void nucleation. *Modeling and Simulation in Materials Science and Engineering, 10*, 63–183.

Shehata, M. T., & Boyd, J. D. (1988). Measurement of spatial distribution of inclusions. In *Inclusions and their influence on materials behaviour* (pp. 123–131). Metals Park: ASM International.

Shima, S., & Oyane, M. (1976). Plasticity theory for porous metals. *International Journal of Mechanical Science, 18*, 285–291.

Simha, C. H. M., Gholipour, J., Bardelcik, A., & Worswick, M. J. (2007). Prediction of necking in tubular hydroforming using an extended stress-based forming limit curve. *Journal of Engineering Materials and Technology, 129*, 36–47.

Siad, L., Ouali, M. O., & Benabbes, A. (2008). Comparison of explicit and implicit finite element simulations of void growth and coalescence in porous ductile materials. *Materials and Design, 29*, 319–329.

Somekawa, H., Tanaka, T., Sasaki, H., Kita, K., Inoue, A., & Higashi, K. (2004). Diffusion bonding in ultra-fine-grained Al-Fe alloy indicating high-strain-rate superplasticity. *Acta Materialia, 52*, 1051–1059.

Sovik, O., & Thaulow, C. (1997). Growth of spheroidal voids in elastic–plastic solids. *Fatigue and Fracture in Engineering Materials and Structures, 20*, 1731–1744.

Stewart, J. B., & Cazacu, O. (2011). Analytical yield criterion for an anisotropic material containing spherical voids and exhibiting tension–compression asymmetry. *International Journal of Solids and Structures, 48*, 357–373.

Smerd, R., Winkler, S., Salisbury, C., Worswick, M., Lloyd, D., & Finn, M. (2005). High strain rate tensile testing of automotive aluminum alloy sheet. *International Journal of Impact Engineering, 32*, 541–560.

Sun, J. (1991). Effect of stress triaxiality on micro-mechanisms of void coalescence and micro-fracture ductility of materials. *Engineering Fracture Mechanics, 39*, 799–805.

Sun, Y. (1995a). Influence of void nucleation and growth on deformation localization in tensile sheet specimen. *Engineering Fracture Mechanics, 51*, 381–389.

Sun, Y. (1995b). Constitutive equations for ductile materials containing large and small voids. *Mechanics of Materials, 19*, 119–127.

Sun, Y., & Wang, D. (1989). A lower bound approach to the yield loci of porous materials. *Acta Mech Sinica, 5*, 237–245.

Sun, D. Z., Sigele, D., Voss, B., & Schmitt, W. (1989). Application of local damage models to the numerical analysis of ductile rupture. *Fatigue and Fracture of Engineering Materials and Structures, 12*, 201–212.

Suquet, P. (1995). Overall properties of non-linear composites: a modified secant moduli theory and its link with Ponte Casteneda"s nonlinear variational procedure. *Comptes rendus de Academi des Sciences, Paris, Serie IIB, 320*, 563–571.

Suquet, P. (1997). Effective properties of nonlinear composites. In *Continuum micromechanics* (CISM Course and Lecture Notes, pp. 197–264, 230).

Tanaka, J. P., Pampillo, C. A., & Low, J. R. (1970a). Review of developments in plane strain fracture toughness testing. *ASTM, STP, 463*, 191.

Tanaka, K., Mori, T., & Nakamura, T. (1970b). Cavity formation at the interface of a spherical inclusion in a plastically deformed matrix. *Philosophical Magazine, 21*, 267–279.

Tandon, G. P., & Weng, G. J. (1986). Average stress in the matrix and effective moduli of randomly oriented composites. *Composites Science and Technology, 27*, 111–132.

Tang, C. Y., Chow, C. L., Shen, W., & Tai, W. H. (1999). Development of a damage-based criterion for ductile fracture prediction in sheet metal forming. *Journal of Materials Processing Technology, 91*, 270–277.

Tani, J., & Kido, H. (2008). Lattice dynamics of Mg2Si and Mg2Ge compounds from first-principles calculations. *Computational Materials Science, 42*, 531–536.

Thomason, P. (1985a). Three-dimensional models for the plastic limit-loads at incipient failure of the intervoid matrix in ductile porous solids. *Acta Metallurgica, 33*, 1079–1085.

Thomason, P. (1985b). A three-dimensional model for ductile fracture by the growth and coalescence of micro-voids. *Acta Metallurgica, 33*, 1087–1095.

Thomason, P. F. (1990). *Ductile fracture of metals*. Oxford: Pergamon Press.

Thomason, P. F. (1993). Ductile fracture by the growth and coalescence of microvoids of non-uniform size and spacing. *Acta Metallurgica, 41*, 2127–2134.

Thomson, C. I. A. (2001). *Modelling the effect of particle clustering on ductile failure*. Ph.D. thesis, Carleton University, Ottawa.

Thomson, C. I. A., Worswick, M. J., Pilkey, A. K., Lloyd, D. J., & Burger, G. (1999). Modeling void nucleation and growth within periodic clusters of particles. *Journal of the Mechanics and Physics of Solids, 47*, 1–26.

Toi, Y., & Kang, S. S. (2005). Mesoscopic natural element analysis of elastic moduli, yield stress and fracture of solids containing a number of voids. *International Journal of Plasticity, 21*, 2277–2296.

Tvergaard, V. (1981). Influence of voids on shear band instabilities under plane strain conditions. *International Journal of Fracture, 17*, 389–407.

Tvergaard, V. (1982). Ductile fracture by cavity nucleation between larger voids. *Journal of the Mechanics and Physics of Solids, 30*, 265–286.

Tvergaard, V. (1990). Material failure by void growth to coalescence. *Advances in Applied Mechanics, 27*, 83–151.

Tvergaard, V., & Needleman, A. (1984). Analysis of the cup-cone fracture in a round tensile test bar. *Acta Metallurgica, 32*, 157–169. 231.

Tvergaard, V., & Needleman, A. (1986). Effect of material rate sensitivity on failure modes in the Charpy V-notched test. *Journal of Mechanics and Physics Solids, 34*, 213–241.

Tvergaard, V., & Needleman, A. (1995). Effects of non-local damage in porous plastic solids. *International Journal of Solids and Structures, 8*, 1063–1077.

Tvergaard, V., & Needleman, A. (1997). Nonlocal effects on localization in a void-sheet. *International Journal of Solids and Structures, 34*, 2221–2238.

Tvergaard, V., & Niordson, C. (2004). Nonlocal plasticity effects on interaction of different size voids. *International Journal of Plasticity, 20*, 107–120.

Voce, E. (1948). The relationship between stress and strain for homogeneous deformation. *Institute of Metals, 74*, 537–562.

Varma, N. S. P., Narasimhan, R., Luo, A. A., & Sachdev, A. K. (2007). An analysis of localized necking in aluminum alloy tubes during hydroforming using a continuum damage model. *International Journal of Mechanical Science, 49*, 200–209.

Wang, D., & Pan, J. (2004). An anisotropic Gurson yield criterion for porous ductile sheet metals with planar anisotropy. *International Journal of Damage Mechanics, 13*, 7–33.

Weck, A. (2006). *The role of coalescence on ductile fracture*. Ph.D. thesis, McMaster University, Hamilton ON, Canada.

Wen, J., Huang, Y., Hwang, K., Liu, C., & Li, M. (2005). The modified Gurson model accounting for the void size effect. *International Journal of Plasticity, 21*, 381–395.

Weng, G. J. (1984). Some elastic properties of reinforced solids, with special reference to isotropic ones containing spherical inclusions. *International Journal of Engineering Science, 22*, 845–856.

Weng, G. J. (1990). The overall elastoplastic stress strain relations of dual-phase metals. *Journal of the Mechanics and Physics of Solids, 38*, 419–441.

Winkler, S., Thompson, A., Salisbury, C., Worswick, M., Van Riemsdijk, I., & Mayer, R. (2008). Strain rate and temperature effects on the formability and damage of advanced high-strength steels. *Metallurgical and Materials Transactions A, 39*, 1350–1358.

Winkler, S. L. (2003). Private Communications.

Worswick, M. J. (1988). *Cavity growth and constitutive softening in ductile solids*. Ph.D. thesis, University of Waterloo, Waterloo, Ontario, Canada.

Worswick, M. J., Pilkey, A. K., Thomson, C. I. A., Lloyd, D. J., & Burger, G. (1998). Percolation damage predictions based on measured second phase particle distributions. *Microstructural Science, 26*, 507–514.

Worswick, M. J., & Pelletier, P. (1998). Numerical simulation of ductile fracture during high strain rate deformation. *European Physical Journal Applied Physics, 4*, 257–267.

Worswick, M. J., Chen, Z. T., Pilkey, A. K., Lloyd, D., & Court, S. (2001). Damage characterization and damage percolation modeling in aluminum alloy sheet. *Acta Materialia, 49*, 2791–2803.

Wu, P. D., Jain, M., Savoie, J., MacEwon, S. R., Tugcu, P., & Neale, K. W. (2003). Evaluation of anisotropic yield function for aluminum sheets. *International Journal of Plasticity, 19*, 121–138.

Xia, W. J., & Chen, Z. T. (2007). A quasi-exact yield criterion for anisotropic porous ductile sheet metals. *Acta Mechanica, 191*(1–2), 93–108.

Xiong, W., Qin, X., & Wang, L. (2007). Densification behaviour of nanocrystalline Mg2Si compact in hot pressing. *Journal of Material Science and Technology, 23*, 595–598.

Xue, L. (2007). Damage accumulation and fracture initiation in uncracked ductile solids subject to triaxial loading. *International Journal of Solids and Structures, 44*, 5163–5181.

Xue, L. (2008). Constitutive modeling of void shearing effect in ductile fracture of porous materials. *Engineering Fracture Mechanics, 75*, 3343–3366.

Yamagiwa, K., Watanabe, Y., Fukui, Y., & Kapranos, P. (2003). Novel recycling system of aluminum and iron wastes – In-situ Al-Al3Fe functionally graded material manufactured by a centrifugal method. *Materials Transactions, 44*, 2461–2467.

Yan, X. Q. (1992). Effect of yield surface curvature on local necking in biaxially stretched sheets in porous materials. *Transactions ASME Journal of Engineering Material and Technology, 114*, 196–200.

Yee, K. C., & Mear, M. E. (1996). Effect of void shape on the macroscopic response of non-linear porous solids. *International Journal of Plasticity, 12*, 45–68.

Yoon, J. H., Stewart, J. B., & Cazacu, O. (2011). Coupled elastic–plastic damage model for a porous aggregate with an incompressible matrix displaying tension–compression asymmetry. *Engineering Fracture Mechanics, 78*, 1407–1423.

Zhang, Z. L. (1996). A sensitivity analysis of material parameters for the Gurson constitutive model. *Fatigue and Fracture of Engineering Materials and Structures, 19*, 561–570.

Zhang, Y., & Chen, Z. (2007). On the effect of stress triaxiality on void coalescence. *International Journal of Fracture, 143*, 105–112.

Zhang, Z., & Niemi, E. (1994a). Studies on the ductility predictions by different local failure criteria. *Engineering Fracture Mechanics, 48*, 529–540.

Zhang, Z., & Niemi, E. (1994b). Analyzing ductile fracture using dual dilational constitutive equations. *Fatigue and Fracture of Engineering Materials and Structures, 17*, 695–707.

Zhang, Z. L., & Skallerud, B. (2010). Void coalescence with and without prestrain history. *International Journal of Damage Mechanics, 19*, 153–174.

Zhang, Z., Thaulow, C., & Odegard, J. (2000). A complete Gurson model approach for ductile fracture. *Engineering Fracture Mechanics, 67*, 155–168.

Zhang, K. S., Bai, J. B., & Francois, D. (2001). Numerical analysis of the influence of the Lode parameter on the void growth. *International Journal of Solids and Structures, 38*, 5847–5856.

Zhang, Z. (1998). A complete Gurson model. In M. H. Aliabadi (Ed.), *Nonlinear fracture and damage mechanics*. Southampton: Computational Mechanics Publications.

Index

Z. Chen and C. Butcher, *Micromechanics Modelling of Ductile Fracture*,
Solid Mechanics and Its Applications 195, DOI 10.1007/978-94-007-6098-1,
© Springer Science+Business Media Dordrecht 2013

Printed by Printforce, the Netherlands